Computational
Microelectronics

Edited by S. Selberherr

Multigrid Methods for Process Simulation

W. Joppich and S. Mijalković

Springer-Verlag Wien New York

Dr. Wolfgang Joppich
Institut für Methodische Grundlagen
Gesellschaft für Mathematik und Datenverarbeitung mbH
Sankt Augustin, Federal Republic of Germany

Dr. Slobodan Mijalković
Faculty of Electronic Engineering
University of Niš
Niš, Yugoslavia

© 1993 Springer-Verlag/Wien
Softcover reprint of the hardcover 1st edition 1993

Typesetting: Thomson Press (India) Ltd., New Delhi 110 001

Printed on acid-free paper

With 126 Figures

ISBN-13: 978-3-7091-9255-9   e-ISBN-13: 978-3-7091-9253-5
DOI: 10.1007/ 978-3-7091-9253-5

To
Jona and Markus

Preface

It was about 1985 when both of the authors started their work using multigrid methods for process simulation problems. This happened independent from each other, with a completely different background and different intentions in mind.

At this time, some important monographs appeared or have been in preparation. There are the three "classical" ones, from our point of view: the so-called "1984 Guide" [12] by Brandt, the "Multi-Grid Methods and Applications" [49] by Hackbusch and the so-called "Fundamentals" [132] by Stüben and Trottenberg.

Stüben and Trottenberg in [132] state a "delayed acceptance, resentments" with respect to multigrid algorithms. They complain: "Nevertheless, even today's situation is still unsatisfactory in several respects. If this is true for the development of standard methods, it applies all the more to the area of really difficult, complex applications."

In spite of all the above mentioned publications and without ignoring important theoretical and practical improvements of multigrid, this situation has not yet changed dramatically. This statement is made under the condition that a numerical principle like multigrid is "accepted", if there exist "professional" programs for research and production purposes. "Professional" in this context stands for "solving complex technical problems in an industrial environment by a large community of users". Such a use demands not only for fast solution methods but also requires a high robustness with respect to the physical parameters of the problem.

The question has to be posed why there is a lack of professional multigrid software? The main reason seems to be that multigrid as well as specific application fields have become individual disciplines. Many multigrid experts work on multigrid problems and not on application problems. Therefore, most of the research will not convince the engineer who looks for the solution of a very concrete problem. For highly specialized applications it is not sufficient to present only solutions of isolated mathematical problems. On the other hand, the complexity and the enormous degree of freedom to choose components complicates the naive use of multigrid. But the main reason for the lack of robust and efficient professional multigrid programs is the separation of disciplines and the fact that "purists" of either side do not work in the deep and dark swamp of details of the other side.

To overcome this principal separation, which is observed, too, with respect to process simulation, both of the authors—the one working in a multigrid research group, the other originating from the engineering discipline—decided to leave their original research fields to approach the difficulties of the respective complementary discipline. By the exchange of experience and ideas, a large collection of possible approaches to be used in multigrid algorithms for process simulation is obtained. These results lead to the following conclusion:

1. Many of the standard multigrid components work well, even in a complex process simulation application.
2. Non-standard approaches have been successfully used in combination with physical models which are "nearly" realistic ones.
3. More sophisticated physical models within a simulator-like approach justify simplifications which still guarantee satisfying results from the numerical point of view.

This is, of course, not yet the elimination of the disappointing situation towards "professional multigrid software for (process) simulation". An improvement of this state is only possible if practical results convince both the research and application community and if developers of simulation software are encouraged to use multigrid. The practical proof of multigrid efficiency, the demonstration that concepts, derived with the aid of model problems, carry over to complex problems and the shown flexibility to use non-standard approaches are, in our opinion, the best arguments to convince. To give the description of the ingredients which are needed for successful multigrid algorithms and to give the compact information why and how multigrid works in a realistic process simulation environment: this is the intention of this book.

ACKNOWLEDGEMENTS: The multigrid group of Professor Dr. U. Trottenberg at the Gesellschaft für Mathematik und Datenverarbeitung mbH, Bonn, is one of the German centers of multigrid development. Their longtime activity on multigrid produced a large set of results which can be considered now to be fundamental. Having so many multigrid experts in an immediate neighbourhood it would have been a bad omission not to use their expertise, experience and originality. This was done by many helpful discussions during the phase of programming and while writing this book. In addition, some of them gave very concrete contributions to this book. We owe thanks to Dr. Barbara Steckel for the Section 2.7.4 on "the fast

adaptive composite grid method, FAC" including comparing remarks to the MLAT-technique, to Dr. Ute Gärtel for her Remarks on "multigrid methods for anisotropic 3D—problems", Remarks 2.17 and Section 2.11.6, to Dr. Anton Schüller for his Section 2.11 on "a concept to parallelize multigrid methods", to Dipl.-Math. H. Schwichtenberg and Dr. G. Winter for their "application of standard multigrid components to the semi-conductor device equations" in Section 2.12 and especially to Dr. Rudolph A. Lorentz who improved the first approaches to use shape-preserving interpolations and who proposed additional interpolation methods. Section 4.6 is a summary of the joint work. For helping out with proof reading we owe thanks to Dr. Guy Lonsdale.
We are also indebted to the process and device simulation group of Professor Dr. N. Stojadinović, at the Department of Microelectronics, Faculty of Electronic Engineering, Niš. Their activities are mainly motivated by the demands of the semiconductor industry and particularly concern the development of the professional process simulation tool MUSIC—based on the adaptive multilevel principles which are described in this book. Special thanks are to Dragan Pantić for writing some critical and difficult portions of MUSIC and for his concrete contribution in Section 5.4.4 on "coupled defects diffusion simulation", to Srdjan Mitrović for writing a code to visualize MUSIC results and for his assistance in producing figures within Chapter 5 and the book cover, and further to Zoran Prijić for helpful suggestions in selecting the realistic processing parameters for the simulation examples of Chapter 5.
Last but not least we have to mention our families which supported us by their patience when we spent a lot of "their" time by writing this book.

Sankt Augustin, Niš, June 1992   W. Joppich and S. Mijalković

Contents

Notation

The following pages contain a brief summary combined with a short explanation of those abbreviations and notations which are used throughout the text of this book. Because two completely different disciplines with their own grown up terminology have to be combined it is not always possible to maintain a uniqueness of the notation. But in any case it is possible to recognize the actual interpretation without difficulties. For instance: in the multigrid context N is used to divide the unit interval into N subintervals having all the same length. N therefore determines the mesh size $h = 1/N$ and the number of grid points, too. Analogously N_t denotes the number of time steps, the number of nodes in time direction. On the other hand, the dopand concentration is denoted by $N = N(x, y, t)$, too. The context makes its actual use obvious. Quantities like $\Omega(t)$ are by the same reasons not explicitly marked as time dependent ones.

Discrete quantities are defined in terms of the general coordinate x and the formal discretization parameter Δx or ΔX to denote some coarser discretization level. Δx thus can be replaced by h, Δt, H, $2h$, ΔT, ... $L_{\Delta x}$ or $u_{\Delta x}$ consequently stand for $L_h, L_H, L_{\Delta t}, \ldots, u_h, \ldots$, and so on.

Equations, tables, figures, ..., are referenced with the chapter number as prefix. The equation (3.5) is therefore found in Chapter 3, and, similarly the equation (1.1) is the first equation of Chapter one.

$\Omega, \bar{\Omega}$	The (space) domain and its closure
Ω_t	The time domain, interval of simulation
T	The simulation time
$\partial\Omega, \partial\Omega_i$	The boundary of Ω and segments of it
$\Omega_{(h)}$	Subdomain of Ω where a resolution of at least mesh size h is required
L	Differential operator
A	Differential operator with only spatial derivates
B	Differential operator for boundary conditions·
$\vec{n} = (n_x, n_y)$	Outer normal unit vector

$\dfrac{\partial}{\partial \vec{n}}$	Derivative into the direction of $\vec{n}$
f, g, u	Continuous functions for the right hand side of both the partial differential equation and the boundary condition; the solution of the continuous problem
$N(x, y, t)$	Concentraton in (x, y) at time $t \geq 0$
n, n_i	Electron concentration and intrinsic carrier concentration
D^x	with x equal to $0, -, +, =$ are intrinsic diffusivities corresponding to different charge states
$D(N)$	Concentration-dependent diffusion coefficient
$\mathscr{H}$	An index set of mesh sizes, for instance $\mathscr{H} = \{h_l \mid h_l = h_1/2^{l-1}, \ l = 1, \ldots, M\}$. Instead of indexing grids with h_l, for sake of simplicity the level index l is used, sometimes
$h/2, h, 2h, H$	Mesh widths for space discretization
$\Delta t/2, \Delta t, 2\Delta t, \Delta T$	Mesh widths for temporal discretization, Δt^n denotes the time-step size at the temporal grid node $t^n = t^{n-1} + \Delta t^n$ for $n \in \mathbb{N}$
N_x, N_y, N_t	Number of grid intervals in x-, respectively y-direction; number of time-steps
$\mathscr{G}_h$	Infinite space grid (lattice) with mesh width h. The points are denoted by (x_i, y_j) or $\vec{p}$, alternatively
G_h	The set of grid points corresponding to the level of discretization $(h, 2h, H, l, \ldots)$. Similar for boundaries
G	A grid structure consisting of several grids, a sequence of grids
$G_{\mathscr{H}}$	A composite mesh
m_g	The number of global grids within a grid sequence
$f_{i,j}, (f)_{\Delta x}$	Evaluation of $f(x, y)$ at (x_i, y_j), injection to the Δx-grid
$f_{\Delta x}$	Grid function on the Δx-grid
$\|f_{\Delta x}\|_\infty, \|f_{\Delta x}\|_2$	Commonly used discrete norms; the discrete maximum (infinite) norm and the discrete L_2-norm, respectively
$N_{\Delta x}^{n+1}$	Grid function $N_{\Delta x}$ at time $t^{n+1} = t^n + \Delta t^{n+1}$
$Id_{\Delta x}$	Identity grid operator on the Δx-grid
$I_{\Delta X}^{\Delta x}, \hat{I}_{\Delta X}^{\Delta x}$	Prolongation operators from the ΔX-grid to the Δx-grid
$\mathbb{I}_{\Delta X}^{\Delta x}$	FMG-interpolation
$I_{\Delta x}^{\Delta X}, \hat{I}_{\Delta x}^{\Delta X}, I^{\Delta x}, \hat{I}^{\Delta x}$	Restriction operators from the Δx-grid to the ΔX-grid or from the continuum to the Δx-grid
$F_{\Delta x}^{\Delta X}, F^{\Delta X}$	Full weighting, a special weighted restriction
$\lambda, \lambda_{l,m}, \lambda^h, \lambda_{l,m}^h$	Eigenvalues
$\varphi_{l,m}, \varphi_{l,m}^h$	Eigenfunctions
$\rho(M)$	Spectral radius of the operator M, asymptotic convergence radius
$\mathbf{I}, \mathbf{A}, \mathbf{D}, \mathbf{L}, \mathbf{P}, \mathbf{U}, \ldots$	Matrices, operators
$\mathbf{f}, \mathbf{u}, \mathbf{u}^{(n)}, \mathbf{u}^{(n+1)}, \ldots$	Vectors

$M_{\Delta x}^{\Delta X}, M_{\Delta x}, S_{\Delta x}$	Two-grid and multilevel operators, smoothing operator
$r_{\Delta x}, r_{\Delta x}^{(n)}$	Residual grid function (after n iterations)
$e_{\Delta x}$	Global error $e_{\Delta x} = (u)_{\Delta x} - u_{\Delta x}$ between the continuous solution u of $Lu = f$ (evaluated on the Δx-grid) and the exact solution $u_{\Delta x}$ of $L_{\Delta x} u_{\Delta x} = f_{\Delta x}$
$l_{\Delta x}$	Local-global error for semi-discrete problems using implicit time discretization schemes
$\tau_{\Delta x}, \tau_{\Delta X}^{\Delta x}$	Local discretization error with respect to the Δx-grid and the relative local discretization error of the ΔX-grid relative to the Δx-grid
$\tilde{e}_{\Delta x}^{(n)}$	Algebraic error $\tilde{e}_{\Delta x}^{(n)} = u_{\Delta x} - w_{\Delta x}^{(n)}$ between $u_{\Delta x}$ and an approximation $w_{\Delta x}^{(n)}$ to it afer n iterations
$v = v_1 + v_2$	Total number of relaxation sweeps, v_1 pre-smoothing steps, v_2 post-smoothing steps
v_{sol}	Number of relaxation sweeps to solve iteratively on the coarsest grid
$\mathcal{A}, \mathcal{B}, \mathcal{L}, \mathcal{C}, \mathcal{D}$	General stencil entries of a discrete five point operator
$\mu(h, \theta)$	Amplification factor for the Fourier mode with frequency θ on a grid with mesh size h
$\tilde{\mu}(h)$	Smoothing rate depending on the grid size
$\mu_{PGS-lex}^*$	h-independent smoothing rate of a special relaxation scheme (here: $PGS - lex$)
RU	Relaxation unit, a measure for the computational work
W_l	Computational work per cycle envolving l levels
$\sum RU$	Total number of RU for a simulation
ρ_{n+1}, ρ_{av}	Reduction of the defect by the $n + 1$-th iteration and the average empirical convergence rate per relaxation sweep
ρ_{ru}, ρ_{nor}	Average convergence factor per relaxation unit and the convergence factor relative (normalized) to a fixed relaxation
lx, ch, eo, oe	Denote the scanning of grid points, columns or lines: lexicographically, chequerboarded (red-black), even-odd and odd-even (ZEBRA), respectively
a, l_x, l_y	Parameters which describe the geometry of the simulation domain
N_d	Implantation dose
R_p	Range of projection
$\dot{U}_{\vec{n}}(x, t)$	Oxide growth rate perpendicular to the moving Si-SiO$_2$-interface
δ_{kl}	Kronecker's symbol
$\text{dist}(z, S)$	$\inf_{w \in S} \| z - w \|$ with the euclidean distance $\| \cdot \|$
$f(x) = O(g(x))$	Landau's symbol: $\| f(x)/g(x) \|_\infty <$ const. for $x \to x_0$ and $\| x - x_0 \|_\infty$ sufficiently small
$f(x) = o(g(x))$	Landau's symbol: $\lim_{x \to x_0} f(x)/g(x) = 0$
$\mathbb{N}, \mathbb{Z}, \mathbb{R}$	Natural numbers, integers, real numbers

Introduction 1

The recent improvements of the VLSI-technology have lead to an enormous demand for integrated circuits which are used in nearly every part of practical life. The increased complexity of the products requires the use of reliable tools for the design and the development of VLSI-devices. Within the design process the simulation of fabrication steps—process simulation— is especially open for numerical simulation.

To establish the numerical simulation as an important and accepted aid to reduce both developing cost and time there are besides other requirements two important tasks to be solved by the exchange of information between engineers, physicists, mathematicians and computer scientists:

1. The investigation of the basic physical mechanism with the aim to get a complete description of the physical principles.
2. The development of efficient and ambitious numerical algorithms which are applicable to the complex physical models.

The simulation has to be "realistic", which is often synonymous with "extremely complex models" depending on a large variety of parameters. And, of course, the algorithm has to provide exact results as fast as possible. This explains the recent search for appropriate algorithms which is not yet finished.

The rapidly changing semiconductor technology defines the requirements which have to be satisfied by simulation tools. This implies that research for process simulation cannot be done as a pure academic discipline and especially not decoupled from the needs of semiconductor industry. From the industrial point of view, the ultimate goal of process simulation is to enable *silicon processing without silicon*, that is, to complete expensive laboratory experiments for technology development by less expensive numerical experiments. The critical obstacle to achieve this goal is the fact that improved models, describing a new technology, usually are not available before the technology itself can be processed and is more or less under control. On the other hand, the improvement of process models becomes possible only by both experimental verification and numerical support by flexible software tools which are robust enough to handle even new models. Process simulation programs in this sense not only serve as a tool for the analysis and the optimization of currently used technology but also serve as a key instrument for the technology development.

To determine adequate physical models for the diffusivities and reaction terms as well as the actual computation of the solution represent formidable tasks in the simulation of multiparticle evolution problems. These problems are related to each other since the definition of physical parameters for the kinetics of point defects is almost impossible without numerical simulation. Multiparticle diffusion in todays process simulation tools already poses heavy demands on computer resources. Moreover, with the continuously growing number of particle profiles whose spatial and temporal redistribution have to be handled simultaneously, a dramatic increase of the computational demands can be expected for the near future. This poses serious problems to the next generation of process simulation programs. Adaptive and parallel algorithms are indisputable.

The general processing sequence includes processes which are not of "evolution" type. Representative examples are ion implantation, deposition and selective etching processes. The results of these processes provide the initial conditions, the geometry and the boundary conditions: they serve as auxiliary processes which are necessary to obtain well-posed initial boundary value problems, both with respect to mathematics and to technology. Classifying these processes as "auxiliary" ones, does not underestimate their importance and complexity. Multilevel treatment of processes *not originating from partial differential equations* leads to so-called non-pde multilevel methods which actually become more and more attractive for multigrid research activities. Nevertheless, there are good reasons to put these methods beyond the scope of this book.

The simulation of evolution processes leads to very large discrete problems. To solve them is the most time consuming phase of process simulation. Typical examples are diffusion and oxidation processes which are responsible for the particle redistribution and for the definition of active regions within the semiconductor devices. The models for transport phenomena in semiconductor fabrication processes are also governed by large scale variations in the coefficients, in the forcing terms and in the boundary conditions. Consequently, the final distributions of relevant physical quantities vary on a large scale over the simulation domain. This problem, together with the concentration-, time- and temperature-dependent physical parameters, poses serious problems upon adaptive simulation tools for process simulation.

The most obvious example of such a behavior is the redistribution of impurities by diffusion. The profiles are characterized by strong variations inside thin volume layers compared to the whole active processing area.

These moving fronts require a higher grid resolution than it is necessary within the remaining simulation domain. The process evolution distributes the active particles within the semiconductor substrate. Consequently, the final distributions cover larger areas than the initial ones. This allows different resolutions in different areas at different moments. For these reasons, an adaptive method, which keeps the error of the discrete solutions for the total simulation below a predetermined tolerance, is a highly desirable feature for process simulation programs, especially if this goal is reached with a minimum of computational work.

The numerical situation of process simulations is characterized by an astonishingly broad spectrum of methods. This already starts with the space discretization technique. As well finite difference (FD) as finite element (FE) approaches are commonly used.

FE-methods are often recommended because of special geometric features of the simulation area. Besides the good approximation of irregular domains the use of irregularly refined meshes is in principle standard. Therefore it is more than surprising that most process simulators need user information to decide where to refine. This means that the refinement areas are determined in advance. Even with FE-approaches really adaptive approaches are not yet standard for process simulation tools. The reasons may be the lack of reliable control mechanisms, the resulting complexity of the data structure and perhaps the missing measure of quality for the created mesh.

With respect to local refinements the situation concerning FD-based simulators is even worse. In the literature there is not the least indication that refinements in the sense of "patched" grids are used. In some cases rectangular tensor product grids with different and varying meshsizes in x- and y-direction are used. This leads to an increased number of points in critical regions where the higher resolution is wanted. On the other hand there are many points even in parts of the domain where they are absolutely not necessary. Caused by the varying meshsize, the second order approximation of centered difference formula may reduce to an only first order one. For a fixed accuracy this requires a larger number of points.

Compared to finite element methods where irregular meshes are well established, the grid refinement for finite difference approaches in combination with multigrid techniques is not yet standard, although it is often quoted to be extremely simple to realize them. This is true in the sense that for multigrid algorithms using finite difference discretizations there are no other questions to answer than for any other approach.

The idea of local refinements within the multigrid algorithm offers some essential advantages compared to singlegrid approaches. The second order discretization is guaranteed by the regular structure of all grids, including the refinement patches. The FAS-algorithm provides a numerical quantity which is well suited to decide where a grid has to be refined during the solution process: this really is adaptivity.

FE-methods allow an excellent approximation of time-dependent geometries with curved boundaries and therefore usually compute the solution on
the original domain. With FD-discretizations such an approach is often
rejected because of the difficulty to discretize boundary conditions at curved
boundaries. In most cases the original domain is mapped onto a rectangle by
conformal mapping or with the aid of a simple variable transformation. This
is absolutely not necessary. It is possible to perform all the computations on
the original domain and using a second order FD-formula on the whole
computational domain. An additional benefit of such an approach is the
maintained original structure of the partial differential equation.

The wide range of methods is continued by the numerical treatment with
respect to time. The space discrete equations may be considered as a stiff
system of ordinary differential equations which can be solved by special
solvers. Explicit methods in time usually require an extreme number of time-
steps because of stability conditions which strongly depend on the meshsize.
With growing importance implicit schemes are used. They allow time
stepping procedures which again introduce adaptivity—now with respect
to time.

Thus adaptivity is the guiding principle which always has to be kept in mind
throughout the following chapters.

A Practical Guide to Standard Multigrid Methods 2

One way to get acquainted with the multigrid idea is to analyze standard problems, "model problems" from the multigrid point of terminology, and to solve them by simple singlegrid and multigrid algorithms. It is natural to restrict the first analysis to such problems where a deep theoretical insight and the knowledge about the structure of the solution is available. This chapter will—as a first goal—help to reduce the time which has to be invested for such an analysis. Starting from well-known examples a basic introduction to multigrid is given and some theory is presented with the aim to explain why and how multigrid works—the main purpose of this chapter.

2.1 Continuous and Discrete Model Problems

With f and g functions,

$$f: \quad \Omega \to \mathbb{R} \qquad g: \quad \partial\Omega \to \mathbb{R}$$
$$(x, y) \mapsto f(x, y) \qquad (x, y) \mapsto g(x, y),$$

L a linear second order elliptic differential operator, B denoting a linear boundary operator, an elliptic boundary value problem on the two-dimensional domain Ω (not necessary a rectangle) with boundary $\partial\Omega$ is formally given by

$$Lu = f \text{ on } \Omega \subset \mathbb{R}^2$$
$$Bu = g \text{ on } \partial\Omega. \tag{2.1}$$

For model problems, B mostly represents either Dirichlet or Neumann boundary conditions. This restriction is not an essential one but it simplifies the representation considerably. The most important step towards the definition of model problems concerns the differential operator L which avoids—in case of the following model problems—first order derivatives and mixed derivatives.

Then the model problem on the unit square in $\mathbb{R}^2$

$$Lu = f \text{ on } \Omega: (0,1) \times (0,1)$$
$$u = g \text{ on } \partial\Omega$$

$$Lu := -\left(a\frac{\partial^2 u}{\partial x^2} + b\frac{\partial^2 u}{\partial y^2} \right) + cu \qquad (2.2)$$

with coefficient functions $a(x,y)$, $b(x,y)$ and $c(x,y)$ leads to well-known representatives of (2.1):

$$a \equiv 1, \quad b \equiv 1, \quad c \equiv 0: \quad Lu = -\left(\frac{\partial^2 u}{\partial x^2} + \frac{\partial^2 u}{\partial y^2} \right) =: -\Delta u \qquad (2.3)$$

$$a \equiv 1, \quad b \equiv 1, \quad c > 0: \quad Lu = -\Delta u + cu \qquad (2.4)$$

$$a \equiv \varepsilon, \quad b \equiv 1, \quad c \equiv 0: \quad Lu = -\left(\varepsilon\frac{\partial^2 u}{\partial x^2} + \frac{\partial^2 u}{\partial y^2} \right), \quad 0 < \varepsilon \ll 1 \qquad (2.5)$$

The corresponding differential equations are often referred to as Poisson equation, Helmholtz equation and the two-dimensional anisotropic model equation.

Because it is not intended to give a systematic introduction to discretization techniques, the finite difference approach which is appropriate for (2.3)–(2.5) is used to derive the discrete equations. With $h = (h_x, h_y)$ as formal discretization quantity the infinite two-dimensional grid

$$\mathcal{G}_h = \{(x,y) \mid x = ih_x, y = jh_y; h_x, h_y > 0; i, j \in \mathbb{Z}\}$$

is introduced. For a square grid h_x is equal to h_y and instead of the multiindex $h = (h_x, h_y)$ the scalar quantity $h(=h_x=h_y)$ is used. In many cases h is given by $h = (1/N_x, 1/N_y)$ where N_x and N_y determine the number of intervals per unit length.

The discrete problem is posed on $G_h := \Omega \cap \mathcal{G}_h$ with discrete boundary conditions on ∂G_h which is in most cases constructed similar to G_h. For rectangular domains $\mathcal{G}_h$ is chosen with grid lines coinciding with $\partial\Omega$. The meshsize h determines the complexity of the discrete problem: the smaller h is the larger the number of points in G_h will be. Grid functions are defined on $\mathcal{G}_h$, G_h or ∂G_h and denoted by the subscript h, for instance $u_h, f_h, g_h, e_h, r_h, \ldots$. The usage of grid functions and grid operators L_h, B_h which act on them is an efficient tool to represent discrete problems.

Assuming at least fourth order differentiability for u and

$$G_h = \{(x_i, y_j) \mid x_i = ih_x, y_j = jh_y; h_x = 1/N_x, h_y = 1/N_y,$$
$$0 \le i \le N_x, 0 \le j \le N_y; N_x, N_y \in \mathbb{N}\}$$

the Taylor series expansion leads to

$$\frac{a}{h_x^2}(-u_{i-1,j} + 2u_{i,j} - u_{i+1,j})$$

$$+ \frac{b}{h_y^2}(-u_{i,j-1} + 2u_{i,j} - u_{i,j+1}) + cu_{i,j} = f_{i,j} + \tau_{i,j} \qquad (2.6)$$

where $\tau_{i,j}$ is an $O(h^2)$-quantity and essentially depends on the fourth order derivative $u^{(iv)}$ which is evaluated within the box $(x_{i-1}, x_{i+1}) \times (y_{j-1}, y_{j+1})$ for $1 \leq i \leq N_x - 1, 1 \leq j \leq N_y - 1$. The discrete boundary conditions are

$$u_{0,j} = g_{0,j}, \quad u_{N_x,j} = g_{N_x,j} \quad \text{for} \quad 0 \leq j \leq N_y \quad \text{and}$$
$$u_{i,0} = g_{i,0}, \quad u_{i,N_y} = g_{i,N_y} \quad \text{for} \quad 1 \leq i \leq N_x - 1$$

Neglecting all the error terms $\tau_{i,j}$ results in a set of difference equations which represents the discrete analogue of the model problem (2.2):

$$L_h u_h = f_h \quad \text{on } G_h$$
$$B_h u_h = g_h \quad \text{on } \partial G_h \qquad (2.7)$$

Example 2.1. *With $L = -\Delta$, $N = N_x = N_y = 4$ (that means $h = \frac{1}{4}$) the discretization of*

$$-\Delta u = f \quad \text{on } \Omega = (0,1)^2$$
$$u = g \quad \text{on } \partial\Omega \qquad (2.8)$$

leads to a set of linear equations for $u_{i,j}$

$$\frac{1}{h^2}(-u_{i-1,j} - u_{i,j-1} + 4u_{i,j} - u_{i,j+1} - u_{i+1,j}) = f_{i,j} \text{ with } (1 \leq i,j \leq 3)$$

$$\text{and} \quad u_{i,j} = g_{i,j} \quad \text{if one of} \quad i,j \in \{0,4\} \qquad (2.9)$$

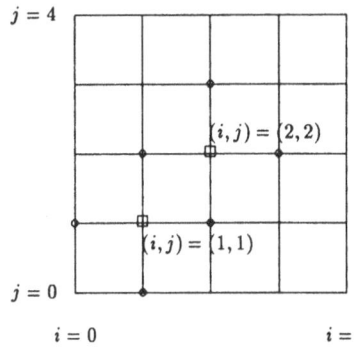

Fig. 2.1 Discretization of $-\Delta u$

The equation for the variable $u_{2,2}$ at (x_2, y_2) is

$$\frac{1}{h^2}(-u_{1,2} - u_{2,1} + 4u_{2,2} - u_{2,3} - u_{3,2}) = f_{2,2}$$

while the equation with respect to $u_{1,1}$ at (x_1, y_1) is given by

$$\frac{1}{h^2}(-u_{0,1} - u_{1,0} + 4u_{1,1} - u_{1,3} - u_{2,1}) = f_{1,1}. \tag{2.10}$$

With the help of the boundary condition this equation can be rewritten as

$$\frac{1}{h^2}(-g_{0,1} - g_{1,0} + 4u_{1,1} - u_{1,3} - u_{2,1}) = f_{1,1} \ or \tag{2.11}$$

$$\frac{1}{h^2}(4u_{1,1} - u_{1,3} - u_{2,1}) = \tilde{f}_{1,1} \tag{2.12}$$

with a new right hand side $\tilde{f}_{1,1} = f_{1,1} + \frac{1}{h^2}(g_{0,1} + g_{1,0})$. Ordering the equations lexicographically by lines of constant i allows the representation of the discrete linear problem (2.9) by the matrix equation $\mathbf{Au} = \mathbf{f}$ with

$$(2.13)$$

The solution vector $\mathbf{u}$ is

$$(u_{0,0} \cdots u_{0,4} \, u_{1,0} \cdots u_{1,4} \, u_{2,0} \, u_{2,1} \, u_{2,2} \, u_{2,3} \, u_{2,4} \, u_{3,0} \cdots u_{3,4} \, u_{4,0} \cdots u_{4,4})^T$$

and the right hand side $\mathbf{f}$ is

$$\left(\frac{1}{h^2}g_{0,0} \cdots \frac{1}{h^2}g_{0,4} \cdots \frac{1}{h^2}g_{2,0} \, f_{2,1} \, f_{2,2} \, f_{2,3} \frac{1}{h^2}g_{2,4} \cdots \frac{1}{h^2}g_{4,0} \cdots \frac{1}{h^2}g_{4,4}\right)^T$$

Performing the above sketched process to eliminate the $g_{i,j}$ and keeping in mind the boundary conditions introduces a "new" right hand side $\tilde{\mathbf{f}}$ of the discrete problem and reduces the linear system to $\tilde{\mathbf{A}}\tilde{\mathbf{u}} = \tilde{\mathbf{f}}$ with

$$\tilde{\mathbf{A}} = \frac{1}{h^2}\begin{pmatrix} 4 & -1 & & -1 & & & & & \\ -1 & 4 & -1 & & -1 & & & & \\ & -1 & 4 & & & -1 & & & \\ -1 & & & 4 & -1 & & -1 & & \\ & -1 & & -1 & 4 & -1 & & -1 & \\ & & -1 & & -1 & 4 & & & -1 \\ & & & -1 & & & 4 & -1 & \\ & & & & -1 & & -1 & 4 & -1 \\ & & & & & -1 & & -1 & 4 \end{pmatrix} \tag{2.14}$$

and $\tilde{\mathbf{f}} = (\tilde{f}_{1,1}\ \tilde{f}_{1,2}\ \tilde{f}_{1,3}\ \tilde{f}_{2,1}\ \tilde{f}_{2,2}\ \tilde{f}_{1,3}\ \tilde{f}_{3,1}\ \tilde{f}_{3,2}\ \tilde{f}_{1,3})^T$. The solution is
$\tilde{\mathbf{u}} = (u_{1,1}\ u_{1,2}\ u_{1,3}\ u_{2,1}\ u_{2,2}\ u_{2,3}\ u_{3,1}\ u_{3,2}\ u_{3,3})^T$.

With column vectors $\mathbf{u}_i := (u_{i,1}\ u_{i,2}\ u_{i,3})^T$ and $\mathbf{f}_i := (f_{i,1}\ f_{i,2}\ f_{i,3})^T$ the discrete system with eliminated boundary conditions is equivalent to the 3×3 block-tridiagonal system

$$\begin{pmatrix} \mathcal{A} & -\mathcal{I} & \\ -\mathcal{I} & \mathcal{A} & -\mathcal{I} \\ & -\mathcal{I} & -\mathcal{A} \end{pmatrix}\begin{pmatrix} \mathbf{u}_1 \\ \mathbf{u}_2 \\ \mathbf{u}_3 \end{pmatrix} = \begin{pmatrix} \mathbf{f}_1 \\ \mathbf{f}_2 \\ \mathbf{f}_3 \end{pmatrix} \tag{2.15}$$

with

$$\mathcal{A} = \begin{pmatrix} 4 & -1 & \\ -1 & 4 & -1 \\ & -1 & 4 \end{pmatrix} \quad and \quad \mathcal{I} = \begin{pmatrix} 1 & & \\ & 1 & \\ & & 1 \end{pmatrix}.$$

If the discretization of the above model problem (2.8) uses a finer meshsize $h = (h_x, h_y)$, corresponding to larger N_x and N_y, the matrix of the problem without eliminated boundary conditions has $((N_x + 1)(N_y + 1))^2$ entries. Eliminating the boundary conditions ends with a $(N_x - 1) \times (N_x - 1)$-block tridiagonal matrix $\tilde{\mathbf{A}}$. $\mathcal{A}$ is a $(N_y - 1) \times (N_y - 1)$-tridiagonal matrix and $\mathcal{I}$ is the corresponding identity matrix. The matrix representation (2.14) is often referred to as model problem (2.8) with "eliminated boundary condition" and is formally written as

$$-\Delta_h u_h = f_h \tag{2.16}$$

with the grid operator Δ_h.

The above sketched elimination process is often used to justify the reformulation of (2.7) to

$$L_h u_h = f_h \text{ including the eliminated boundary conditions.} \qquad (2.17)$$

The possibility of such an elimination for other types of boundary conditions is demonstrated in Section 4.2 for homogeneous Neumann boundary conditions.

The discretization of general elliptic problems either leads to matrix equations (2.13) or (2.14) which correspond to the formal descriptions (2.7) and (2.17), alternatively. The description in terms of matrices is not very pleasing, while the terminology of grid operators and grid functions is easy to use. From the practical point of view, the representation of discrete operators L_h with the aid of "difference stars" or "stencils" is helpful. A more pronounced form to write the left hand side of (2.9) is

$$\frac{1}{h^2}(- 1u(x - 1h, y - 0h) - 1u(x - 0h, y - 1h) + 4u(x - 0h, y - 0h)$$

$$- 1u(x + 0h, y + 1h) - 1u(x + 1h, y + 0h)).$$

Looking at Figure 2.1, where those unknowns which contribute to the equation for $u_{i,j}$ ($\square$) are marked by $\circ$, suggests to replace this expression by

$$\frac{1}{h^2} \begin{bmatrix} & -1_{0,1} & \\ -1_{-1,0} & 4_{0,0} & -1_{1,0} \\ & -1_{0,-1} & \end{bmatrix}_h u(x, y)$$

which is the stencil corresponding to the second-order difference approximation (2.9) of $-\Delta u$.

Remark 2.1. *A general difference formula at a certain grid point* $(x, y) \in G_h$, $h = (h_x, h_y)$, *is*

$$L_h u_h(x, y) = \sum_{\substack{\kappa \in \mathbb{I} \times \mathbb{I} \\ \kappa = (\kappa_1, \kappa_2)}} s_\kappa u_h(x + \kappa_1 h_x, y + \kappa_2 h_y)$$

where $\mathbb{I}$ *is an index set* $\mathbb{I} \subset \mathbb{Z}$ *and* $0 \in \mathbb{I}$. *The coefficients* s_κ *are the entries of the stencil which is, for* $L_h u_h$

$$[s_\kappa]_h u_h(x, y) := \begin{bmatrix} & & \vdots & & \\ & s_{-1,1} & s_{0,1} & s_{1,1} & \\ \cdots & s_{-1,0} & s_{0,0} & s_{1,0} & \cdots \\ & s_{-1,-1} & s_{0,-1} & s_{1,-1} & \\ & & \vdots & & \end{bmatrix}_h u_h(x, y). \qquad (2.18)$$

The number of non-vanishing entries in $[s_\kappa]_h$ *depends on the operator L and its discretization. They additionally depend on the space position* (x, y) *and for*

nonlinear operators L on the solution, too. The index h marks the fact that the
discrete operator which is identified with the stencil is applied on a grid with
meshsize h and that the entries usually depend on h. If the actually used h is
obvious, the index is omitted.

The discretization of second order differential operators using standard
second order finite difference formula usually leads to difference equations
which can be represented by compact nine-point or five-point stencils:

$$
\begin{bmatrix} s_{-1,1} & s_{0,1} & s_{1,1} \\ s_{-1,0} & s_{0,0} & s_{1,0} \\ s_{-1,-1} & s_{0,-1} & s_{1,-1} \end{bmatrix}_h
\qquad
\begin{bmatrix} & s_{0,1} & \\ s_{-1,0} & s_{0,0} & s_{1,0} \\ & s_{0,-1} & \end{bmatrix}_h
\tag{2.19}
$$

The stencil for the operator L of the model problem (2.2) after second order
approximation on a quadratic grid is

$$
\frac{1}{h^2}\begin{bmatrix} & -b & \\ -a & 2(a+b)+h^2c & -a \\ & -b & \end{bmatrix}.
\tag{2.20}
$$

The special selection of a, b and c automatically provides the stencils for the
discrete counterparts of the differential operators (2.3)–(2.5):

$$
-\Delta:\ \frac{1}{h^2}\begin{bmatrix} & -1 & \\ -1 & 4 & -1 \\ & -1 & \end{bmatrix}
\tag{2.21}
$$

$$
-\Delta+c:\ \frac{1}{h^2}\begin{bmatrix} & -1 & \\ -1 & 4+ch^2 & -1 \\ & -1 & \end{bmatrix}
\tag{2.22}
$$

$$
-\left(\varepsilon\frac{\partial^2 u}{\partial x^2}+\frac{\partial^2 u}{\partial y^2}\right):\ \frac{1}{h^2}\begin{bmatrix} & -1 & \\ -\varepsilon & 2(\varepsilon+1) & -\varepsilon \\ & -1 & \end{bmatrix}
\tag{2.23}
$$

Due to the elimination of boundary conditions these stencils reduce to four-
and three-point formulas near the edges, respectively near the corners of the
unit square.

Conclusion 2.1. *Up till now several formulations to describe the discrete*
problem have been introduced. They are applicable on non-rectangular grids,
too, and they are extendable to more general problems. To simplify and to unify
the notation, the general discrete boundary value problem is given from now by
$L_h u_h = f_h$ *including the (eliminated) boundary conditions.*

2.1.1 Basic Iterative Schemes

Having set up the discrete problem, the corresponding linear system of equations has to be solved. For the model problem $-\Delta_h u_h = f_h$ with coarse grids (large h) this is not extremely difficult to do and can be done either directly or with the aid of iterative methods. The smaller the meshsize becomes the larger the linear system of equations will be. The resulting matrix is, although very sparse, a matrix of $((N_x - 1)(N_y - 1))^2$ entries after eliminating the boundary conditions, and the number $N_{eq} = (N_x - 1)(N_y - 1)$ of unknowns may be too large for a direct solver. Another argument for iterative approaches is a more general one and is not restricted to a large number of unknowns. Approximating the continuous problem whose solution is the prime task by a discrete one, introduces a certain error, the discretization or truncation error τ_h. With an approximation w_h to the solution of the discrete problem u_h, the total error then consists of the discretization error τ_h plus the algebraic error: $u - w_h = u - u_h + u_h - w_h = \tau_h + \tilde{e}_h$. The question is why to solve the discrete problem exactly when the total error will never be below the discretization error. For practical reasons it is sufficient to solve only upto the level of truncation error—this may be hard enough for fine grids.

A natural choice is to use basic iterative methods like Gauss-Seidel, Jacobi, weighted Jacobi, successive overrelaxation (SOR), alternating directions (ADI) and conjugate gradient (CG) methods. Most of the algorithms are easily formulated, easy to program and do not require a lot of memory. A deep theoretical investigation is left to the standard literature on matrix iterative analysis. Because of their direct relevance for both singlegrid and multigrid methods the two basic relaxation schemes "immediate and simultaneous replacement procedure" together with "weighted" versions of them (weighted Jacobi, SOR) are presented in some details. They provide not only with very effective iterative solvers but also with highly efficient "smoothers" in the multigrid environment. The corresponding generalization to block-iterative techniques is not described formally, but such schemes will be introduced and used later on, because of their practical importance for many multigrid applications.

The following descriptions refer to the linear system of equations

$$\mathbf{Au} = \mathbf{f}.$$

The $N_{eq} \times N_{eq}$-matrix $\mathbf{A} = (a_{ij})_{(N_{eq} \times N_{eq})}$ is splitted into

$$\mathbf{A} = \mathbf{D} - \mathbf{L} - \mathbf{U}$$

with $\mathbf{D}$ the diagonal part of $\mathbf{A}$, while $\mathbf{L}$, respectively $\mathbf{U}$ are the strictly lower and upper nondiagonal part of $\mathbf{A}$ with reverse sign.

Then $\mathbf{Au} = \mathbf{f}$ becomes

$$(\mathbf{D} - \mathbf{L} - \mathbf{U})\mathbf{u} = \mathbf{f} \text{ or}$$

$$\mathbf{Du} = (\mathbf{L} + \mathbf{U})\mathbf{u} + \mathbf{f} \text{ and consequently}$$
$$\mathbf{u} = \mathbf{D}^{-1}(\mathbf{L} + \mathbf{U})\mathbf{u} + \mathbf{D}^{-1}\mathbf{f}.$$

This helps to establish the simultaneous replacement procedure or total step method in vector representation for the $k + 1$-st iteration. Another representation of this algorithm shows in a more practical way how to compute the components of $\mathbf{u}^{(n+1)} = (u_1^{(n+1)}, u_2^{(n+1)}, \ldots, u_{N_{eq}}^{(n+1)})^T$ using only those of the previous iteration step.

Algorithm 2.1. (*Jacobi*)

$$\forall_{n=0,\ldots} \quad \mathbf{u}^{(n+1)} = \mathbf{P}_J \mathbf{u}^{(n)} + \mathbf{D}^{-1}\mathbf{f} \tag{2.24}$$

$$\text{with} \quad \mathbf{P}_J := \mathbf{D}^{-1}(\mathbf{L} + \mathbf{U}) \tag{2.25}$$

$$\forall_{n=0,\ldots} \quad \forall_{i=1}^{i=N_{eq}} \quad u_i^{(n+1)} = -\frac{1}{a_{ii}} \left(\sum_{\substack{j=1 \\ j \neq i}}^{j=N_{eq}} a_{ij} u_j^{(n)} - f_i \right) \tag{2.26}$$

Using exclusively the old vector $\mathbf{u}^{(n)}$ to compute the new one, $\mathbf{u}^{(n+1)}$, and replacing the new approximation as a whole ("simultaneously") when all new components are available is the name giving procedure for this algorithm.

A simple transition to the single step procedure starts with rewriting (2.26) as

$$\forall_{n=0,\ldots} \quad \forall_{i=1}^{i=N_{eq}} \quad u_i^{(n+1)} = -\frac{1}{a_{ii}} \left(\sum_{j=1}^{j=i-1} a_{ij} u_j^{(n)} + \sum_{j=i+1}^{j=N_{eq}} a_{ij} u_j^{(n)} - f_i \right) \tag{2.27}$$

and to replace $\mathbf{u}^{(n)}$ within the first sum by $\mathbf{u}^{(n+1)}$. This procedure to use new components as soon as they are available, replacing them after each single step, is the origin to name the Gauss-Seidel method also single step or immediate replacement procedure.

The different sums in (2.27) correspond to the matrix product of $\mathbf{L}$ and $\mathbf{U}$ with $\mathbf{u}^{(n)}$, respectively $\mathbf{u}^{(n+1)}$ in the Gauss-Seidel variant. Similar to Algorithm 2.1 a vector formulation is available. With the above splitting of $\mathbf{A}$ the equation to solve is

$$(\mathbf{D} - \mathbf{L} - \mathbf{U})\mathbf{u} = \mathbf{f} \quad \text{or}$$
$$(\mathbf{D} - \mathbf{L})\mathbf{u} = \mathbf{U}\mathbf{u} + \mathbf{f} \quad \text{and}$$
$$\mathbf{u} = (\mathbf{D} - \mathbf{L})^{-1}(\mathbf{U}\mathbf{u} + \mathbf{f}).$$

The complete description of the Gauss-Seidel method both in terms of matrices and in component representation is

Algorithm 2.2. (*Gauss-Seidel*)

$$\forall_{n=0,\ldots} \quad \mathbf{u}^{(n+1)} = \mathbf{P}_{GS} u^{(n)} + (\mathbf{D} - \mathbf{L})^{-1}\mathbf{f} \tag{2.28}$$

$$\text{with} \quad \mathbf{P}_{GS} := (\mathbf{D} - \mathbf{L})^{-1}\mathbf{U} \tag{2.29}$$

$$\forall_{n=0,\dots} \ \forall_{i=1}^{i=N_{eq}} \ u_i^{(n+1)} = -\frac{1}{a_{ii}} \left(\sum_{j=1}^{j=i-1} a_{ij}u_j^{(n+1)} + \sum_{j=i+1}^{j=N_{eq}} a_{ij}u_j^{(n)} - f_i \right) \tag{2.30}$$

To complete the basic collection of iterative methods so-called weighted variants of the above algorithms have to be mentioned. Having a new (intermediate) iterate of the Jacobi method

$$\mathbf{D}\tilde{\mathbf{u}}^{(n+1)} = (\mathbf{L} + \mathbf{U})\mathbf{u}^{(n)} + \mathbf{f}$$

the corresponding correction of $\mathbf{u}^{(n)}$ is

$$\tilde{\mathbf{u}}^{(n+1)} - \mathbf{u}^{(n)} = (\mathbf{D}^{-1}(\mathbf{L} + \mathbf{U}) - \mathbf{I})\mathbf{u}^{(n)} + \mathbf{D}^{-1}\mathbf{f}.$$

Instead of using this correction to get $\mathbf{u}^{(n+1)}$ the really added term is weighted by a factor $\omega > 0$, the relaxation parameter, so leading to

$$\mathbf{D}\mathbf{u}^{(n+1)} = \mathbf{D}\mathbf{u}^{(n)} + \omega((\mathbf{L} + \mathbf{U}) - \mathbf{D})\mathbf{u}^{(n)} + \omega\mathbf{f}$$
$$= (\omega(\mathbf{L} + \mathbf{U}) + (1 - \omega)\mathbf{D})\mathbf{u}^{(n)} + \omega\mathbf{f}.$$

The weighted or damped Jacobi-method then is the following algorithm.

Algorithm 2.3. (*Weighted Jacobi*)

$$\forall_{n=0,\dots} \ \mathbf{u}^{(n+1)} = \mathbf{P}_J(\omega)\mathbf{u}^{(n)} + \omega\mathbf{D}^{-1}\mathbf{f} \tag{2.31}$$

$$\text{with} \quad \mathbf{P}_J(\omega) := \omega\mathbf{D}^{-1}(\mathbf{L} + \mathbf{U}) + (1 - \omega)\mathbf{I} \tag{2.32}$$

$$\forall_{n=0,\dots} \ \forall_{i=1}^{i=N_{eq}} \ u_i^{(n+1)} = u_i^{(n)} - \frac{\omega}{a_{ii}} \left(\sum_{\substack{j=1 \\ j \neq i}}^{j=N_{eq}} a_{ij}u_j^{(n)} - f_i \right) \tag{2.33}$$

Analogous steps applied to the Gauss-Seidel method establish

$$(\mathbf{D} - \omega\mathbf{L})\mathbf{u}^{(n+1)} = ((1 - \omega)\mathbf{D} + \omega\mathbf{U})\mathbf{u}^{(n)} + \omega\mathbf{f} \quad \text{or}$$
$$\mathbf{u}^{(n+1)} = (\mathbf{D} - \omega\mathbf{L})^{-1}((1 - \omega)\mathbf{D} + \omega\mathbf{U})\mathbf{u}^{(n)} + \omega(\mathbf{D} - \omega\mathbf{L})^{-1}\mathbf{f}$$

which defines the successive overrelaxation method (SOR). Because the above description is somewhat long-winded and SOR is an often used iterative solver, the description in component form is added:

Algorithm 2.4. (*SOR*)

$$\forall_{n=0,\dots} \ \mathbf{u}^{(n+1)} = \mathbf{P}_{SOR}(\omega)\mathbf{u}^{(n)} + \omega(\mathbf{D} - \omega\mathbf{L})^{-1}\mathbf{f} \tag{2.34}$$

$$\text{with} \quad \mathbf{P}_{SOR}(\omega) := (\mathbf{D} - \omega\mathbf{L})^{-1}((1 - \omega)\mathbf{D} + \omega\mathbf{U}) \tag{2.35}$$

$$\forall_{n=0,\dots} \ \forall_{i=1}^{i=N_{eq}} \ u_i^{(n+1)} = u_i^{(n)} - \frac{\omega}{a_{ii}} \left(\sum_{j=1}^{j=i-1} a_{ij}u_j^{(n+1)} + \sum_{j=i}^{j=N_{eq}} a_{ij}u_j^{(n)} - f_i \right) \tag{2.36}$$

Example 2.2. *Applying the Gauss-Seidel method to the system* $\tilde{A}\tilde{u} = \tilde{f}$ *of Example 2.1 leads to the iteration for each unknown due to*

$$\tilde{u}_{i,j}^{(n+1)} = \tfrac{1}{4}(h^2\tilde{f}_{i,j} + \tilde{u}_{i-1,j}^{(n+1)} + \tilde{u}_{i,j-1}^{(n+1)} + \tilde{u}_{i,j+1}^{(n)} + \tilde{u}_{i+1,j}^{(n)})$$

for all $1 \le i,j \le N-1$.

Some remarks on the above point oriented schemes conclude the abstract description of the iteration procedures.

Remark 2.2.

1. *The algorithms are well-defined if* $\forall_{i=1}^{i=N_{eq}}\ a_{ii} \neq 0$.
2. *The Jacobi method requires additional storage for the new solution* $\mathbf{u}^{(n+1)}$ *because the components are not immediately updated. The method does not depend on the ordering of the index. This allows both an efficient vectorization and parallelization.*
3. *Gauss-Seidel relaxation immediately replaces the components of the solution vector and therefore needs no memory for intermediate results. This method depends on the ordering of points (comparing for instance properties of the Gauss-Seidel scheme with lexicographic and red-black ordering of points, respectively). A different ordering of points in combination with Gauss-Seidel improves the capabilities of parallelization. For the discrete model problem, the red-black ordering of points allows an update of black points in any order, because only red points are necessary to determine the new black point values: the calculation of any black point value is independent from any other black point value.*
4. *The Gauss-Seidel iteration of Algorithm 2.2 is equivalent to*

$$\forall_{n=0,\dots}\ \forall_{i=1}^{i=N_{eq}}\ u_i^{(n+1)} = u_i^{(n)} - \frac{1}{a_{ii}}\left(\sum_{j=1}^{j=i-1} a_{ij}u_j^{(n+1)} + \sum_{j=i}^{j=N_{eq}} a_{ij}u_j^{(n)} - f_i\right),$$

$$(2.37)$$

   which shows the similarity to SOR.
5. *The weighted methods with* $0 < \omega < 1$ *are called damped or underrelaxed. For* $\omega = 1$ *they coincide with the original scheme and in case of* $\omega > 1$ *they are called extrapolated or overrelaxed.*
6. *Both the Jacobi- and the Gauss-Seidel scheme simply resolve the i-th equation for the unknown* u_i. *The difference origins in the moment when already computed values are used for further calculations: Gauss-Seidel incorporates new values as soon as possible, while the Jacobi scheme waits until all values of the n + 1-th step are available. One expects Gauss-Seidel to converge faster because new information is transported faster.*

2.1.2 Convergence Analysis in Practice

One disadvantage of the above schemes is their poor convergence. A first impression thereof is given in Figure 2.2 with the aid of the convergence history both for the residual and the global error of the model problem (2.8) with certain f and g on a square grid with $h = \frac{1}{128}$ and $h = \frac{1}{256}$. The Gauss-Seidel method is applied four hundred times. This figure motivates the discussion of some important quantities and equations which have to be kept in mind for the consistent development and presentation of the multigrid principle. The discretization error τ_h of the chosen example is neglectable and therefore the error $u - w_h$ is approximately given by the algebraic error $\tilde{e}_h$ (2.39).

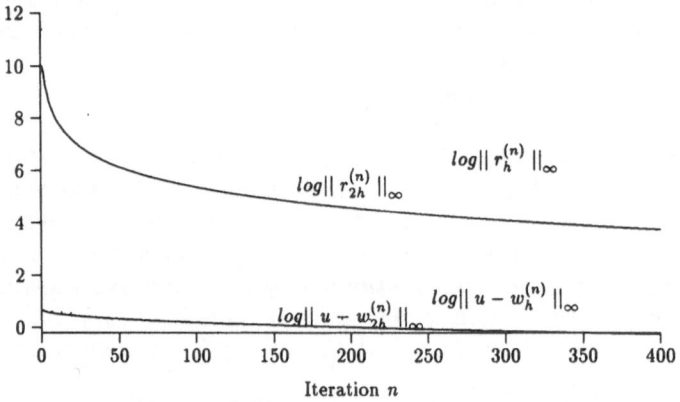

Fig. 2.2 Convergence history of residuals and errors, $h = \frac{1}{256}$

Using the notation with eliminated boundary conditions the exact solution u_h of the discrete problem satisfies

$$L_h u_h = f_h. \tag{2.38}$$

Unfortunately, the exact solution is not available, but in general there is an approximation w_h. This approximate solution usually depends on the number of iterations to obtain it: $w_h = w_h^{(n)}$. The quality of this approximation to the solution u_h is of interest and the most natural measure is the algebraic error

$$\tilde{e}_h = u_h - w_h, \quad \text{or} \quad \tilde{e}_h^{(n)} = u_h - w_h^{(n)} \tag{2.39}$$

after n iterations of the applied iterative method. An approximation to u_h will not satisfy (2.38) exactly. The failure is the residual or the defect

$$r_h := f_h - L_h w_h, \quad \text{respectively} \quad r_h^{(n)} := f_h - L_h w_h^{(n)}. \tag{2.40}$$

All these quantities are vectors or grid functions, too, and are measured in terms of appropriately defined discrete norms.

Rearranging (2.39) to $w_h = u_h - \tilde{e}_h$ and inserting this into (2.40) leads to

$$r_h = f_h - L_h w_h = f_h - L_h(u_h - \tilde{e}_h) = f_h - L_h u_h + L_h \tilde{e}_h.$$

In combination with (2.38) this establishes the defect equation

$$L_h \tilde{e}_h = r_h \tag{2.41}$$

which is an important relationship between $\tilde{e}_h$ and r_h. It is important to point out that the linearity of L_h is necessary to derive (2.41).

Remark 2.3. *Assuming the invertability of L_h and using appropriate norms produces an inequality*

$$\| \tilde{e}_h \| \leq \| L_h^{-1} \| \, \| r_h \|$$

which shows a relation "upto a factor" between residual and error. This is one effect which one might guess with the aid of Figure 2.2. The assumed factor will certainly depend on h but may be bounded by a constant, especially for $h \to 0$. A detailed discussion would lead to stability investigations which are omitted here. Another observation is the initially good reduction of both residual and error norms which becomes worse with an increasing number of iterations. Both degrade as h tends to zero (compare the solid and dotted lines for 2h and h in Figure 2.2). It is especially this fact which demands for multigrid by practice. The theoretical investigation and the explanation automatically establishes tools like the model problem analysis and the smoothing analysis.
Comparing the equations (2.38) and (2.41) tells that the solution u_h with right hand side f_h satisfies the same system of linear equations as the error $\tilde{e}_h$ with the right hand side r_h does.
Assuming a unique solution, the error is equal to zero if and only if the residual is zero, too: $\tilde{e}_h = 0 \Leftrightarrow r_h = 0$.

The connection between residual and error raises the idea to interprete an iterative process to solve (2.38) as an iterative method to solve (2.41). With an approximation $w_h^{(0)}$ to u_h, the defect is given by $r_h^{(0)} = f_h - L_h w_h^{(0)}$. Solving the corresponding defect equation

$$L_h \tilde{e}_h^{(0)} = r_h^{(0)}$$

approximately, can be interpreted to solve a "similar" problem

$$\hat{L}_h \widehat{\tilde{e}_h^{(0)}} = r_h^{(0)}$$

with an operator $\hat{L}_h$ similar to L_h. Thus assuming $\hat{L}_h^{-1}$ to exist, the solution

$$\widehat{\tilde{e}_h^{(0)}} = \hat{L}_h^{-1} r_h^{(0)} \tag{2.42}$$

is used to improve the first approximation $w_h^{(0)}$, giving a better approximation

$$w_h^{(1)} = w_h^{(0)} + \widehat{\tilde{e}_h^{(0)}}. \tag{2.43}$$

This procedure is called residual correction because of (2.42) and (2.43). Independent from the actually used approximate solver $\hat{L}_h^{-1}$ this defines an iteration method to solve the original problem

$$w_h^{(1)} = w_h^{(0)} + \widehat{\tilde{e}_h^{(0)}} = w_h^{(0)} + \hat{L}_h^{-1} r_h^{(0)} = w_h^{(0)} + \hat{L}_h^{-1} f_h - \hat{L}_h^{-1} L_h w_h^{(0)}$$

$$= (Id_h - \hat{L}_h^{-1} L_h) w_h^{(0)} + \hat{L}_h^{-1} f_h \tag{2.44}$$

with an operator $M_h := Id_h - \hat{L}_h^{-1} L_h$ where Id_h stands for the identity operator. Applying this scheme iteratively, it is not more than an exercise to show

$$\tilde{e}_h^{(n)} = (Id_h - \hat{L}_h^{-1} L_h)^{(n)} \tilde{e}_h^{(0)} \tag{2.45}$$

$$r_h^{(n)} = (Id_h - L_h \hat{L}_h^{-1})^{(n)} r_h^{(0)}. \tag{2.46}$$

The asymptotic convergence property of an iterative method is characterized by the spectral radius of its corresponding operator

$$\rho(M_h) = \max \left\{ |\lambda^h| \,|\, \lambda^h \text{ is eigenvalue of } M_h \right\}$$

which is also called (asymptotic) convergence factor. The iterative process converges if and only if $\rho(M_h)$ is smaller than one. This is by (2.45) equivalent to $\tilde{e}_h^{(n)} \to 0$ as $n \to \infty$ and due to the connection between residual and error this also requires $r_h^{(n)} \to 0$ as $n \to \infty$.

With appropriate norms (2.45) and (2.46) lead to the definition of the error reducing factor and the defect reducing factor per iteration step, respectively:

$$\| Id_h - \hat{L}_h^{-1} L_h \| \quad \text{and} \quad \| Id_h - L_h \hat{L}_h^{-1} \| \tag{2.47}$$

If the norm is chosen to be the spectral norm it is an immediate consequence that both factors have to be smaller than one for convergence. Because of $\rho(M_h) = \lim_{n \to \infty} \| M_h^n \|^{1/n}$ it can be expected to find relations like

$$\| \tilde{e}_h^{(n)} \| \leq \| Id_h - \hat{L}_h^{-1} L_h \| \, \| \tilde{e}_h^{(n-1)} \|$$

with

$$\| Id_h - \hat{L}_h^{-1} L_h \| \approx \rho(Id_h - \hat{L}_h^{-1} L_h) < 1$$

and similar for the defect reduction.

Example 2.3. *The convergence history of Figure 2.2 shows the curves both for error norms and residual norms with an increasing number of iterations. Neglecting the initially strong decrease, the development of error and residual is, after a certain number of iterations, very similar, although still different. Empirical values of the two reduction factors are obtained by the measurement of* $\dfrac{\| \tilde{e}_h^{(n)} \|}{\| \tilde{e}_h^{(n-1)} \|} \approx \rho(Id_h - \hat{L}_h^{-1} L_h)$ *and* $\dfrac{\| r_h^{(n)} \|}{\| r_h^{(n-1)} \|} \approx \rho(Id_h - L_h \hat{L}_h^{-1})$ *with* n *very large.*

Table 2.1. *Error reduction and defect reduction for a model problem*

Gauss-Seidel	error reduction	defect reduction
lexicographically	$\dfrac{\|\tilde{e}_h^{(n)}\|}{\|\tilde{e}_h^{(n-1)}\|}$	$\dfrac{\|r_h^{(n)}\|}{\|r_h^{(n-1)}\|}$
1000 iterations		
$h = \frac{1}{128}$	0.9993	0.9987
$h = \frac{1}{256}$	0.9996	0.9988

For the model problem $-\Delta u = f$ on the unit square with the solution $u(x, y) = x^3 + y^2$ one thousand Gauss-Seidel iterations with lexicographical ordering of the grid points yield very impressive values which are shown in the following table for two different meshsizes.
Although they are in good accordance with theoretical predictions, the error reduction factor as well as the defect reduction factor are still different and have not yet reached those values which can be expected by theory. This is demonstrated later.

With the aid of the asymptotic convergence factor (error reduction factor) an estimation of the iterative work to reduce the error by a given factor is obtained. Starting with an initial error of the size $\|\tilde{e}_h^{(0)}\|$ the number N_0 of iterations is of interest when

$$\|\tilde{e}_h^{(N_0)}\| \le \varepsilon \|\tilde{e}_h^{(0)}\| \quad \text{with} \quad \varepsilon \ll 1$$

is valid first.
In the decimal system ε is chosen to be 10^{-d} where d is interpreted to be the gain of correct digits. Then the number N_0 of iterations to improve d digits of the solution is estimated by

$$N_0 \ge -\frac{d}{\log_{10}\rho(Id_h - \hat{L}_h^{-1}L_h)} \tag{2.48}$$

Example 2.4. *The estimation (2.48) with $d = 1$ applied to the values of Table 2.1 states: the lexicographic Gauss-Seidel iteration for the above model problem needs with $h = \frac{1}{256}$ about 5750 iterations to gain one digit. On a coarser grid with $h = \frac{1}{128}$ approximately 2300 iterations are needed for the same error reduction.*

2.1.3 Convergence Analysis for Gauss-Seidel and Jacobi Methods

The poor convergence of iterative methods as well as its dependence on h has been demonstrated for the discrete model problem $-\Delta_h u_h = f_h$ in the previous section. The theoretical explanation follows in this section.

The convergence properties of iteration schemes depend on the spectral radius of their iteration matrices as they are given, for instance in the Algorithms 2.1, 2.2, 2.3 and 2.4 by

$$\rho(\mathbf{P}_J) = \rho(\mathbf{D}^{-1}(\mathbf{L} + \mathbf{U}))$$
$$\rho(\mathbf{P}_{GS}) = \rho((\mathbf{D} - \mathbf{L})^{-1}\mathbf{U})$$
$$\rho(\mathbf{P}_J(\omega)) = \rho(\omega\mathbf{D}^{-1}(\mathbf{L} + \mathbf{U}) + (1 - \omega)\mathbf{I})$$
$$\rho(\mathbf{P}_{SOR}(\omega)) = \rho((\mathbf{D} - \omega\mathbf{L})^{-1}((1 - \omega)\mathbf{D} + \omega\mathbf{U}))$$

An analytical determination of the absolutely largest eigenvalue of the iteration operator is not always possible, but for the discrete model problem this analysis can be done and offers helpful insight into the discrete problem and to the iterative solution method.

It is a simple calculation to verify that

$$\varphi_{l,m}(x, y) = \sin l\pi x \cdot \sin m\pi y \quad (x, y)\in\Omega \quad \text{and}$$
$$\lambda_{l,m} = (l^2 + m^2) \qquad \text{for all } l, m = 1, 2, \ldots$$

are eigenfunctions, respectively eigenvalues of the Laplacian. More precisely, the above set of functions and values satisfies the eigenvalue problem

$$-\Delta u = \lambda u \quad \text{in} \quad \Omega = (0, 1)^2$$
$$u = 0 \qquad \text{at } \partial\Omega.$$

Analogous to the continuous case the corresponding discrete eigenvalue problem

$$-\Delta_h u_h = \lambda^h u_h \quad \text{in} \quad G_h = (0, 1)^2 \cap \mathscr{G}_h$$
$$u_h = 0 \qquad \text{at } \partial G_h$$

on a square grid with $h = 1/N$ leads to the same eigenfunctions

$$\varphi^h_{l,m}(x, y) = \sin l\pi x \cdot \sin m\pi y \quad (x, y)\in\bar{G}_h \quad \text{or}$$

$$\varphi^h_{l,m}(ih, jh) = \sin l\pi ih \cdot \sin m\pi jh = \sin l\pi\frac{i}{N} \cdot \sin m\pi\frac{j}{N} \tag{2.49}$$

with $0 \le i, j \le N$ and $1 \le l, m \le N - 1$. It is not necessary to consider $l, m \ge N$, because the corresponding eigenfunctions on G_h contribute no more information. Due to the discretization, the eigenvalues $\lambda^h_{l,m}$ to the above eigenfunctions differ from the eigenvalues $\lambda_{l,m}$ of the continuous problem. Exploiting trigonometric formula shows for all $1 \le l, m \le N - 1$

$$\lambda^h_{l,m} = \frac{1}{h^2}[4 - 2\cos l\pi h - 2\cos m\pi h] = \frac{4}{h^2}\left[\sin^2\frac{l\pi h}{2} + \sin^2\frac{m\pi h}{2}\right]$$

Remark 2.4. *For the discrete model problem this means*

1. $\lambda^h_{l,m} > 0$ for all l, m.

2. $\lambda_{min}^h = \lambda_{1,1} = \dfrac{4}{h^2}(1 - \cos \pi h) \approx 2\pi^2 - \frac{1}{6}\pi^4 h^2 + O(h^4)$.

3. $\lambda_{max}^h = \lambda_{N-1,N-1} = \dfrac{4}{h^2}(1 + \cos \pi h) \approx \dfrac{8}{h^2} - 2\pi^2 + \frac{1}{6}\pi^4 h^2 + O(h^4)$.

Comparing $\lambda_{min}^h = \lambda_{1,1}^h$ of the discrete problem with the smallest eigenvalue of the continuous operator shows the consequence of the discretization: the perturbation term $-\frac{1}{6}\pi^4 h^2 + O(h^4)$ is introduced.

Usually, the theoretical analysis of relaxation schemes for general problems is rather complicated and the eigenvalues of the corresponding operators are mostly difficult to determine. But the special structure of the Jacobi scheme, Algorithm 2.1, allows a direct analysis for the model problem. Because of

$$\mathbf{P}_J = \mathbf{D}^{-1}(\mathbf{L} + \mathbf{U}) = \mathbf{D}^{-1}(\mathbf{D} - \mathbf{A}) = \mathbf{I} - \mathbf{D}^{-1}\mathbf{A}$$

the corresponding eigenvalues are

$$\lambda_{l,m}^h(\mathbf{P}_J) = 1 - \frac{h^2}{4}\lambda_{l,m}^h = 1 - (1 - \tfrac{1}{2}\cos l\pi h - \tfrac{1}{2}\cos m\pi h)$$

$$= \tfrac{1}{2}(\cos l\pi h + \cos m\pi h).$$

The absolutely largest eigenvalue is the asymptotic convergence factor of the Jacobi scheme for the model problem.

$$\rho(\mathbf{P}_J) = \cos \pi h = 1 - \tfrac{1}{2}h^2\pi^2 + O(h^4) = 1 - O(h^2)$$

The following remark gives a short summary of valuable results with respect to model problems.

Remark 2.5.

1. *If the matrix $\mathbf{A}$ is diagonaldominant, that means $|a_{ii}| > \sum_{i \neq j}|a_{ij}|\ \forall_{i=1}^{i=N}$ then $\|\mathbf{P}_{GS}\|_\infty \leq \|\mathbf{P}_J\|_\infty < 1$. $\|\ \|_\infty$ is the operatornorm corresponding to the discrete maximum vectornorm.*
2. *If $\mathbf{A} = \mathbf{D} - \mathbf{L} - \mathbf{U}$ is irreducible and $\mathbf{L} + \mathbf{U}$ is positive definite then*

$$either \quad \rho(\mathbf{P}_{GS}) < \rho(\mathbf{P}_J) < 1$$
$$or \quad \rho(\mathbf{P}_{GS}) = \rho(\mathbf{P}_J) = 1$$
$$or \quad \rho(\mathbf{P}_{GS}) > \rho(\mathbf{P}_J) > 1$$

The two first results simply state that "if the Jacobi method converges, the Gauss-Seidel scheme converges faster". A more quantitative description of "faster" is given for matrices $\mathbf{A}$ which have "property A", that means where $\mathbf{A}$ is a block matrix with diagonal diagonal blocks:
3. *If $\mathbf{A}$ has "property A" then $\rho(\mathbf{P}_{GS}) = (\rho(\mathbf{P}_J))^2$.*
In other words: under the above condition, which for the model problem

depends on the ordering of points, the asymptotic convergence of the Gauss-Seidel method is twice that of the Jacobi scheme. For the model problem this leads to $\rho(\mathbf{P}_{GS}) = (\cos \pi h)^2 = 1 - \pi^2 h^2 + O(h^4) = 1 - O(h^2)$.

4. *If $\mathbf{A}$ is symmetric and positive definite then $\rho(\mathbf{P}_{GS}) < 1$.*
   The spectral radii of the damped methods depend on the relaxation parameter ω. It is natural to ask for an optimal ω_{opt} which minimizes $\rho(\mathbf{P}_{SOR}(\omega))$, respectively $\rho(\mathbf{P}_J(\omega))$.

5. *If $\lambda := \rho(\mathbf{P}_J)$ and $-\lambda$ are eigenvalues of $\mathbf{P}_J$, then $\omega = 1$ is optimal in the sense that $\rho(\mathbf{P}_J(1)) = \inf_{\omega > 0} \rho(\mathbf{P}_J(\omega))$.*
   For the model problem this leads to the conclusion that no value $\omega \neq 1$ will improve the convergence. Therefore, using the Jacobi method to solve iteratively, it is not necessary to introduce a relaxation parameter. SOR shows a different behavior.

6. *For any matrix $\mathbf{A}$ and $\omega > 1$ $\rho(\mathbf{P}_{SOR}(\omega)) \geq |\omega - 1|$ holds.*
   Consequently only $\omega \in (0, 2)$ guarantees convergent schemes. A complete description of the ω-dependence is given by the following result (Young's theorem):

7. *For a consistently ordered $\mathbf{A}$ and real valued eigenvalues of $\mathbf{P}_J$ there is*

$$\omega_{opt} = \frac{2}{1 + \sqrt{1 - \rho(\mathbf{P}_J)}} \quad \text{and}$$

$$\rho(\mathbf{P}_{SOR}(\omega_{opt})) = \omega_{opt} - 1 = \left(\frac{\rho(\mathbf{P}_J)}{1 + \sqrt{1 - \rho(\mathbf{P}_J)}} \right)^2.$$

Other values of ω lead to ($\mu := \rho(\mathbf{P}_J) < 1$):

$$\rho(\mathbf{P}_{SOR}(\omega))$$
$$= \begin{cases} 1 - \omega + \frac{1}{2}\omega^2\mu^2 + \omega\mu\sqrt{1 - \omega + \frac{1}{4}\omega^2\mu^2} & \text{for} \quad 0 \leq \omega \leq \omega_{opt} \\ \omega - 1 & \text{for} \quad \omega_{opt} \leq \omega \leq 2 \end{cases}$$

With respect to the model problem the optimal relaxation parameter is

$$\omega_{opt} = \frac{2}{1 + \sqrt{\cos^2 \pi h}} = \frac{2}{1 + \sin \pi h}$$

and so providing with

$$\rho(\mathbf{P}_{SOR}(\omega_{opt})) = \omega_{opt} - 1 = \frac{1 - \sin \pi h}{1 + \sin \pi h} = 1 - \pi h + O(h^2) = 1 - O(h).$$

Obviously $\omega_{opt} \to 2$ and $\rho(\mathbf{P}_{SOR}(\omega_{opt})) \to 1$ as h tends to zero, but the asymptotic behavior with respect to h is improved if SOR is compared to the unweighted method – considered as singlegrid iterative solver.

Remark 2.6. *The above information allows a verification of the data, given in Table 2.1 for the Gauss-Seidel relaxation. Due to point 3 of Remark 2.5 the*

spectral radius for the Jacobi procedure applied to the model problem is
$\rho(\mathbf{P}_J) = \cos \pi h$ *and* $\rho(\mathbf{P}_{GS}) = \cos^2 \pi h$. *With* $h = \frac{1}{128}$ *the asymptotic convergence rate is 0.9994 and the theoretical value is 0.9998 for* $h = \frac{1}{256}$. *These values are in good accordance with those of Table 2.1 after one thousand relaxations.*

2.1.4 Practical Analysis of Iterative Schemes

Similar to the Jacobi scheme the weighted variant allows a direct analysis for the discrete model problem, too. This offers a valuable insight into the structure of both the solution and the error, which is helpful for the further understanding of multigrid details. The analysis is simplified because the exact eigenvalues $\lambda_{l,m}^h(\omega)$ of the iteration matrix $\mathbf{P}_J(\omega)$ are determined with the aid of the known eigenvalues $\lambda_{l,m}^h$ of the discrete operator $-\Delta_h$:

$$\lambda_{l,m}^h(\omega) = 1 - \omega \frac{h^2}{4} \lambda_{l,m}^h = 1 - \omega \left(\sin^2 \frac{l\pi h}{2} + \sin^2 \frac{m\pi h}{2} \right) \qquad (2.50)$$

for all $1 \leq l, m \leq N - 1$. The second reason, and a very important one, too, is the fact that the eigenfunctions of the operators $-\Delta_h$ and $\mathbf{P}_J(\omega)$ coincide and that they are given by (2.49). These eigenfunctions $\varphi_{l,m}^h$ are often referred to as "modes", "Fourier modes" or "frequencies".
For all $0 < \omega \leq 1$ and $1 \leq l, m \leq N - 1$ (2.50) implies $|\lambda_{l,m}^h(\omega)| < 1$. On the other hand, for all ω which guarantee convergence, the eigenvalue

$$\lambda_{1,1}^h(\omega) = 1 - 2\omega \sin^2 \frac{\pi h}{2} \approx 1 - \omega \frac{\pi^2 h^2}{2} \qquad (2.51)$$

is very close to one if h is small. This again shows that no value of ω provides really good convergence of the damped Jacobi method. Especially (2.51) explains the often stated "h-dependent convergence": the smaller h becomes the worse the convergence will be.
With respect to this singlegrid analysis the Fourier modes with wave numbers (l, m) in the range of $1 \leq l, m < N/2$ are called low-frequent or low-frequencies. The others, those with $\max(l, m) \geq N/2$, are high-frequent or high-frequencies. The importance of this distinction becomes transparent in the following. The wave number which represents the frequency is connected to the wave length. A mode l represents $l/2$ full sine waves on the unit interval and consequently is of wave length $2/l$.

Example 2.5. *With* $l = 2$ *there is only one full sine wave on* $(0, 1)$ *and the wave length is one. Similarly* $l = N/2$ *implies a wave length of* $2/l = 4/N = 4h$, *while there are* $N/4$ *full sine waves on the unit interval. In two dimensions one has to consider the product of two one-dimensional modes with wave numbers* l *and* m. *For instance, the 2D-frequency* $\varphi_{l,m}^h$ *with* $(l, m) = (1, 6)$ *is the product function of* *those 1D-modes which are represented by a full sine wave into x-direction and three full sine waves into y-direction.*

One important phenomenon of Fourier modes is the "aliasing" effect which is often meant by saying that a frequency is "visible" or not. Especially frequencies with $2N > l$, $m \geq N$ cannot be represented on the h-grid. Their wave lengths are smaller than $2h$ and they coincide with frequencies corresponding to the wave numbers $(2N - l, 2N - m)$. The explanation is the simple trigonometric identity

$$\sin((2N - l)\pi x) = -\sin l\pi x$$

for all h-grid places x. In other words: h-grid modes with wave length smaller than $2h$ are on the h-grid identified with frequencies whose wave lengths are larger than $2h$. The denotation of visibility will be used again in connection with two consecutive grids and meshizes h and H, respectively. If $H = 2h$ is chosen, the above definition of "high" and "low" remains correct. For other ratios H/h it has to be modified [12] slightly.

Example 2.6. *Figure 2.3 explains the aliasing for a one-dimensional frequency with $N = 8, l_1 = N + \tilde{l} = 13$ and $2N - l_1 = 3$.*

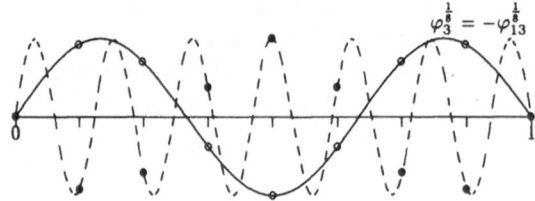

Fig. 2.3 Aliasing of 1D-Fourier modes, $N = 8, l_1 = 13$ and $l_2 = 2N - l_1$

Because the eigenfunctions $\varphi_{l,m}^h$ of the iteration operator build a basis, any h-grid function is a linear combination of just these $\varphi_{l,m}^h$. Especially the initial error $\tilde{e}_h^{(0)}$ may be given by the series

$$\tilde{e}_h^{(0)} = \sum_{l,m=1}^{l,m=N-1} c_{l,m} \varphi_{l,m}^h$$

with some coefficients $c_{l,m}$. For an iterative method like the Jacobi scheme (2.45) holds and the error $\tilde{e}_h^{(n)}$ after n relaxation sweeps is related to the initial error by

$$\tilde{e}_h^{(n)} = (\mathbf{P}_J(\omega))^n \tilde{e}_h^{(0)}.$$

Inserting the eigenfunction expansion of $\tilde{e}_h^{(0)}$ this shows

$$\tilde{e}_h^{(n)} = \sum_{l,m=1}^{l,m=N-1} c_{l,m}(\mathbf{P}_J(\omega))^n \varphi_{l,m}^h = \sum_{l,m=1}^{l,m=N-1} c_{l,m}(\lambda_{l,m}^h(\omega))^n \varphi_{l,m}^h. \qquad (2.52)$$

The last equality essentially depends on the identity of eigenfunctions of both $\mathbf{P}_J(\omega)$ and $-\Delta_h$. It states an eigenvector expansion of $\tilde{e}_h^{(n)}$ with coefficients which are the damped coefficients $c_{l,m}$, where the amplification factor is the n-th power of the corresponding eigenvalue. It is important to note that there are no modes mixed by the Jacobi method. This simply means that the application of $\mathbf{P}_J(\omega)$ to a single mode only damps the amplitude of exactly that component and does not introduce new modes.

Remark 2.7. *Especially for the lexicographic Gauss-Seidel relaxation, but also for other schemes, the relaxation operator has eigenfunctions which differ from those of the difference operator. Using the eigenfunctions of the discrete operator as a basis therefore mixes modes and leads to rather complicated representations. The error $\tilde{e}_h^{(n)}$ then, in general, is expressed by terms like*

$$\tilde{e}_h^{(n)} = \sum_{l,m=1}^{l,m=N-1} a_{l,m} \mathbf{P}(\omega)^n \varphi_{l,m}^h = \sum_{l,m=1}^{l,m=N-1} a_{l,m} \left(\sum_{\kappa,l=1}^{\kappa,l=N-1} b_{\kappa,l} \varphi_{\kappa,l}^h \right).$$

But in principle, the basic relaxation methods behave similar on the set of frequencies. This is demonstrated by the following experiments. Selected frequencies with wave numbers (l, m) are used as initial guesses to show how the basic iterative schemes for $L_h = -\Delta_h$ and homogeneous Dirichlet boundary conditions act on them.

The Fourier modes $\varphi_{l,m}^h$ with $(l, m) \in \{(1, 1); (3, 3); (6, 6)\}$ on a grid with $h = \frac{1}{128}$ show, that the error norm after each relaxation sweep decreases the faster the higher the wave numbers are, that means the more oscillatory the modes are. This is true and demonstrated in Figure 2.4 both for the weighted Jacobi method with $\omega = 0.8$ and for the Gauss-Seidel relaxation with either lexicographic and red-black ordering of points (compare (2.50) and (2.52) for the weighted Jacobi method). The choice of $\omega = 0.8$ is justified below, see also Section 2.10.

In realistic situations, the initial guess will not consist of just a single mode but it will be a combination of several frequencies. Performing the above experiment with an initial guess composed of $\varphi_{1,1}^h$, $\varphi_{3,3}^h$ and $\varphi_{64,64}^h$ shows a somewhat different result. For the first sweeps of an initial relaxation phase an extreme reduction of the error is observed. This situation is similar to that of Figure 2.2, but more pronounced. The quantitative behavior of the weighted Jacobi ($\omega = 0.8$) and the Gauss-Seidel relaxation is plotted in Figure 2.5.

After the initial phase with the rapid error decrease, the error reduction becomes worse because the high-frequent components of the error are damped out and the iteration schemes have to work hard for a further reduction of the error, now being dominated by the low-frequent components. This observation is, for the weighted Jacobi method, explained again by (2.50) and (2.52).

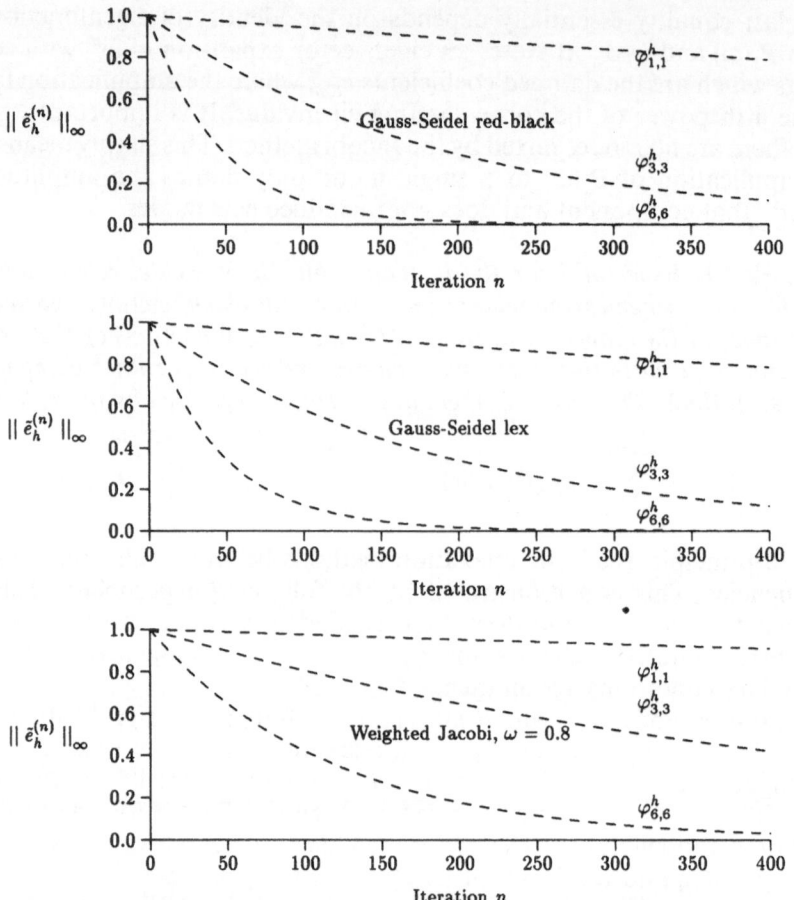

Fig. 2.4 Convergence history of initial guesses $\varphi_{1,1}^h$, $\varphi_{3,3}^h$ and $\varphi_{6,6}^h$

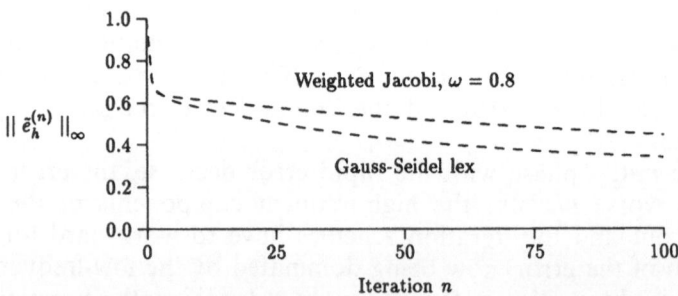

Fig. 2.5 Convergence history, initial guess $\frac{1}{3}(\varphi_{1,1}^h + \varphi_{6,6}^h + \varphi_{64,64}^h)$

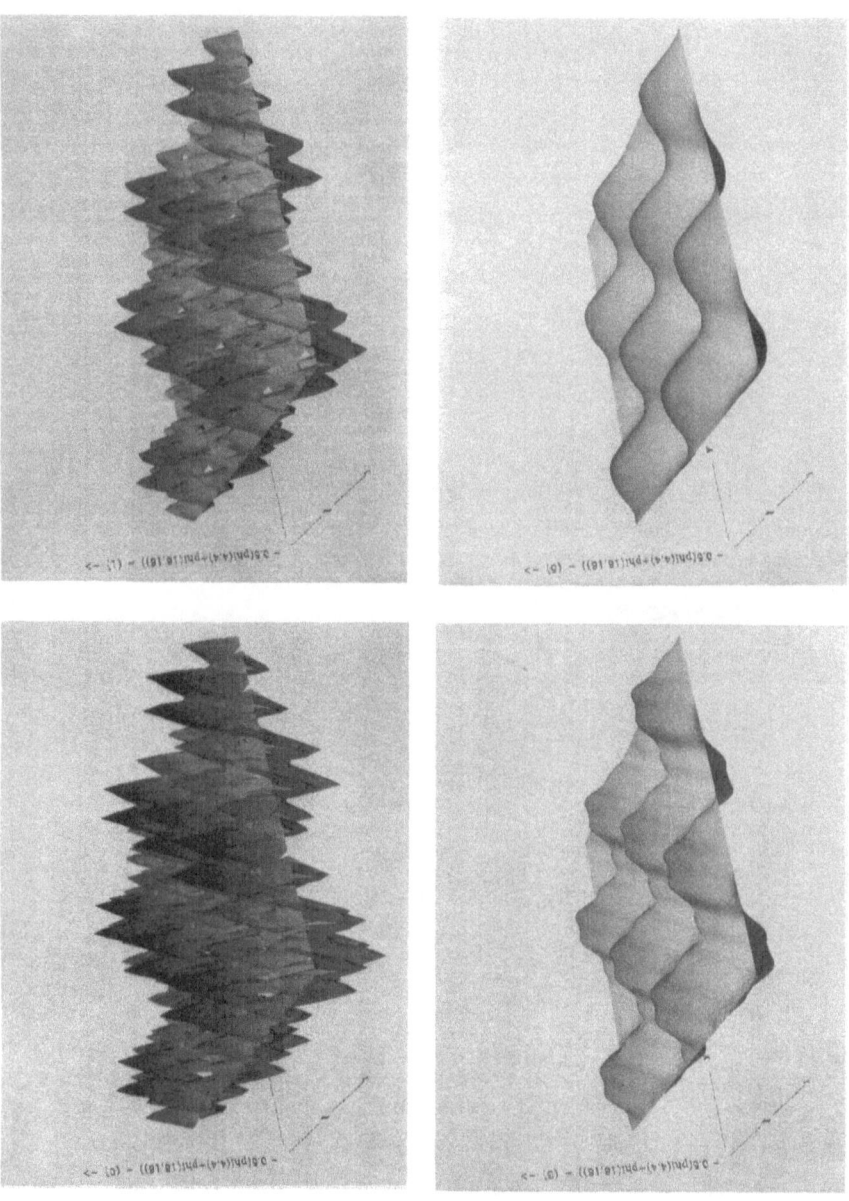

Fig. 2.6 Initial phase of relaxation, initial guess $\frac{1}{2}(\varphi_{4,4}^h + \varphi_{16,16}^h)$

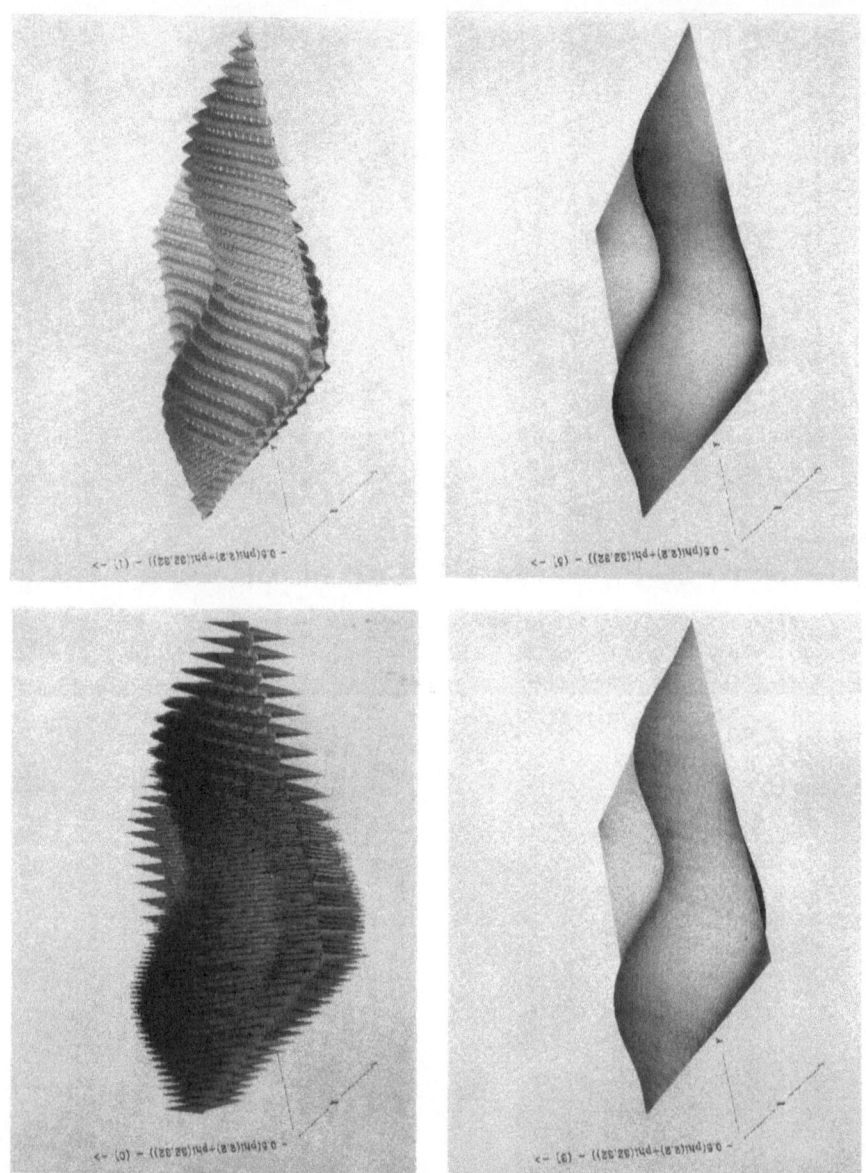

Fig. 2.7 Initial phase of relaxation, initial guess $\frac{1}{2}(\varphi^h_{2,2} + \varphi^h_{32,32})$

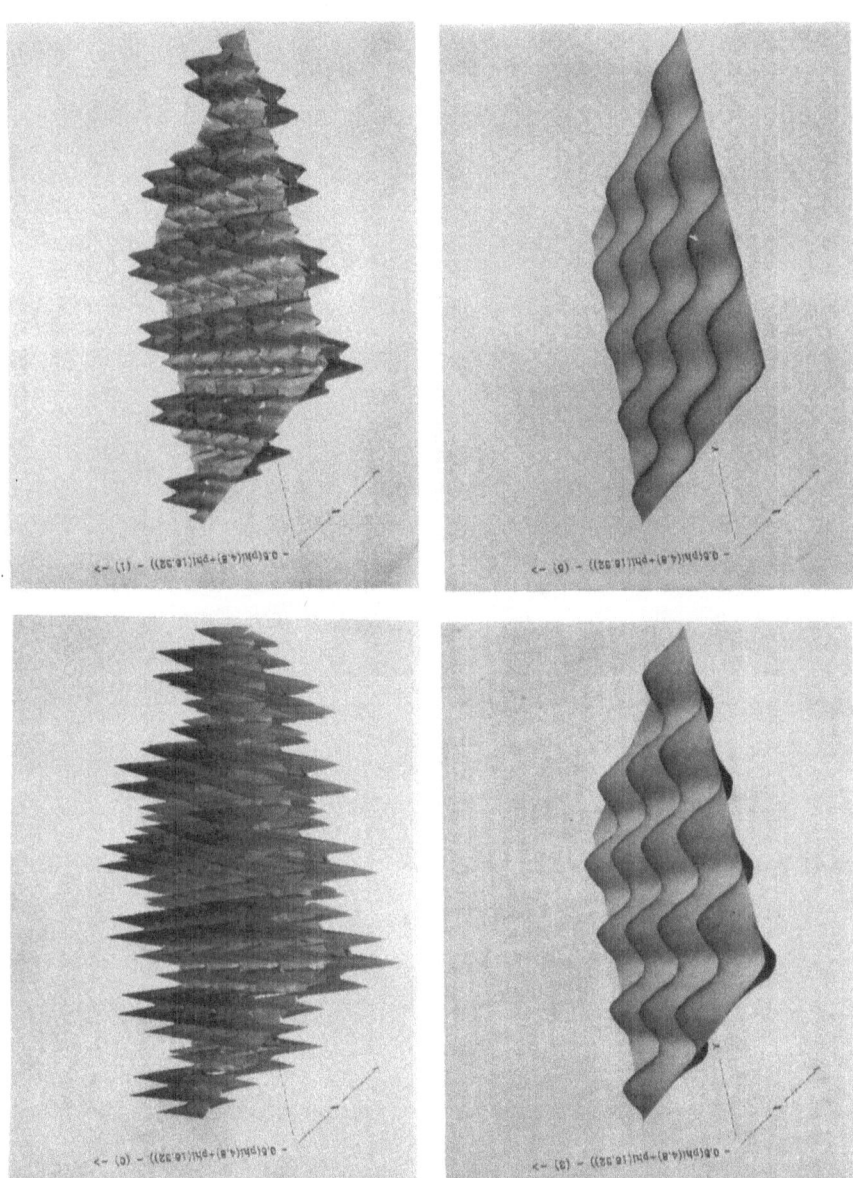

Fig. 2.8 Initial phase of relaxation, initial guess $\frac{1}{2}(\varphi^h_{4,8} + \varphi^h_{16,32})$

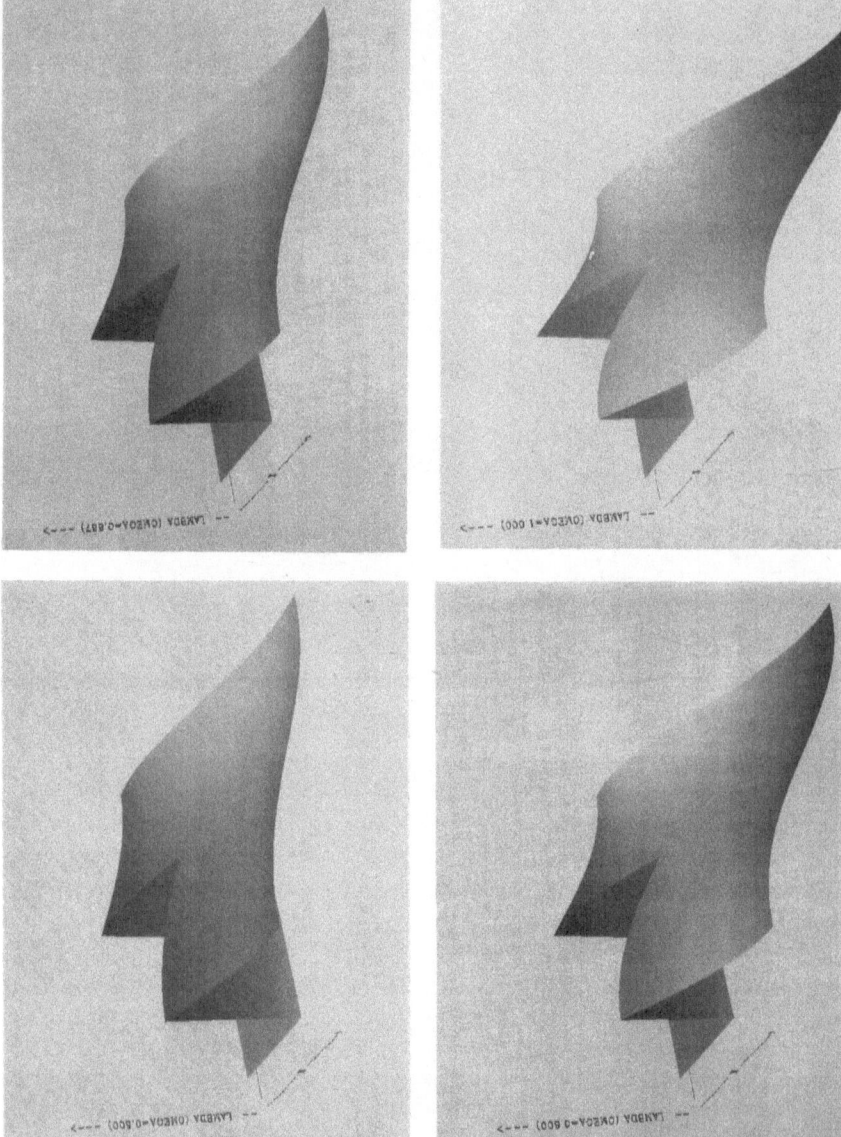

Fig. 2.9 Eigenvalues $\lambda_{l,m}^h(\omega)$ of the weighted Jacobi method for $\omega \in \{0.5, 0.67, 0.8, 1.0\}$ for high-frequent modes

Conclusion 2.2. *The bad convergence properties of the iteration schemes depend on the bad damping of the Fourier modes with small wave numbers (l, m), the low frequencies. This is expressed very impressively by the pictures of Figures 2.6–2.8 which present the initial convergence phase of the lexicographic Gauss-Seidel relaxation for $L_h = -\Delta_h$ applied to different initial guesses on a grid with $h = \frac{1}{64}$. Starting with an initial guess, the result after one, three and five iteration sweeps are shown. All the conclusions, derived by the analysis of the weighted Jacobi scheme can be seen, in principle, for the Gauss-Seidel method, too. The higher the frequencies the faster they disappear while lower frequencies are reduced extremely slow.*

From the above result that low-frequent components of the error are not damped satisfactorily by the ω-Jacobi-iteration for any value of ω, it is natural to ask for that value of ω which provides the best damping of all high-frequent modes, that means for all $\varphi_{l,m}^h$ with $N/2 \leq \max(l, m) \leq N - 1$. The question then is to find

$$
\min_{0 \leq \omega \leq 1} \left(\max_{N/2 \leq \max(l,m) \leq N - 1} |\lambda_{l,m}^h(\omega)| \right)
$$

and to look for an upper bound which should be independent from the mesh-size h and smaller than one. Using the equivalent representation of the eigenvalues $\lambda_{l,m}^h(\omega) = 1 - \omega/2(2 - \cos l\pi h - \cos m\pi h)$ it is shown in [132] that $\omega = 0.8$ is the optimal choice. This value gives a damping factor of 0.6 for all high-frequent Fourier modes and for all small h.

Conclusion 2.3. *In contrast to the convergence of the weighted Jacobi method where $\omega = 1$ is the best choice, the optimal damping of all high-frequent modes is guaranteed by the choice $\omega = 0.8$. Different values of ω work out the ω-dependence of the damping property well (Figure 2.9). To conclude this analysis of the weighted Jacobi iteration method with respect to the model problem, it turns out that this procedure has the smoothing property: the high-frequent modes are damped out very fast but the smooth components converge extremely slow. Many iterative schemes possess this smoothing property, too.*

2.2 The Multigrid Principle

The first basic observation which leads to multigrid—the smoothing property of relaxation schemes—has been explained previously for basic iterative schemes. It is worth to recall that many relaxation schemes—and not only the presented basic ones—possess this smoothing property. This means that many relaxation methods reduce high-frequent modes effectively but let the smooth frequencies change slowly. Additionally, the damping of high-frequent modes does in many cases not depend on the meshsize h.

The h-dependent convergence factor, for instance $\rho = 1 - O(h^2)$, would naturally lead to the idea of using an $H > h$ to improve the asymptotic convergence factor in order to decrease the numerical cost of the computation. But one has to keep in mind the original "h"-problem where the mesh-size h has been dictated by accuracy requirements of the boundary value problem. So it is not allowed to forget the h-problem and to look only for the H-problem because of the improved convergence. But both problems can be combined in a proper way.

2.2.1 From the Coarse Grid Correction Idea to the Multigrid Correction Scheme

In Section 2.1.2 the residual correction has been introduced to improve w_h, the approximation to the solution of the discrete problem. This correction is obtained by solving a problem $\hat{L}_h \hat{e}_h = \hat{r}_h$ which is "similar" to the defect equation of the given discrete problem. This creates an iterative procedure with the corresponding operator

$$M_h = Id_h - \hat{L}_h^{-1} L_h.$$

Remark 2.8. *Thinking in terms of matrices and choosing $\hat{L}_h^{-1}$ to be the diagonal part of L_h results in $M_h = \mathbf{P}_J$ which leads to all the known disadvantages with respect to the convergence behavior.*

Introducing a coarser grid G_H $(H > h)$ and an appropriate approximation L_H to L_h on G_H the h-defect equation $L_h \tilde{e}_h = r_h$ is supplemented by $L_H \tilde{e}_H = r_H$ which can be solved fast if the number of gridpoints in G_H is small compared to that of G_h. Unfortunately, both equations have nothing in common, except perhaps that L_H is "similar" to L_h. There is not yet a relation between $\tilde{e}_h$ and $\tilde{e}_H$, respectively between r_h and r_H. The necessary coupling of the two h- and H-problems becomes possible with the help of linear transfer operators I_h^H and I_H^h which map fine grid functions onto coarse grid functions and vice versa. I_h^H denotes a restriction operator and I_H^h is the prolongation or interpolation operator.

Algorithm 2.5. *Coarse grid correction (CGC)*
The iterative scheme to compute a new approximation $w_h^{(n)}$ of the fine grid problem then proceeds by the following steps

$$\begin{array}{lll}
(1)\ \textit{Computation of residuals} & r_h^{(n)} := f_h - L_h w_h^{(n-1)} & \\
(2)\ \textit{Restriction of residuals} & r_H^{(n)} := I_h^H r_h^{(n)} & \\
(3)\ \textit{Exact solution of the} & & \\
\quad\ \ \textit{coarse grid problem} & L_H \tilde{e}_H^{(n)} = r_H^{(n)} & (2.53) \\
(4)\ \textit{Transfer of the correction} & \tilde{e}_h^{(n)} := I_H^h \tilde{e}_H^{(n)} & \\
(5)\ \textit{Correction} & w_h^{(n)} := w_h^{(n-1)} + \tilde{e}_h^{(n)} &
\end{array}$$

Because of $w_h^{(n)} = w_h^{(n-1)} + I_H^h L_H^{-1} I_h^H (f_h - L_h w_h^{(n-1)})$ the connected "coarse grid correction (CGC)" iteration operator is given by

$$M_h^H(CGC) = Id_h - I_H^h L_H^{-1} I_h^H L_h. \tag{2.54}$$

Defining $\hat{L}_h^{-1} := I_H^h L_H^{-1} I_h^H$ immediately links the idea of coarse grid correction with the residual correction consideration of Section 2.1.2.

Remark 2.9. *The coarse grid correction procedure is not convergent if it is used as an iterative solution scheme. The spectral radius $\rho(M_h^H(CGC))$ is larger or equal to one, because there exist h-grid functions w_h with $I_h^H L_h w_h = 0$ which are reproduced by $M_h^H(CGC)$, see Example 2.8.*

The analysis of the basic iterative schemes has shown that after some relaxation sweeps essentially low-frequent Fourier modes dominate. Because it is necessary for the coarse grid correction to map h-grid functions by some restriction operator I_h^H into the space of H-grid functions it is a self-suggesting question to ask how the low-frequent h-modes behave on the coarser grid. For simplicity it is assumed to have $H = 2h$ and to inject fine grid values onto the coarse grid (Figure 2.10).

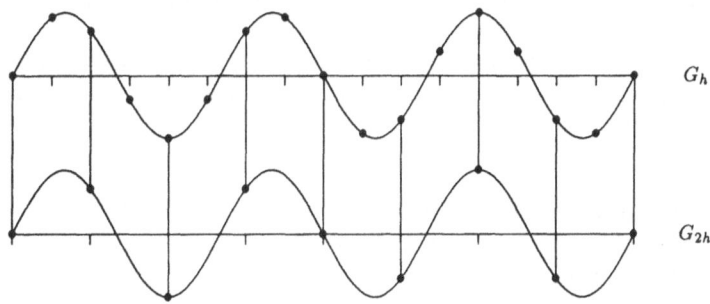

Fig. 2.10 Injection of a 1D h-grid frequency, $l = 6, N = 16$

The coarse grid

$$\mathcal{G}_H = \{(x, y) | x = iH, y = jH; i, j \in \mathbb{Z}\} = \{(x, y) | x = 2ih, y = 2jh; i, j \in \mathbb{Z}\}$$

consists of all even-numbered points of $\mathcal{G}_h$. For the model problem with $h = 1/N$ and $H = 2h = 1/(N/2)$ the injection of grid functions stands for the evaluation of these functions at coarse grid places, which in the present situation are the even-numbered fine grid points. The evaluation of a low-frequent fine grid frequency $\varphi_{l,m}^h$ at the node $(x, y) = (2ih, 2jh) \in G_H$ gives

$$\varphi_{l,m}^h(x, y) = \sin l\pi 2ih \cdot \sin m\pi 2jh = \sin l\pi iH \cdot \sin m\pi jH = \varphi_{l,m}^H(x, y)$$

and shows that these modes $(l, m \in \{1, \ldots, N/2 - 1\})$ of the h-grid maintain their wave number with respect to the H-grid. But compared to that wave number which defines "high-frequent", $N/4$ on G_H, the injected low-frequent h-grid mode appears more oscillatory on the coarse grid.

Example 2.7. *For $N = 16$ the wave number $l = 6$ represents a low-frequent mode. On $G_H = G_{2h} = G_{1/8}$ this wave number stands for a high-frequent one, because the high frequencies on grid G_H are those with wave numbers between 4 and 7.*

Now it is easy to determine those low-frequencies with respect to the h-grid which stay low-frequent on the grid $G_H = G_{2h}$ and those which become high-frequent. On the other hand, it is of interest to know how the high-frequent modes of the h-grid are represented on the coarse grid.

Example 2.8. *An h-grid high-frequent mode with l (or m) equal to $N/2$ coincides with the zero function on the H-grid:*

$$\varphi_{N/2,m}^h(iH, jH) = \varphi_{N/2,m}^h(2ih, 2jh) = \sin \pi i \cdot \sin 2jm\pi h = 0$$

for all nodes of G_H. This explains Remark 2.9, because such functions are reproduced by the iteration operator of the coarse grid correction process (2.53). $M_h^H(CGC)$ has an eigenvalue with absolute value larger or equal to one.

High-frequent fine grid modes cannot be represented on the coarse grid. They coincide with coarse grid modes having the same wave numbers. Due to the comments concerning aliasing functions, applied to a grid with now $H = 1/(N/2)$, such a coarse grid mode with, for instance, $N > l > N/2$ aliases with the H-grid $N - l$ frequency.

Example 2.9. *Let $N = 16$, $h = \frac{1}{16}$ and $(l, m)_h = (13, 2)_h$ a high-frequent mode. On the coarse grid the wave numbers still are $(l, m)_H = (13, 2)_H$, but with respect to $H = \frac{1}{8}$. The aliasing phenomenon lets this function appear on the H-grid as a mode with $(l, m)_H = (3, 2)_H$ which is obviously low-frequent, even on the coarse grid.*

Remark 2.10. *Wave numbers correspond to a certain wave length, and the above discussion of representing fine grid modes on a coarse grid can be repeated with respect to the wave length instead of wave numbers. On the h-grid the low-frequent h-grid modes have wave lengths larger than $4h$. A high-frequent fine grid mode contains at least one component with wave length between $2h$ and $4h$. If the modes are even more oscillatory than h-grid high-frequent ones, then the wave length is smaller than or equal to $2h$ and this "very high-frequent" mode is not visible on the h-grid. It aliases with a mode having a wave length larger than $2h$.*
The h-grid modes with wave length larger than $4h = 2H$ can be represented by H-grid frequencies, they are "visible" on the H-grid. On the other hand, all h-grid frequencies with a wave length smaller than or equal to $4h = 2H$ cannot be represented on the H-grid, they are invisible.

These discussions should provide the insight that "visibility" strongly depends on the relative structure of the coarse grid with respect to the fine grid. The choice of the grids decides what modes are "high-frequent" and which not.

Remembering Algorithm 2.5 and taking into account the representation of fine grid modes with respect to the coarse grid it is an obvious conclusion to avoid the transfer of h-grid modes which are invisible on the H-grid. They cannot be reduced on this grid. Fortunately, it is already known how to avoid the transfer of such frequencies from the fine grid to the coarse grid: performing several relaxation sweeps over the fine grid points reduces the high-frequent components of the error. The resulting error is smooth and can be approximated well on the H-grid by the solution of the coarse grid problem, the correction, which has to be transferred back to the fine grid. Combining the ideas of smoothing and coarse grid correction within an iterative process provides a new two grid iteration procedure. With respect to the convergence properties the two components, smoothing and coarse grid correction, can be interchanged. More general, it is allowed to distribute the smoothing part into a pre-smoothing and into a post-smoothing step. If the total number of relaxation sweeps is v, the convergence property of the corresponding iterative scheme only depends on v and not on the numbers v_1 and v_2 of pre-, respectively post-smoothing steps ($v = v_1 + v_2$). Such a splitting offers a certain degree of symmetry as well as it guarantees that high-frequent components as they might be introduced by the transfer of the correction from the coarse grid to the fine grid are damped out after the coarse grid correction step.

The computation of some grid function $\bar{w}_h^{(n)}$ by $v_1 \geq 0$ steps of a given relaxation method from an initial guess $w_h^{(n-1)}$ is written as

$$\bar{w}_h^{(n)} := RELAX^{v_1}(w_h^{(n-1)}, L_h, f_h)$$

or in terms of grid operators $\bar{w}_h^{(n)} := \mathscr{S}_h^{v_1} w_h^{(n-1)}$. This extends the coarse grid correction scheme (Algorithm 2.5) to a two-grid method which computes the new approximation $w_h^{(n)}$ from $w_h^{(n-1)}$.

Algorithm 2.6. *Correction scheme (CS)*

(1) *Pre-smoothing* $\bar{w}_h^{(n)} := RELAX^{v_1}(w_h^{(n-1)}, L_h, f_h)$

(2) *Computation of residuals* $r_h^{(n)} := f_h - L_h \bar{w}_h^{(n)}$

(3) *Restriction of residuals* $r_H^{(n)} := I_h^H r_h^{(n)}$

(4) *Exact solution of the coarse grid problem* $L_H \tilde{e}_H^{(n)} = r_H^{(n)}$

(5) *Transfer of the correction* $\tilde{e}_h^{(n)} := I_H^h \tilde{e}_H^{(n)}$

(6) *Correction* $\tilde{w}_h^{(n)} := \bar{w}_h^{(n)} + \tilde{e}_h^{(n)}$

(7) *Post-smoothing* $w_h^{(n)} = RELAX^{v_2}(\tilde{w}_h^{(n)}, L_h, f_h)$

$$(2.55)$$

The iteration operator of this two-grid method is

$$M_h^H(CS) = \mathscr{S}_h^{\nu_2} M_h^H(CGC) \mathscr{S}_h^{\nu_1}$$

with $M_h^H(CGC)$ due to (2.54) $M_h^H(CGC) = Id_h - I_H^h L_H^{-1} I_h^H L_h$. Applied to $\tilde{e}_h^{(n)} = u_h - w_h^{(n)}$ the iteration leads (compare (2.45)) to

$$\tilde{e}_h^{(n+1)} = M_h^H(CS)\tilde{e}_h^{(n)}.$$

Although the different components of the two-grid method have not yet been specified completely, the idea to extend the two-grid method to a real multigrid method suggests itself: the exact solution of the coarse grid problem (step (4) of (2.55)) often is too time consuming, because the problem size may be still very large. Therefore an iterative method is attractive again. Carrying over all the ideas developed for the h-grid problem now to the H-grid problem leads to the recursive definition of a multigrid method using more than two grids. The creation of additional coarser grids is stopped when a direct solution of the "coarsest grid problem" is possible. For many model problems the coarsest grid may consist of only one interior point.

Remark 2.11. *The recursive application of a $h - 2h$-method to fine grid frequencies explains the effect of this idea. On the h-grid the high-frequent components are damped out very fast. Essentially the low-frequencies remain and dominate. They are smooth on G_h and they are approximated well on G_{2h} by corresponding coarse grid modes which appear more oscillatory on the $2h$-grid. The subset of $2h$-high-frequent modes is smoothed on this grid. The remaining $2h$-grid low-frequencies are transferred to the $4h$-grid where they appear more oscillatory, again. The repeated application of all these steps shows that it can be expected to get all the fine grid components of the error damped. The damping takes place on grids "where the low-frequent h-grid mode seems to be high-frequent". Assuming an "ideal" coarse-to-fine transfer which produces no new high-frequent error components by the correction interpolation from a coarse grid to the next finer one, the total procedure will lead to an effective reduction of the total error on the fine grid.*

The transition from the $h - H$-two-grid method to a multigrid method is due to the above explanations characterized by the use of $\gamma \geq 1$ recursive iterations of the two-grid method applied to the coarse grid defect equation, using a sequence of coarser grids and the zero grid function as initial guess. For a grid sequence the description of the multigrid procedure is facilitated by using the "level" number to mark the grid equation instead of the mesh-size (see for example Algorithm 5.5): The algorithmic representation of multigrid methods is either given in terms of two-grid schemes or with the aid of the complete grid sequence. Because Multigrid algorithms are based upon two-grid methods and this representation form contains all required components, the further descriptions exclusively use the two-grid formulation. Additionally, the proofs of several theorems on multigrid (for instance the h-independent convergence, see Section 2.10) essentially make use of it.

Example 2.10. *Without discussing details of the algorithmic components at this point, it is of interest to apply a multigrid algorithm and to look for its convergence behavior. The problem of Example 2.3 (Figure 2.2) is solved now by a multigrid algorithm on a grid sequence with a finest grid corresponding to the grids of the previously performed singlegrid experiments. The multigrid convergence history of both the errors and the residuals is given in Figure 2.11 for different meshsizes $h \in \{\frac{1}{64}, \frac{1}{128}, \frac{1}{256}\}$. The coarsest grid always consists of only one inner point. The most significant differences which will be explained or proven later on are:*

1. *Both quantities, $r_h^{(n)}$ and $u - w_h^{(n)}$ (which is again the algebraic error $\tilde{e}_h^{(n)}$, approximately) are reduced extremely fast by each iteration step and not only during the initial phase of cycling. The reduction factors of approximately 0.05 are reached within every cycle until the machine accuracy is reached. This situation occurs after roughly nine cycles for the given problem.*
2. *The reduction factors do not depend on the meshsize (the different curves are parallel).*
3. *The error $\| u - w_h^{(n)} \|_\infty$ after one multigrid cycle is considerably smaller than the corresponding error of the singlegrid method after four hundred iterations. Because one cycle of the algorithm used is not more expensive than four singlegrid (fine grid) relaxation sweeps, the multigrid method is a very lucrative one.*

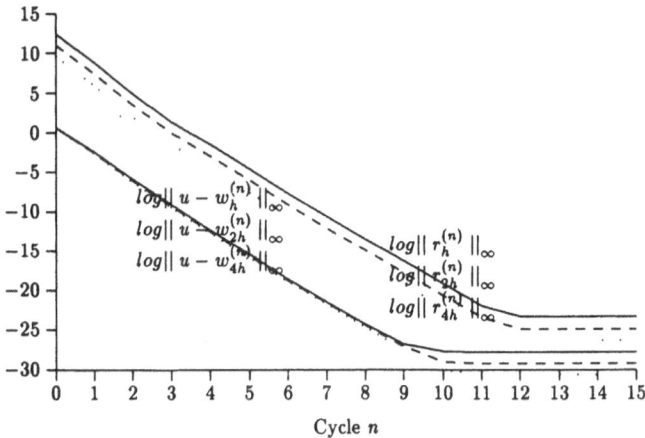

Fig. 2.11 Multigrid convergence, residuals and errors, $h = \frac{1}{256}$

2.3 The Components of the Algorithm

So far, the correction scheme of Algorithm 2.6 is more a general algorithmic description rather than a well-defined iterative method. There are many details—components—open for discussion and decision. To choose them properly and putting them together to a multigrid strategy which depends on the problem to solve, is a very demanding task. For the beginning it is adequate to look at each component of the algorithm separately.

The first decision to be made concerns the discretization technique and the finest grid. Both do strongly depend on the differential operators and on the domain (L, B and Ω). Most approaches exploit the underlying geometric information to set up the discrete problem on grids covering the domain Ω. This approach is common to the well-known discretization techniques like finite differences, finite volumes and finite elements. Many applications permit the discretization with knot-oriented schemes. In such cases the unknowns are placed at the intersections of grid lines. In other cases, for instance with systems of partial differential equations for flow-problems, cell oriented schemes are favoured (see Section 2.9 and Figure 2.12-(b)). In this case certain unknowns are positioned within a control volume and others are placed at the border. In any way, the grid plays an essential role. It has to allow a stable and consistent discretization of the continuous problem. The resulting data structure should pose no or only minor restrictions to the chosen algorithm and should be flexible enough to permit an extension of both the problem and the solution technique.

Figure 2.12 shows several regular grids which are often used to compute discrete approximations. Grids like the one of Figure 2.12-(a) are used throughout this chapter for the model problems and in many other applications. For Neumann boundary conditions grids like that in Figure 2.12-(c) allow an easy second order discretization with step size h. For local refinements, grids like that in Figure 2.12-(d) are a first approach to refine grids at places where it becomes necessary, but they have to be used carefully with respect to the discrete problem. Mesh sizes of adjacent cells have to obey certain relations if a desired order of approximation has to be reached globally.

On general domains the discretization near the boundary becomes more difficult. One approach is to introduce local or global "boundary fitted" coordinates which is, in practice, a coordinate transformation or the change of the computational domain. Unfortunately, the partial differential equation and the boundary conditions are transformed, too. This may cause new difficulties for the numerical procedure. As a benefit of this approach the discretization near the boundary is simplified.

Another strategy could be to use a regular mesh for "standard" discretization at most grid places and to modify the discretization only at places where it is necessary, for instance near the boundary. For "Poisson-like"

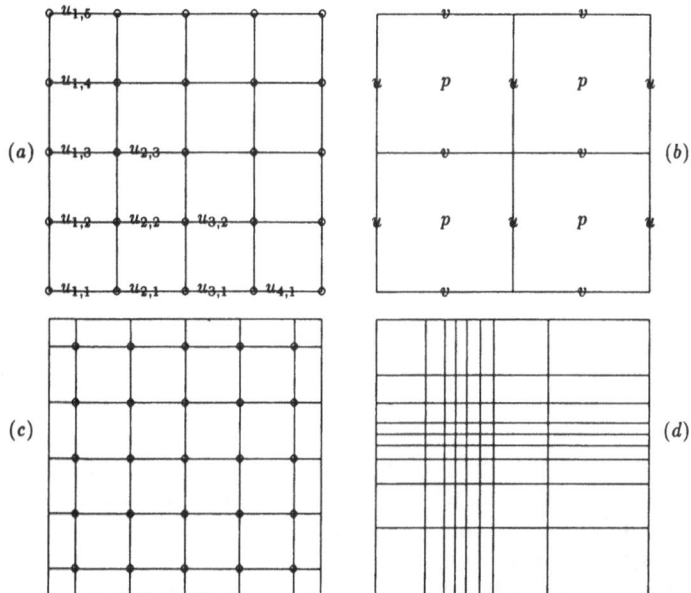

Fig. 2.12 Different grids

problems with Dirichlet boundary conditions the Shortley-Weller-approximation is well-known. Although the global approximation is only $O(h)$, the convergence order with respect to $h \to 0$ can be $O(h^2)$ if the second order approximation is violated only at points near the boundary. This is true for instance on regular square grids. The Shortley-Weller stencil near a curved boundary $\partial\Omega$ is, for the model problem (2.2) with $c = 0$, given by

$$2 \left[\begin{array}{ccc} & -\dfrac{b}{h_N(h_N + h_S)} & \\[2mm] -\dfrac{a}{h_W(h_W + h_E)} & \dfrac{a}{h_W h_E} + \dfrac{b}{h_S h_N} & -\dfrac{b}{h_E(h_E + h_W)} \\[2mm] & -\dfrac{b}{h_S(h_S + h_N)} & \end{array} \right]. \qquad (2.56)$$

Second order approximations for Neumann-type boundary conditions on curved boundary segments can be derived, in principle, with the finite difference technique, too (see Section 4.2.3 and [6]).

Assuming a discrete version $L_h u_h = f_h$ of the continuous problem on G_h, the iterative solution—not yet thinking about multigrid and smoothing— requires a numbering of points to determine the sequence of resolving the equations which are assigned to the unknowns associated with some grid node.

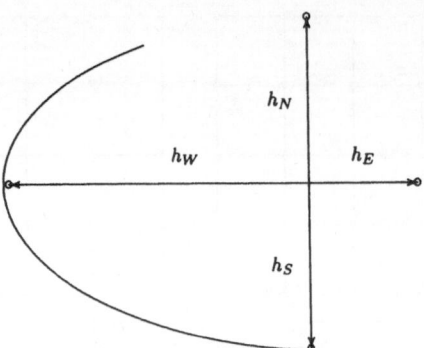

Fig. 2.13 Shortley-Weller approximation

As already mentioned, the results of Jacobi-like relaxations do not depend on the ordering of points, while the Gauss-Seidel smoothing gives different results for a different scanning of the equations. This is the reason for the often appended information "lexicographic", "chequered(red-black)", "even-odd" (lex, ch(rb), eo) or ZEBRA, and so on. Additional variants are produced by interchanging the role of x- and y-direction or by scanning the points in

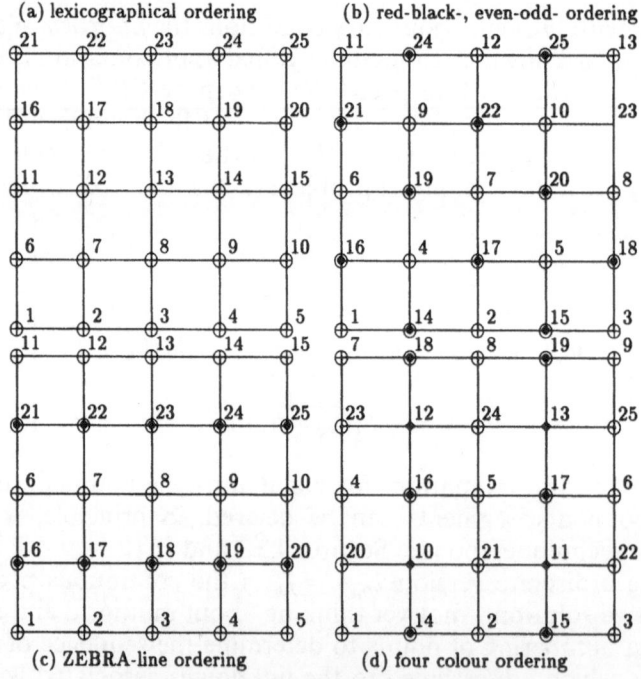

(a) lexicographical ordering   (b) red-black-, even-odd- ordering

(c) ZEBRA-line ordering   (d) four colour ordering

Fig. 2.14 Strategies to enumerate grid nodes

reverse order. The mostly used ordering schemes are the lexicographic ordering (Figure 2.14-(a), marching from left to the right and from the bottom to the top), the chequered ordering (Figure 2.14-(b)), the ZEBRA line ordering (Figure 2.14-(c)) and the four colour ordering (Figure 2.14-(d)). The lexicographical ordering assigns to the point (ih, jh) a smaller number than to the point $(i'h, j'h)$ if $j \le j'$ and in case of $j = j'$ if $i < i'$.

The chequerboard ordering corresponds to a splitting of the grid points into the set of "red" points ($i + j$ is an even number)

$$G_h^{red} = \{(ih, jh) | i + j \in 2\mathbb{Z}\} \quad \text{and into}$$
$$G_h^{black} = \{(ih, jh) | i + j \in \mathbb{Z} - 2\mathbb{Z}\} = G_h - G_h^{red},$$

or equivalently saying: "red $- i + j$ is an even number" while "black $- i + j$ is an odd number".

The zebra line ordering collects at first those points which belong to "red" (even) lines

$$G_h^{red} = \{(ih, jh) | j \in 2\mathbb{Z}\}$$

and then those of "black" (odd) ones

$$G_h^{black} = \{(ih, jh) | j \in \mathbb{Z} - 2\mathbb{Z}\} = G_h - G_h^{red}.$$

The four colour ordering splits G_h into four disjoint subsets

$$G_h^\kappa = \{(ih, jh) | i, j \in \mathbb{Z} \text{ and } (i, j) = \begin{cases} (even, \ even) & \kappa = 1 \\ (odd, \ odd) & \kappa = 2 \\ (even, \ odd) & \kappa = 3 \\ (odd, \ even) & \kappa = 4 \end{cases} \}.$$

Within every "colour" the points are numbered lexicographically, again.

Example 2.11. *In Example 2.1 the matrix $\tilde{A}$ (2.14) corresponds to a lexicographic ordering of points. As an effect of the red-black ordering the resulting matrix for the same problem changes to*

$$\tilde{A} = \frac{1}{h^2} \begin{pmatrix} 4 & & & & & -1 & -1 & & \\ & 4 & & & & -1 & & -1 & \\ & & 4 & & & -1 & -1 & -1 & -1 \\ & & & 4 & & -1 & & & -1 \\ & & & & 4 & & -1 & & -1 \\ -1 & -1 & -1 & & & 4 & & & \\ -1 & & -1 & -1 & & & 4 & & \\ & -1 & -1 & & -1 & & & 4 & \\ & & -1 & -1 & -1 & & & & 4 \end{pmatrix}.$$

$$(2.57)$$

Within this block matrix $\tilde{\mathbf{A}} = \dfrac{1}{h^2} \begin{pmatrix} 4 \mathscr{I}^{\text{red}} & -\mathscr{I} \\ -\mathscr{I}^T & 4 \mathscr{I}^{\text{black}} \end{pmatrix}$ *the size of the blocks*
depends on the number of red, respectively, black points.

Both the previous analysis of the basic relaxation schemes and the development process of the two-grid—and consequently of the multigrid methods—
show the fundamental role of the component "relaxation".

From the early stages of multigrid development up till now, the number of
investigated smoothing procedures has increased continuously. Methods of
ADI- and CG-type have to be mentioned as well as the incomplete (line)
LU-decomposition (ILU, ILLU), distributive relaxations (for instance the
Kaczmarz-relaxation), collective relaxations—this list could be extended.
Here the "classical" schemes are favoured. They have turned out to be very
effective smoothers not only for model problems, but also for many serious
applications. This justifies the restriction to smoothing methods which are
extensions of the already discussed basic relaxation schemes. They differ
from the standard procedures by a specific patterning of the grid points or by
building blocks of equations corresponding to certain blocks of points,
where the blocks may be treated again with some patterning scheme. The
resulting schemes will be denoted by the same terminology as the numbering
of grid points, although it should be mentioned that there is no general
correspondence between "ordering" strategy and "relaxation" strategy.

The Jacobi relaxation method for the discrete model problem is analyzed
easily because the eigenfunctions of the difference operator and of the
corresponding iteration operator are the same. A similar situation holds for
more general linear Poisson-like problems. This is the reason for the intensive
use of Jacobi-type relaxations for analysis purposes, although they are not
really efficient smoothers, even if "optimal" parameters are used.

The first variant of the standard Jacobi iteration is the so-called red-black
relaxation. This smoothing method consists of two half-steps, each of them
applied to a different set of grid points. The first partial step determines new
values in Jacobi manner for G_h^{red}. The second step uses this result to calculate
new values for G_h^{black}. It has to be pointed out, that in case of compact nine-
point stencils as they may appear in case of second order discretization of
general second order differential operators or even with the "Mehrstellen"
discretization of the Laplacian, this type of relaxation is not equivalent to
the Gauss-Seidel relaxation with red-black ordering of points: within every
half-step the application of the nine-point formula incorporates points which
belong to the same colour. But with five-point stencils similar to that of Δ_h,
the Jacobi-type red-black relaxation is equivalent to the Gauss-Seidel
relaxation with red-black ordering of points. A straight-forward generalization is the four-colour relaxation.

Gauss-Seidel relaxation with lexicographic ordering of grid points is the
most basic smoother and is widely used. It works with satisfactory results—

but not yet in the optimal way—for Poisson-like problems. The Gauss-Seidel relaxation with red-black ordering is in such cases superior. Point oriented relaxation techniques loose their smoothing properties as soon as anisotropies appear. Then block oriented schemes become attractive, especially if they are combined with a problem-dependent ordering of the blocks. The scanning of blocks and the manner of updating characterizes the block-oriented smoothing procedures, too.

A column (line) relaxation collects all equations corresponding to the grid points of a certain column (line) together and solves this subsystem for the respective unknowns. If the columns (lines) are passed through one after another lexicographically and if the values are updated immediately, this procedure is called lexicographic column (line) Gauss-Seidel relaxation (CGS-lex, LGS-lex). The "even-odd (ZEBRA)" relaxations scan the even-numbered blocks first and the odd-numbered afterwards. The denotation CGS-eo (LGS-eo) for the even-odd column (line) Gauss-Seidel type relaxation is self-explaining.

Highly recommended for anisotropic problems, especially with an $\varepsilon(x, y)$ depending on the space position and varying extremely, the "alternating direction (AD) ZEBRA" relaxation combines the four different steps CGS for even-numbered columns, CGS for odd-numbered columns, LGS for even-numbered lines and LGS for odd-numbered lines to a very efficient smoothing procedure (see Section 2.4).

Remark 2.12. *Experience shows the superiority of block oriented schemes to point oriented schemes. For the model problem it is possible to show that the spectral radius of the optimal block SOR is related to that of the optimal point SOR by $\rho(\mathbf{P}_{block-SOR}(\omega_{opt})) \approx \rho(\mathbf{P}_{point-SOR}(\omega_{opt}))^{\sqrt{2}}$. This reduces the number of required iterations for a fixed accuracy compared to the point relaxation method.*

Coarsening strategies have to be defined for every chosen grid structure if the use of multigrid methods is intended. It is not only a feeling that the adequate coarsening strategy is dictated by the underlying problem, too. The mostly used grids are regular square grids. They permit a variety of easy to use coarsening rules to set up the grid sequence.

The method to double the meshsize from G_h to G_H, setting $H = 2h$, is called "standard coarsening". Two consecutive grids, related to each other by standard coarsening are shown in Figure 2.15.

Another uniform coarsening is the "red-black" coarsening (Figure 2.16), omitting every other point and so generating a coarse grid which corresponds to a rotated grid with a $h - H$ relation of $H = \sqrt{2}h$. This type of coarsening is typically for special multigrid reduction methods [137].

Nonuniform coarsening is of importance with respect to anisotropic and singularly perturbed problems. A basic procedure is to double the meshsize

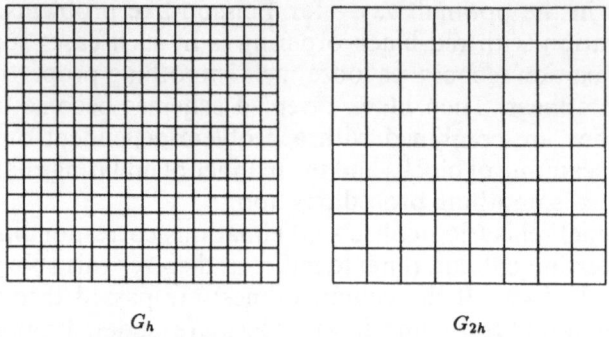

G_h G_{2h}

Fig. 2.15 G_{2h} created from G_h by standard coarsening

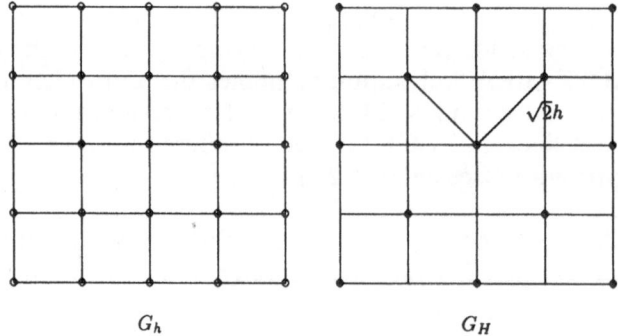

G_h G_H

Fig. 2.16 G_H created from G_h by red-black coarsening

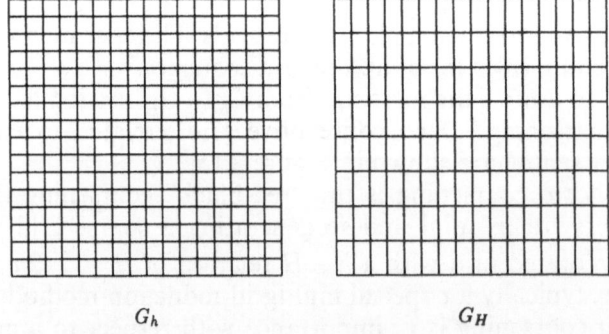

G_h G_H

Fig. 2.17 G_H created from G_h by semi coarsening

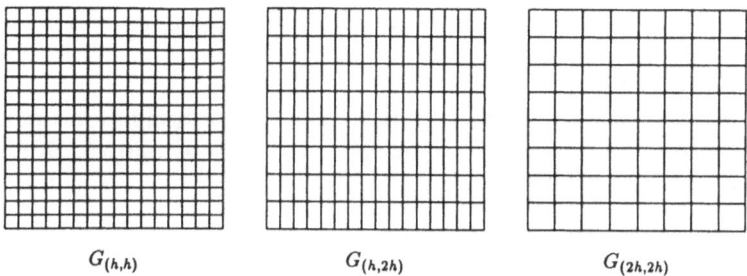

$$G_{(h,h)} \qquad\qquad G_{(h,2h)} \qquad\qquad G_{(2h,2h)}$$

Fig. 2.18 Two steps of the alternating semi coarsening applied to $G_{(h,h)}$

only into one direction (of two): $(H_x, H_y) = (h, 2H)$. One step of semi coarsening provides the grid sequence shown in Figure 2.17.

The semi coarsening may be applied several times into the same direction, but it may be applied, too, into different directions for each coarsening step. This is the "alternating semi coarsening" (Figure 2.18).

Remark 2.13. *Creating regular square grids for the model problem by standard coarsening allows an easy calculation of characteristic grid quantities. Assuming a fixed h_1, a grid sequence consisting of M grids is determined by the meshsizes $h_1, h_1/2, h_1/4, \ldots, h_1/2^{M-1}$. For model problems on the unit square it is often sufficient to choose $h_1 = \frac{1}{2}$. The coarsest grid then consists of only one inner grid point and the step sizes are $\frac{1}{2}, \frac{1}{4}, \frac{1}{8}, \frac{1}{16}, \frac{1}{32}, \frac{1}{64}, \frac{1}{128}, \frac{1}{256}, \ldots$. The number of gridpoints, including those at the boundary, are $\mathcal{N}_1 = 9, \mathcal{N}_2 = 25, \mathcal{N}_3 = 81, \mathcal{N}_4 = 289, \mathcal{N}_5 = 1089, \mathcal{N}_6 = 4225, \mathcal{N}_7 = 16641, \mathcal{N}_8 = 66049, \ldots$. An elementary estimation shows, that the total number of points within a sequence of M grids is about $\frac{4}{3}\mathcal{N}_M$ with $\mathcal{N}_M = (1 + 1/h_M)^2$.*

A component which has been used only formally and without any specification is L_H, the coarse grid operator. Approximating a partial differential operator by a difference operator L_h with respect to G_h establishes the discrete problem $L_h u_h = f_h$ on G_h. The most natural way to define the operator L_H is to proceed analogously with respect to G_H. The definition of all "coarse" grid operators within a complete sequence for multigrid is simplified by this natural choice which is used throughout this book unless it is explicitly indicated to be different.

Example 2.12. *The stencils corresponding to Δ_h and Δ_{2h} differ only by the preceding factor of $\dfrac{1}{h^2}$, respectively $\dfrac{1}{(2h)^2}$.*

In general, for the coarse grid correction, L_H should be a reasonable approximation to L_h to guarantee the overall consistency of the discrete problem. An alternative choice of L_H originates from the variational

(Galerkin-type) approach. Then L_H is determined due to

$$L_H = I_h^H L_h I_H^h,$$

which is the "discrete Galerkin condition". The prolongation operator I_H^h and the restriction operator I_h^H, interpreted in terms of matrices, are assumed to be adjoint to each other:

$$I_h^H = c(I_H^h)^T, \quad c = \text{constant}.$$

Both conditions are the "discrete variational properties".

Having a consistent approximation L_h to L, the overall consistency of the coarse grid correction approximation depends on the relative consistency of L_H with respect to L_h. Due to the construction of the Galerkin-type coarse grid operator with properly chosen restriction and interpolation, the overall consistency is provided automatically.

As a disadvantage of the Galerkin approach one may regard the dependence of L_H from L_h. Determining $L_H = I_h^H L_h I_H^h$, with L_h a five-point stencil, the coarse grid operator L_H will not always correspond to a five-point stencil. But nine-point approximations on the fine grid, will remain nine-point formula on the coarse grid.

Example 2.13. *The determination of a Galerkin-type coarse grid operator is demonstrated for $L_h = \Delta_h$, due to (2.21), using bilinear interpolation and the full weighting restriction (both of them are discussed in detail later). Applying $I_h^H L_h I_H^h$ step by step to a coarse grid function within a nine-point surrounding of $u_{i,j}$ explains the precalculation of L_H.*

1. *Interpolation of the coarse grid values to the next finer grid.*
2. *Evaluating the fine grid operator $L_h = \Delta_h$ within the fine grid nine-point surrounding of $u_{i,j}$ to the interpolated values produces a fine grid pattern like*

$$
\begin{array}{ccc}
0 & d & 0 \\
a & z & c \\
0 & b & 0
\end{array}
$$

with, omitting the factor $\dfrac{1}{h^2}$,

$$z = -2u_{i,j} + \tfrac{1}{2}(u_{i-1,j} + u_{i,j-1} + u_{i+1,j} + u_{i,j+1}) \quad \text{and}$$

$$a = -\tfrac{1}{2}(u_{i-1,j} + u_{i,j}) + \tfrac{1}{4}(u_{i-1,j-1} + u_{i,j-1} + u_{i,j+1} + u_{i-1,j+1}),$$

b, c, and d similar.

3. *The new coarse grid value $\tilde{u}_{i,j}$ is obtained by an averaging process (full weighting) which takes into account all fine grid points within the previously*

used fine grid nine-point surrounding of $u_{i,j}$:

$$\tilde{u}_{i,j} = 4\{-3u_{i,j} + \tfrac{1}{2}(u_{i-1,j} + u_{i,j-1} + u_{i,j+1} + u_{i+1,j})$$
$$+ \tfrac{1}{4}(u_{i-1,j-1} + u_{i-1,j+1} + u_{i+1,j-1} + u_{i+1,j+1})\}$$

The expression within the brackets corresponds to the stencil

$$\begin{bmatrix} \tfrac{1}{4} & \tfrac{1}{2} & \tfrac{1}{4} \\ \tfrac{1}{2} & -3 & \tfrac{1}{2} \\ \tfrac{1}{4} & \tfrac{1}{2} & \tfrac{1}{4} \end{bmatrix},$$

which is upto omitted factors the Galerkin representation of $L_H = I_h^H L_h I_H^h$ and which shows the expansion of the five-point formula into a nine-point formula.

The visit of the different levels during the multigrid iteration depends on the recursion parameter γ which is a number greater or equal to one. This number may depend on the grid level, for instance

$$\gamma = \begin{cases} \gamma_1 & \text{for an "odd-numbered" level,} \\ \gamma_2 & \text{if the level is "even-numbered".} \end{cases}$$

Usually γ is fixed and chosen to be smaller than four. With $\gamma = 1$ the so-called V-cycle is generated, while $\gamma = 2$ leads to the W-cycle. These names are obvious, when looking to the scheme in Figure 2.19 where four levels are used for demonstration purposes.

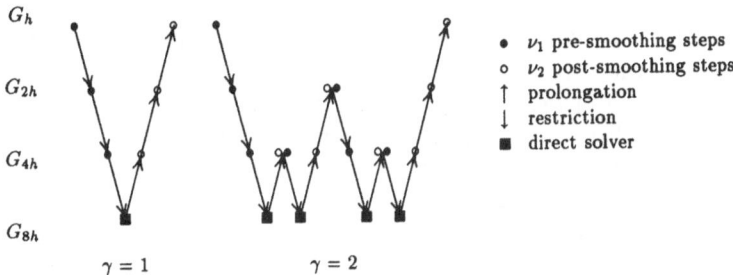

Fig. 2.19 The structure of V- and W-cycles

$\gamma = 1$ or $\gamma = 2$ are the mostly used values. As one may imagine, algorithms with $\gamma = 2$ are more expensive than those using $\gamma = 1$. On the other hand, experience shows that W-cycles are more robust than V-cycles because γ determines in a certain sense the accuracy of the coarse grid correction.

The fixed parameter γ allows the recursive definition of the multigrid operator. As explained in [132], the multigrid iteration is given by the recursion

$$M_2 = \mathscr{S}_2^{\nu_2}(Id_2 - I_1^2 L_1^{-1} I_2^1 L_2)\mathscr{S}_2^{\nu_1} \quad \text{and}$$
$$M_{l+1} = \mathscr{S}_{l+1}^{\nu_2}(Id_{l+1} - I_k^{l+1}(Id_l - M_l^\gamma)L_l^{-1} I_{l+1}^l L_{l+1})\mathscr{S}_{l+1}^{\nu_1}$$

for $l = 2, \ldots, M - 1$. The term $(Id_l - M_l^\gamma)$ reflects the approximate solution of the G_l equation by γM_l multigrid iterations. Such a representation is used to derive theoretical results.

Similar to γ the number $v = v_1 + v_2$ of relaxations, respectively the number of pre- and post-smoothing steps on the current level may be predefined or be chosen in dependence on a certain condition. With fixed values v_1 and v_2 the cycles are denoted as $V(v_1, v_2)$, $W(v_1, v_2)$ and similar for other types. In most practical applications $v \le 3$ is a good choice. The number of post-relaxations may be zero. But as already mentioned and as it will become obvious by the subsequently following analysis of the bilinear interpolation operator it is recommended to perform at least one post-smoothing step. Therefore the mostly used combinations of (v_1, v_2) are $(1, 1)$ and $(2, 1)$ for $v = 2$ and $v = 3$, respectively. Such an approach is very easy to program and works in most situations.

With the aid of the already given information and remembering Remark 2.11 another choice suggests itself: relaxation has to damp high-frequent modes on the respective grid level. So it is appropriate to relax as long as there exist significant contributions of such components which have to be damped. The condition to continue relaxation might be a convergence factor which is smaller than the worst damping factor of all high-frequent modes (the so-called "smoothing factor") because this indicates to still existing high-frequent components. Of course, one has to know this "smoothing factor"; but this quantity is in many cases not known in advance.

Remark 2.14. *In addition to preassigned cycling and iteration parameters which define when to stop the cycling or when to switch from one grid to another one in [12] accomodative algorithms are proposed. They often use residual norms of two consecutive grids for switching purposes. In Section 5.3.1 a strategy is proposed which stops the iteration as soon as the level of the truncation error is reached. Such a flexible cycling structure is desirable in many applications but it is not suitable for a detailed analysis of the method.*

Grid transfer operators have opened the way from the coarse grid correction idea to the multigrid correction scheme. The proper choice of both the residual restriction and the prolongation plays a key-role within the successful development of multigrid algorithms. The transfer operators depend, of course, on the chosen grids and the interpolation near boundaries is not a trivial task.

If $G_H \subset G_h$ the first restriction operator which comes into mind is the injection (INJ). With standard coarsening the coarse grid function then is defined by

$$I_h^H v_h = v_h|_{G_H}.$$

Multiplying this operator by $\frac{1}{2}$ leads to the sometimes used "half injection"

(HI). According to the basic idea of multigrid which requires a good transfer of low-frequent components, the injection possesses the property that $I_h^H \varphi_{l,m}^h$ is a good approximation of the fine grid low-frequency $\varphi_{l,m}^h$. Although easy to program the injection is of practical importance only for special applications.

Transfer operators which define a coarse grid value by a certain averaging of neighbouring fine grid values provide robust algorithms for more general problems. Even in the case of difference operators L_h with substantially changing coefficients the weighted restrictions are successfully applied. Half weighting (HW) and full weighting (FW) are the mostly recommended restrictions. The name "half weighting" points to the weighting which is something between (INJ) and (FW). The corresponding stencils are

$$\frac{1}{16}\begin{bmatrix} 1 & 2 & 1 \\ 2 & 4 & 2 \\ 1 & 2 & 1 \end{bmatrix}_h^{2h} \qquad (F_h^{2h}) \text{ for full weighting and} \qquad (2.58)$$

$$\frac{1}{8}\begin{bmatrix} & 1 & \\ 1 & 4 & 1 \\ & 1 & \end{bmatrix}_h^{2h} \qquad \text{for half weighting} \qquad (2.59)$$

and they have to be interpreted in the same way as (2.18). Assuming standard coarsening, the coarse grid function $v_{2h} = F_h^{2h} v_h$ for nodes in $G_{2h} \cap G_h$ is defined by

$$\begin{aligned} v_{2h}(x, y) = \tfrac{1}{16} [& v_h(x - h, y - h) + v_h(x - h, y + h) + v_h(x + h, y - h) \\ & + v_h(x + h, y + h) + 2(v_h(x - h, y) + v_h(x, y - h) \\ & + v_h(x, y + h) + v_h(x + h, y)) + 4v_h(x, y)] \end{aligned}$$

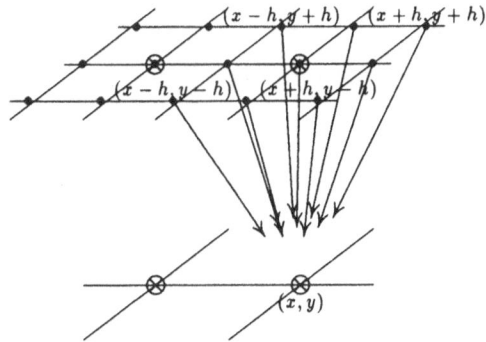

Fig. 2.20 Full weighting, schematic representation

It is not necessary to limit the restriction formula to nine-point neighbour-hoods of the coarse grid point. Higher order weightings which incorporate more than nine fine grid points have been defined but they are of minor practical importance.

Remark 2.15. *The idea behind the weighted restrictions is to represent fine grid functions correctly on the coarse grid. Then the weighting should satisfy a conservation property like*

$$H^2 \sum_{x \in G_H} I_h^H r_h(x) = h^2 \sum_{x' \in G_h} r_h(x')$$

which means that (discrete) integrals are preserved. This "exact representa-tion" is violated if injection (INJ) is used for the Poisson equation, discretized by the five-point formula, in combination with red-black relaxation. If smooth-ing ends with a red point relaxation sweep then all red points have zero residuals. Because of $G_{2h} \subset G_h^{\mathrm{red}}$ the transferred residual $r_{2h} = I_h^H r_h$ is zero, although the fine grid residual is not zero at all. The coarse grid correction will in general not improve the fine grid solution.
If, on the other hand, smoothing terminates with a black relaxation sweep the fine grid residuals in black points vanish but are nonzero in red points. By the same argument as above, now a nonzero defect is transferred from the fine grid to the coarse one. Because this defect is too large, half injection is used and provides improved coarse grid values. Half weighting is, in this case, a good choice for residual restriction. That red-black relaxation and INJ is not a good combination for equations similar to the Poisson or the Helmholtz equation is impressively demonstrated by Table 4.8 in Chapter 4.

Applying the half weighting operator to a low-frequent fine grid Fourier component $\varphi_{l,m}^h$ with wave numbers $1 \le l, m \le N/2 - 1$ shows for $(x, y) \in G_{2h}$

$$[I_h^{2h} \varphi_{l,m}^h](x, y) = \tfrac{1}{4}(2 + \cos \pi lh + \cos \pi mh)\varphi_{l,m}^{2h}.$$

For FW a similar relation holds:

$$[F_h^{2h} \varphi_{l,m}^h](x, y) = \tfrac{1}{4}(1 + \cos \pi lh)(1 + \cos \pi mh)\varphi_{l,m}^{2h}.$$

The preceding factor can be reformulated to

$$\left(1 - \sin^2 \frac{l\pi h}{2}\right)\left(1 - \sin^2 \frac{m\pi h}{2}\right).$$

Performing the same calculations for other wave numbers l, m shows the same result: the full weighting operator, applied to fine grid modes, produces coarse grid modes multiplied by a factor which depends on the wave numbers. Similar to the aliasing phenomenon high-frequent fine grid fre-quencies are mapped onto "some" coarse grid frequencies. In more detail, for FW the following is easy to show:

$$
\left.\begin{aligned}
F_h^{2h}\varphi_{l,m}^h &= \left(1 - \sin^2\frac{l\pi h}{2}\right)\left(1 - \sin^2\frac{m\pi h}{2}\right)\varphi_{l,m}^{2h} \\
F_h^{2h}\varphi_{N-l,m}^h &= -\sin^2\frac{l\pi h}{2}\left(1 - \sin^2\frac{m\pi h}{2}\right)\varphi_{l,m}^{2h} \\
F_h^{2h}\varphi_{l,N-m}^h &= -\left(1 - \sin^2\frac{l\pi h}{2}\right)\sin^2\frac{m\pi h}{2}\,\varphi_{l,m}^{2h} \\
F_h^{2h}\varphi_{N-l,N-m}^h &= \sin^2\frac{l\pi h}{2}\sin^2\frac{m\pi h}{2}\,\varphi_{l,m}^{2h}
\end{aligned}\right\} \qquad (2.60)
$$

These relations are helpful to derive a Fourier representation of the full weighting operator. The above property may be interpreted as a damping of certain frequencies by the restriction operator. The averaging weightings act on frequencies like filters.

The transfer of information from a coarse grid to the next finer one, prolongating the correction, is usually done by interpolation. The mostly used interpolation rule for corrections is due to the smoothness of this quantity the (bi)linear interpolation. Nevertheless, depending on the problem and on the chosen solution strategy more sophisticated interpolation methods are sometimes helpful. Here only higher order formula, nonlinear interpolation schemes, shape preserving methods or the interpolation with the aid of the difference equation are mentioned (see Sections 2.6, 2.7.1 and 4.6).

For a two-dimensional problem on square grids connected by standard coarsening the bilinear interpolation is performed in the following way:

1. for $(x_i, y_j) \in G_h \cap G_{2h}$ take $u_h(x_{2i}, y_{2j}) = u_{2h}(x_i, y_j)$
2. for fine grid points (x_{2i+1}, y_{2j}) and (x_{2i}, y_{2j+1}) which are no coarse grid points but whose fine grid lines coincide with coarse grid lines, the fine grid values are determined by

$$
u_h(x_{2i+1}, y_{2j}) = \tfrac{1}{2}(u_{2h}(x_i, y_j) + u_{2h}(x_{i+1}, y_j)) \quad \text{and}
$$
$$
u_h(x_{2i}, y_{2j+1}) = \tfrac{1}{2}(u_{2h}(x_i, y_j) + u_{2h}(x_i, y_{j+1})), \text{ respectively}
$$

3. the fine grid value in (x_{2i+1}, y_{2j+1}) is calculated by

$$
u_h(x_{2i+1}, y_{2j+1})
$$
$$
= \tfrac{1}{4}(u_{2h}(x_i, y_j) + u_{2h}(x_{i+1}, y_j) + u_{2h}(x_{i+1}, y_{j+1}) + u_{2h}(x_i, y_{j+1})),
$$

always using $1 \le i,j \le N/2 - 1$. To describe interpolation methods, a notation similar to stencils is introduced [132]. In this terminology the above sketched bilinear interpolation is given by

$$
\frac{1}{4}\begin{bmatrix} 1 & 2 & 1 \\ 2 & 4 & 2 \\ 1 & 2 & 1 \end{bmatrix}_{2h}^{h}
$$

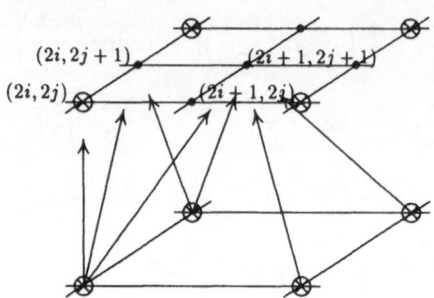

$(2i, 2j + 1)$ $(2i + 1, 2j + 1)$

$(2i, 2j)$ $(2i + 1, 2j)$

Fig. 2.21 Bilinear interpolation, schematic representation

This representation is suggestive, because the interpolation of a δ-like unit function being 1 at $(x, y) \in G_{2h}$ and zero elsewhere generates a pattern like the stencil's one on G_h. The brackets indicate the distribution of coarse grid values and the h, respectively $2h$, represent the step size of the involved grids. The general interpretation of such interpolation stencils is given in [132].

Example 2.14. *For the model problem with standard coarsening the restriction operator may be considered as an operator transferring a vector (grid function) from an* $(N - 1) \times (N - 1)$*-dimensional space to an* $(N/2 - 1) \times (N/2 - 1)$*-dimensional one. Such an operator is easily described in terms of matrices. To illustrate this, the fine grid is constructed with* $N = 8$*, thus consisting of 49 inner points. Standard coarsening leads to a next coarser grid with 9 inner points. Writing down all the equations to calculate the components of the coarse grid residual by full weighting establishes a linear system of equations*

$$\mathbf{A}_{FW}\mathbf{r}_h = \mathbf{r}_{2h}$$

with $\mathbf{r}_h = (r_{i,j}^h)^T$ *for* $1 \le i,j \le 7$ *and* $\mathbf{r}_{2h} = (r_{i,j}^{2h})^T$ *for* $1 \le i,j \le 3$ *and* $\mathbf{A}_{FW}$ *a* 9×49*-matrix which is equal to*

$$\frac{1}{16}\begin{pmatrix}
121 & 242 & 121 \\
 & 121 & 242 & 121 \\
 & & 121 & 242 & 121 \\
 & & & 121 & 242 & 121 \\
 & & & & 121 & 242 & 121 \\
 & & & & & 121 & 242 & 121 \\
 & & & & & & 121 & 242 & 121 \\
 & & & & & & & 121 & 242 & 121 \\
 & & & & & & & & 121 & 242 & 121
\end{pmatrix}_{9 \times 49}$$

The bilinear interpolation operator (BL) can be interpreted as an operator mapping vectors (grid functions) from an $(N/2 - 1) \times (N/2 - 1)$*-dimensional* *space to an* $(N-1) \times (N-1)$-*dimensional one. Writing this mapping in terms of matrices leads to*

$$\mathbf{A}_{BL}\mathbf{u}_{2h} = \mathbf{r}_h \quad \text{with} \quad \mathbf{A}_{BL} = 4\mathbf{A}_{FW}^T.$$

Thus the bilinear interpolation and the full weighting satisfy one of the mentioned discrete variational conditions. They are, up to a constant adjoint to each other.

It is of interest to see how prolongation methods act on the frequencies. The available information concerning the aliasing and the restriction of grid functions leads to the assumption that interpolating a coarse grid frequency will not only transfer this frequency to the fine grid but will also activate those fine grid modes which alias with the original frequency on the coarse grid. Therefore the approach

$$I_{2h}^h \varphi_{l,m}^{2h}(x, y) = a_{l,m} \varphi_{l,m}^h(x, y) + b_{l,m} \varphi_{N-l,m}^h(x, y)$$
$$+ c_{l,m} \varphi_{l,N-m}^h(x, y) + d_{l,m} \varphi_{N-l,N-m}^h(x, y) \qquad (2.61)$$

is a natural one. Using the information that for $(x, y) \in G_{2h} \cap G_h$ the coarse grid value is $I_{2h}^h \varphi_{l,m}^{2h}(x, y) = \sin l\pi x \cdot \sin m\pi y = \sin l\pi 2ih \cdot \sin m\pi 2jh$ leads to a first condition for the coefficients $a_{l,m}$, $b_{l,m}$, $c_{l,m}$ and $d_{l,m}$. Three additional equations for the nodes with the indices $(2i+1, 2j)$, $(2i, 2j+1)$ and $(2i+1, 2j+1)$ set up a linear system of equations. The unique solution which determines the coefficients of the above expansion is

$$\left.\begin{aligned}
a_{l,m} &= \tfrac{1}{4}(1 + \cos l\pi h)(1 + \cos m\pi h) \\
b_{l,m} &= \tfrac{1}{4}(\cos l\pi h - 1)(1 + \cos m\pi h) \\
c_{l,m} &= \tfrac{1}{4}(1 + \cos l\pi h)(\cos m\pi h - 1) \\
d_{l,m} &= \tfrac{1}{4}(\cos l\pi h - 1)(\cos m\pi h - 1)
\end{aligned}\right\} \qquad (2.62)$$

for all wave numbers l, m. Due to [12] restriction and prolongation operators are classified by certain "orders". In this terminology (INJ) is of order zero, (FW) has order two while bilinear and bicubic interpolations are of second respectively fourth order. Then it is shown that the sum of restriction and interpolation order has to be larger than the order of the differential operator. Consequently, bilinear interpolation together with full weighting is a sufficient choice for second order differential problems.

Example 2.15. *The introduction of high-frequent fine grid modes by interpolation is not only a theoretical statement. The introduced high-frequent modes may contribute considerably, as the values in the following table demonstrate. Some coefficients of the previously given expansion for the interpolated mode are given in dependence on different meshsizes ($2h = \frac{1}{32}$ and $2h = \frac{1}{64}$) and on varying wave numbers (l, m).*

Table 2.2. *Coefficients of fine grid modes contributing to* $I^h_{2h}\varphi^{2h}_{l,m}$

$(2h = \frac{1}{32}, h = \frac{1}{64})$	(l,m):	$a_{l,m}$	$b_{l,m}$	$c_{l,m}$	$d_{l,m}$
	$(1,1)$:	0.9988	-0.0006	-0.0006	$0.3 \cdot 10^{-6}$
	$(6,6)$:	0.9574	-0.0210	-0.0210	0.0004
	$(8,16)$:	0.8210	-0.0324	-0.1408	0.0055
	$(15,15)$:	0.7577	-0.1127	-0.1127	0.0167
	$(31,31)$:	0.2751	-0.2494	-0.2494	0.2260
$(2h = \frac{1}{64}, h = \frac{1}{128})$	$(1,1)$:	0.9997	-0.0001	-0.0001	$0.2 \cdot 10^{-7}$
	$(6,6)$:	0.9892	-0.0053	-0.0053	$0.2 \cdot 10^{-4}$
	$(8,16)$:	0.9527	-0.0092	-0.0376	0.0003
	$(15,15)$:	0.9341	-0.0323	-0.0323	0.0011
	$(31,31)$:	0.7432	-0.1188	-0.1188	0.0190

The contribution of the coarse grid mode to the final fine grid frequency is determined by the coefficient $a_{l,m}$ which varies extremely with respect to (l, m). For $l, m \ll N/2$ and h small, the coefficient $a_{l,m}$ will be close to one while $b_{l,m}, c_{l,m}$ and $d_{l,m}$ are approximately zero.
These values show that the bilinear interpolation maps a coarse grid frequency onto a fine grid function which is "more or less" the corresponding coarse grid mode with some "perturbations" by higher frequencies. The contributing modes for $I^h_{2h}\varphi^{2h}_{6,6}$ are shown in Figure 2.22 with $2h = \frac{1}{32}$.

Remark 2.16. *Relations like (2.60) and (2.61) together with (2.62) are valid in more general situations and are intensively stressed by the two-level and model problem analysis. This important property of the different operators—to have a Fourier representation—is the basis of multigrid analysis tools, for instance the smoothing analysis of relaxation schemes and the two-level analysis (see Section 2.10). Performing the calculations with respect to the set of basis functions consisting of sine products is a quite natural choice for model problems with Dirichlet boundary conditions because the error at the boundary is zero. This condition is easily satisfied by an expansion of the error in terms of sine functions. For the analysis of problems including Neumann boundary conditions the base of cosine products is an appropriate choice. It provides similar results as presented with the sine products.*
A more general approach is to use Fourier modes $e^{i\theta(x/h)}$ with $\theta \in (-\pi, \pi]$. Their generalization to the 2D-case simply uses $\theta := (\theta_1, \theta_2)$, $\mathbf{x} := (x, y)$, $h := (h_x, h_y)$ and $\theta \cdot \frac{\mathbf{x}}{h} := \theta_1 \frac{x}{h_x} + \theta_2 \frac{y}{h_y}$ which leads to $e^{i\theta \cdot \mathbf{x}/h} := e^{i\theta_1(x/h_x)} e^{i\theta_2(y/h_y)}$ with $\theta_1, \theta_2 \in (-\pi, \pi]$. The link between frequencies in terms of θ and those which are identified by wave numbers (l, m) is given by $\theta_1 = -\pi + 2\pi l/N$, respectively $\theta_2 = -\pi + 2\pi m/N$, and considering only the imaginary part. The normalization of θ to the interval $(-\pi, \pi]$ is justified by the properties of sine and cosine.

Fig. 2.22 $I^h_{2h} \varphi^{2h}_{6,6}$ —contributing Fourier modes

For an operator with constant coefficients and periodic boundary conditions on the unit square the $e^{i\theta \cdot x/h}$ are eigenfunctions of this operator. For the discrete operator L_h which is given by (2.23) it is easy to show that $L_h e^{i\theta \cdot x/h} = \frac{2}{h^2}(1 + \varepsilon - \varepsilon \cos \theta_1 - \cos \theta_2)e^{i\theta \cdot x/h}$. The factor is the "symbol" or the "formal eigenvalue" of L_h.

This approach can be extended to systems of partial differential equations (Section 2.9) and it is used in Section 2.10 to present the mode analysis in a more general framework.

2.4 Multigrid Strategies for Model Problems—Complete Algorithms

Using a grid sequence and treating the problem on different levels, especially separating the damping of high- and low-frequent modes, combined with appropriate inter-grid transfers yields a multigrid convergence which is independent of the step size h. To guarantee this property, the different components have to be chosen properly. It is not a trivial task at all, to decide which component is best used for a problem with certain characteristica like strongly varying coefficients, discontinuities, singularities or solution-dependent coefficients. One might guess that Fourier analysis is a natural tool to provide some realistic a priori information about the expected behavior of concrete algorithms. This is in fact true, and the concrete choice of algorithmic components can be justified by this tool. Within this section it is intended to demonstrate the composition of algorithms in practice and to show the effect of different components, even for classes of problems. This is possible because coefficients, right hand sides, boundary conditions or the shape of domains do not detoriate the performance of multigrid algorithms—as long as they are not "extreme". Having this in mind, the class of "isotropic" problems, given for instance by (2.2) with coefficient functions $a(x, y)$, $b(x, y) > 0$, $c(x, y) = 0$ and $a(x, y) \approx b(x, y)$ for all $(x, y) \in \Omega$ is fairly well represented by the Poisson equation. A reasonable multigrid algorithm then is composed of:

discretization	: symmetric second order finite differences
finest grid	: uniform grid of meshsize h
coarser grid	: created by standard coarsening
smoothing	: Gauss-Seidel with lexicographic ordering
residual transfer	: full weighting
correction transfer	: bilinear interpolation
coarse grid operator	: analogous to the discretization on the finest grid
cycle type	: V, that means $\gamma = 1$
smoothing steps	: $\nu = \nu_1 + \nu_2 = 2 + 1 = 3$

This choice of components results in an algorithm with a convergence factor of approximately 0.1 per cycle thus reducing the error by one order of magnitude within one single multigrid iteration step.

Replacing the lexicographically ordered Gauss-Seidel relaxation by red-black relaxation and using half weighting instead of full weighting leads to a multigrid algorithm which is one of the most efficient for the Poisson equation. Its convergence speed per multigrid iteration is still increased to approximately 0.06. These empirical (measured) values differ only slightly from the predictions by the two level analysis (for the above methods 0.119 and 0.034, respectively). Consequently, Gauss-Seidel relaxations on uniform grids and creating coarser grids by standard coarsening are reasonable selections for the class of isotropic problems. Combined with bilinear interpolation of the correction and full weighting, respectively half weighting in case of red-black relaxation, very efficient algorithms result.

Example 2.16. *A more detailed analysis of the above "very efficient" algorithms comes out by varying the meshsize of the finest grid (using regular meshes with $h_1 = \frac{1}{2}, h_2 = \frac{1}{4}, \ldots,$ this is equivalent to a different number of levels used) and evaluating the algorithms with $\gamma = 1$ and $\gamma = 2$ (V-, respectively W-cycle). $v = v_1 + v_2$ is equal to two or three with v_2 always equal to one. Solving the standard problem (see Example 2.3) the following table gives empirical defect- and error-reduction rates for cycles using red-black relaxation for smoothing and half injection for residual transfer.*

It is worth mentioning the following facts which are implicitly contained in Table 2.3:

1. *As expected, defect- and error-reduction rates are identical (within experimental accuracy) if the more expensive cycles are considered. Nevertheless, due to the higher numerical effort the $W(2, 1)$-cycle converges faster than the $V(2, 1)$-cycle.*
2. *Again, for the more costly algorithms, the reduction rates are independent from the meshsize of the finest grid used, confirming by this way the theorem of the h-independent multigrid convergence. However, this property should*

Table 2.3. *Asymptotic error reduction and defect reduction for a model problem*

	V(1, 1)		V(2, 1)		W(2, 1)	
L_2-norm	$\dfrac{\|\tilde{e}_h^{(n)}\|}{\|\tilde{e}_h^{(n-1)}\|}$	$\dfrac{\|r_h^{(n)}\|}{\|r_h^{(n-1)}\|}$	$\dfrac{\|\tilde{e}_h^{(n)}\|}{\|\tilde{e}_h^{(n-1)}\|}$	$\dfrac{\|r_h^{(n)}\|}{\|r_h^{(n-1)}\|}$	$\dfrac{\|\tilde{e}_h^{(n)}\|}{\|\tilde{e}_h^{(n-1)}\|}$	$\dfrac{\|r_h^{(n)}\|}{\|r_h^{(n-1)}\|}$
$h = \frac{1}{64}$	0.151	0.150	0.050	0.051	0.032	0.032
$h = \frac{1}{128}$	0.170	0.152	0.050	0.052	0.032	0.033
$h = \frac{1}{256}$	0.183	0.155	0.052	0.053	0.033	0.033

---

be interpreted carefully. The theory only states that the asymptotic $(h \to 0)$ multigrid convergence rate ρ is bounded away from one $0 \le \rho \le c < 1$ by a constant c independent from h. This does not imply that a certain algorithm has to converge with the same speed for all admissable meshsizes.

The recursive character of the multigrid cycling and the consequences of relaxing and transferring information between different levels is demonstrated by the output of a multigrid program solving again the problem which was already discussed in the Examples 2.3 and 2.10. A sequence of seven grids is used. The smallest meshsize is $h = \frac{1}{128}$. Standard coarsening ends with a coarsest grid consisting of only one inner point. The initial guess for inner points is zero and one $V(2,1)$-cycle with red-black relaxation is performed. The L_2-Norm of the defect is evaluated on each level before and after the smoothing steps.

```
GRID  7 L2-DEF-OLD   0.28581D+04 the defect at the beginning of
        L2-DEF-NEW   0.19171D+04 the first cycle and after
        L2-DEF-NEW   0.11921D+04 two pre-smoothing steps;
GRID  6 L2-DEF-NEW   0.82105D+03 the defect on the next coarser
        L2-DEF-NEW   0.60983D+03 level after residual restriction
        L2-DEF-NEW   0.39481D+03 and two pre-smoothing steps;
GRID  5 L2-DEF-NEW   0.27406D+03 ...
        L2-DEF-NEW   0.20692D+03
        L2-DEF-NEW   0.13432D+03
GRID  4 L2-DEF-NEW   0.93478D+02
        L2-DEF-NEW   0.69012D+02
        L2-DEF-NEW   0.43731D+02

              . . . . .
                  .
              . . . . .

GRID  4 L2-DEF-NEW   0.40472D+02
        L2-DEF-NEW   0.37875D+01
GRID  5 L2-DEF-NEW   0.12820D+03
        L2-DEF-NEW   0.12780D+02
GRID  6 L2-DEF-NEW   0.38392D+03
        L2-DEF-NEW   0.40145D+02
GRID  7 L2-DEF-NEW   0.11715D+04
        L2-DEF-NEW   0.13063D+03
        L2-DEF-RED   0.45704D-01 reduction of the defect by cycle 1
```

The above situation changes and the multigrid strategy has to be modified as soon as a certain anisotropy is introduced. Discretizing the operator of the model problem (2.5) by second order finite differences on an equispaced grid with meshsize h, see stencil (2.23), and determining the approximation w_h to the discrete solution u_h by Gauss-Seidel point relaxation the new value at

(x_i, y_j) is

$$w_{i,j} = \frac{1}{2(1 + \varepsilon)} (\varepsilon w_{i-1,j} + w_{i,j-1} + w_{i,j+1} + \varepsilon w_{i+1,j} + h^2 f_{i,j}),$$

where the Gauss-Seidel relaxation indices have been omitted. In terms of the error $\tilde{e}_h = u_h - w_h$ this becomes

$$\tilde{e}_{i,j} = \frac{1}{2(1 + \varepsilon)} (\varepsilon \tilde{e}_{i-1,j} + \tilde{e}_{i,j-1} + \tilde{e}_{i,j+1} + \varepsilon \tilde{e}_{i+1,j}). \tag{2.63}$$

With $\varepsilon = 1$ (Poisson equation) the error-averaging, that means the Gauss-Seidel smoothing effect into each direction is explained immediately by

$$\tilde{e}_{i,j} = \tfrac{1}{4}(\tilde{e}_{i-1,j} + \tilde{e}_{i,j-1} + \tilde{e}_{i,j+1} + \tilde{e}_{i+1,j}).$$

The anisotropic model equation, ε very small, guarantees this averaging process only into the dominant direction which is also called the direction of strong coupling, in the present case the y-direction. Performing only a few relaxation sweeps the error will not become small into the x-direction. Even for high-frequent modes the amplification factor of one PGS-lex relaxation sweep becomes worse as ε tends towards zero:

$$\varepsilon = 1 \qquad \text{amplification} \quad 0.5$$
$$\varepsilon = 0.1 \qquad\qquad\qquad\quad\; 0.83$$
$$\varepsilon = 0.01 \qquad\qquad\qquad\quad\; 0.98$$

Instead of discretizing (2.5) on an equispaced two-dimensional h-grid, this can be done with respect to a grid characterized by $(h_x, h_y) = (h, 2h)$. The corresponding stencil

$$\frac{1}{4h^2} \begin{bmatrix} & -1 & \\ -4\varepsilon & 2(1 + 4\varepsilon) & -4\varepsilon \\ & -1 & \end{bmatrix}$$

shows a reduced anisotropy. The improved averaging of the Gauss-Seidel scheme in x-direction appears in the formula

$$\tilde{e}_{i,j} = \frac{1}{2(1 + 4\varepsilon)} (4\varepsilon \tilde{e}_{i-1,j} + \tilde{e}_{i,j-1} + \tilde{e}_{i,j+1} + 4\varepsilon \tilde{e}_{i+1,j}) \tag{2.64}$$

for the corresponding error.

Having multigrid in mind and knowing about the treatment of frequencies on different scales, an omitted coarsening into the non-dominant direction is equivalent with leaving visible components of the fine grid visible on the next coarser grid. From the multigrid idea it is not necessary to treat components on fine grids whenever they are represented on coarser ones where they are reduced at lower cost. Therefore it does not matter that pointwise relaxation does not smooth with respect to the non-dominant direction—as long as the coarsening strategy makes the high-frequent components with respect to this direction appear on the coarser scales. Repeated semi coarsening into the dominant direction makes the direction of strong coupling disappear because of the growing stencil entries (introducing additional factors "4"). When all stencil entries, perhaps after several semi coarsening steps, are of approximately the same magnitude, the smoothing with respect to all directions and all remaining frequencies is achieved by point Gauss-Seidel relaxation and standard coarsening.

This is the basic idea underlying the first reasonable multigrid approach for anisotropic equations: the lexicographic point Gauss-Seidel relaxation is still used, but the standard coarsening strategy is replaced by semi coarsening into the y-direction (y-line coarsening, coarsening into the direction of strong coupling) until all the stencil entries are of approximately the same size. Then standard coarsening may be applied further (but still maintaining different meshsizes h_x, respectively h_y).

With constant coefficient problems the corresponding multigrid algorithm works efficiently. But if coefficient functions change their relative magnitude, the coarsening has to be done in an adaptive way depending on this local variation. This leads to rather complicated algorithms which can be avoided by changing the relaxation procedure and keeping the coarsening strategy fixed.

The idea is to use a relaxation scheme which produces smooth errors both into the x- and y-direction. This is achieved by solving for all strongly coupled unknowns. For the given problem, $\varepsilon \ll 1$, Gauss-Seidel y-line (column Gauss-Seidel, CGS) relaxation is recommended. The occurring tridiagonal systems of equations are solved directly by solvers whose numerical work is proportional to the number of unknowns within the currently considered grid line. This does not disturb the overall multigrid optimality. Usually, the lastmentioned approach is used because a simple extension of this idea, alternatingly relaxing into each of the coordinate direction (alternating direction relaxation, AD), maintains the simplicity of the algorithm and is an easily realized and highly efficient modification, especially if the size of the coefficients varies over the domain. For such a situation the alternating direction relaxation may be combined with the even-odd scanning of lines and rows (alternating direction ZEBRA-relaxation, AD-ZEBRA). Such components considerably improve the damping of oscillatory Fourier modes as the corresponding amplification factors for different relaxation schemes and ε in Table 2.4 show.

Table 2.4. *Damping of high-frequent components, standard coarsening*

ε	PGS-ch	CGS-lex	CGS-ZEBRA	AD-lex	AD-ZEBRA
1	0.25	0.447	0.250	0.149	0.048
0.1	0.83	0.447	0.125	0.373	0.102
0.01	0.98	0.447	0.125	0.438	0.122

Table 2.5. *Empirical convergence rates for the anisotropic model problem*

	PGS-ch		CGS-ZEBRA		AD-ZEBRA
ε	V(2, 1)	W(2, 1)	V(2, 1)	W(2, 1)	W(1, 1)
1	0.077	0.049	0.020	0.016	0.008
0.1	0.056	0.054	0.041	0.029	0.036
0.01	0.92	0.92	0.044	0.029	0.047

This table summarizes the above explained phenomena and demonstrates the superiority of the line-, respectively column-oriented relaxation schemes, especially if a chequered scanning of the blocks is applied. Having already pointed out the principal importance of good smoothing properties for an efficient multigrid cycling, there should be left no doubt that a satisfactory smoothing of the relaxation scheme used carries over to the convergence of concrete algorithms. This has been shown in practice for the anisotropic model problem in [78, 132]. The components are chosen similar to the algorithm for Poisson-like problems but different relaxation schemes are used within V- and W-cycles. For $v = v_1 + v_2$ as specified and full weighting for residual transfer empirical (asymptotic) convergence rates are computed on a grid sequence with finest meshsize $h = \frac{1}{64}$ (Table 2.5). The resulting values prove the robustness of those algorithms using y-line or alternating direction ZEBRA relaxation.

Remark 2.17. *For three-dimensional problems the situation becomes even more complicated. Consider, for example, the model equation*

$$-\frac{\partial}{\partial x}\left(a\frac{\partial u}{\partial x}\right) - \frac{\partial}{\partial y}\left(b\frac{\partial u}{\partial y}\right) - \frac{\partial}{\partial z}\left(c\frac{\partial u}{\partial z}\right) = f \tag{2.65}$$

with (smooth) coefficients that may depend on the three space variables.
The standard multigrid approach for Poisson-like equations, characterized by the components 'standard coarsening, point relaxation, (tri-) linear interpolation, full weighting' leads to an efficient solution method only if the coefficients a, b, c are of the same order of magnitude. If this is not the case, more sophisticated smoothing methods have to be choosen. (For a detailed analysis see [134].) Again, block relaxation strategies (while keeping standard coarsening) yield efficient smoothing methods also for problems with strong anisotropies. The underlying principle: "Solve simultaneously for all strongly coupled unknowns" here means that line or even plane relaxation steps have to be performed. Carrying out such block relaxations in a suitable way, for example, line relaxations using special direct solvers for tridiagonal systems and plane relaxations by appropriate 2D multigrid methods, preserves the O(N) complexity of the resulting multigrid method. In cases when the (variable) coefficients are allowed to have changing anisotropies in different parts of Ω, alternating block methods are used.

In order to achieve a good balance of efficiency and robustness, for a concrete problem the 'cheapest' among the suggested methods with sufficient smoothing properties should be chosen. In the following an appropriate choice of the smoothing method by distinguishing several typical cases for (2.65) is proposed. (The relations '~' and '≫' between the coefficients have the meaning 'of same order of magnitude as' and 'essentially much larger than', respectively).
Smoothing method for (2.65):

1. $a \sim b \sim c$: point relaxation;
2. $a \gg b \sim c$: x-line relaxation;
3. $a \sim b \gg c$: (x, y)-plane relaxation, *performed by a 2D multigrid method, which uses point relaxation for error smoothing;*
4. $a \gg b \gg c$: (x, y)-plane relaxation, *performed by a 2D multigrid method, which uses x-line relaxation for error smoothing;*
5. $a \gg b$, $a \gg c$: (x, y)-plane relaxation, *performed by a 2D multigrid method, which uses alternating line relaxation for error smoothing;*
6. general case: alternating plane relaxation, *performed by 2D multigrid methods, which use alternating line relaxation for error smoothing.*

When plane relaxation *is used, it is sufficient to carry out each relaxation step by one step of a "cheap" 2D multigrid method, described by the components 'V(1, 1)-cycle, full weighting, linear interpolation, Gauss-Seidel relaxation (point- or linewise, depending on the kind of anisotropy of the 2D problems in the planes)'.*
Table 2.6 contains convergence factors ρ *that have been measured for concrete problems of the form (2.65), covering the 6 cases cited above, using in each case the mentioned 'cheapest, safe' relaxation type. Here,* Ω *is the 3D unit-cube and* $\tilde{a}$, $\tilde{b}$, $\tilde{c}$ *are polynomial coefficients of highly anisotropic behavior:*

$$\tilde{a} = a_1(x)a_2(y)a_3(z) \quad \text{with} \quad a_i(x) = 1 + 4(\gamma_i - 1)(x - x^2),$$
$$\tilde{b} = b_1(x)b_2(y)b_3(z) \quad \text{with} \quad b_i(x) = 1 + (\gamma_i - 1)x,$$
$$\tilde{c} = c_1(x)c_2(y)c_3(z) \quad \text{with} \quad c_i(x) = \gamma_i + (1 - \gamma_i)x,$$
$$\gamma_1 = 10, \ \gamma_2 = 2, \ \gamma_3 = 5.$$

Table 2.6. *Convergence factors for the 3D anisotropic model problem*

case	a	b	c	ρ^*	ρ
1.	1	1	1	.20	.19
2.	100	1	1	.074	.045
3.	100	100	1	.052	.030
4.	10000	100	1	.052	.0039
5.	10000	$\tilde{b}$	$\tilde{c}$		.036
6.	$\tilde{a}$	$\tilde{b}$	$\tilde{c}$		.10

The numbers ρ^* are 3D two-grid-convergence factors, calculated by model problem analysis (see [134] and Section 2.10). The numbers ρ have been computed with a multigrid program using the indicated relaxation method (and the remaining components being standard), performing $W(1,1)$-cycles and using a finest grid G_h with $32 \times 32 \times 32$ intervals. (For further details see [38].)

It is very convenient to construct extremely "fast converging" algorithms — if this is considered to be equivalent with having a small spectral radius. Even if the convergence is h-independent and fast in this sense the computational work to achieve these properties has to be taken into account if the interest is laid upon "efficient" algorithms.

By the direct operation count for model problems it has been proven that the number of operations per multigrid cycle is directly proportional to the number of unknowns belonging to the finest grid. The concrete relation, that means the constant of proportionality, naturally depends on the components of the algorithms (L_h, L_H, coarsening, cycle, smoothing,...), but the main result, the efficiency of multigrid turns out — and not only for model problems.

In many important applications an operations count is impossible or at least very tedious. Therefore an easy to determine measure is asked for. The first step to this goal is to assume standard coarsening. This assumption is not a necessary one but it simplifies the following representation. The second one is to assume γ to be independent from any grid level and the last one is to assume that the work to solve the problem on the coarsest grid is neglectable — an assumption which is justified as the model problems with only one inner point on the coarsest grid show.

W_l^{l-1} denotes the computational work of the two-grid method on grid level l excluding the work to solve the $l-1$-level coarse grid problem. W_l^{l-1} thus includes the work of v relaxation sweeps, of the residual calculation and its transfer and of the interpolation of corrections and the correction of the old approximation. The recursive definition of the multigrid cycle using a set of M grids then provides a formula to determine the total work W_M for one cycle by

$$W_M = \sum_{l=2}^{l=M} \gamma^{M-l} W_l^{l-1} \tag{2.66}$$

The individual contributions on each level are proportional to the number $\mathcal{N}_l$ of unknowns on the corresponding grid. Formula (2.66) thus reduces to

$$W_M \approx \sum_{l=2}^{l=M} \gamma^{M-l} (v W_l(\mathscr{S}) + W_l(R) + W_l(P)) \mathcal{N}_l \tag{2.67}$$

where $W_l(\mathscr{S})$, $W_l(R)$ and $W_l(P)$ measure the computational work per grid point for smoothing, residual restriction and prolongation of corrections, respectively. Constructing the different components identically for all grids makes the work per grid point independent of the respective level. Exploiting the facts that standard coarsening reduces the number of points from one level to the next one by a factor of $(1/2^d)$, d the space dimension of the underlying problem, and that by experience $W_l(\mathscr{S})$ is approximately as large as $W_l(R)$ and that $W_l(P)$ is neglectable, (2.67) changes to

$$W_M \approx (v+1)W_M(\mathscr{S})\mathscr{N}_M \sum_{l=2}^{l=M} \left(\frac{\gamma}{2^d}\right)^{M-l}. \tag{2.68}$$

The product $W_M(\mathscr{S})\mathscr{N}_M$ is often called "work unit (WU)". Here the most expensive contribution is used as name giving component for the "relaxation unit (RU)" because it denotes the "work" per relaxation on level M. As a consequence, in two dimensions, a complete multigrid cycle costs approximately

$$\tfrac{4}{3}(v+1)RU \quad \text{for} \quad \gamma = 1$$
$$2(v+1)RU \quad \text{for} \quad \gamma = 2$$
$$4(v+1)RU \quad \text{for} \quad \gamma = 3$$

which shows for $\gamma \le 3$ the $O(\mathscr{N}_M)$ property—which is one essential part of the "multigrid optimality". In general, the $O(\mathscr{N}_M)$ dependence is true with standard coarsening for all γ satisfying $1 \le \gamma \le 2^d - 1$. For larger values the linear dependence on the number of unknowns is no longer valid. But this is a more theoretical consideration because relevant applications in a natural way limitate the allowed computational work and therefore restrict γ to be either one or two. Although the W-cycle is more "robust" and advantageous from the theoretical point of view (h-independent convergence has been proven under quite general assumptions) the V-cycle is in many applications sufficient: the numerical cost are low and the convergence is acceptable.

Example 2.17. *The numerical work for a $V(2,1)$-cycle is approximately $5\tfrac{1}{3}RU$. Returning to the algorithm for the Poisson equation which showed an asymptotic convergence behavior of about 0.06, the numerical work which is invested to achieve this convergence has to be taken into account. Computing a "convergence per RU" (ρ^{1/W_M}) close to 0.59 tells that the numerical work corresponding to one relaxation sweep contributes within the cycling by an error reduction of at least this value. The convergence of iterative schemes on grids with comparable resolution lead to convergence rates of 0.98 and worse. This comparison explains what "efficiency of the multigrid cycling" stands for. Due to the different empirical convergence rates of the $V(2,1)$- and $W(2,1)$-cycle in Table 2.5 (0.053, respectively 0.033) the corresponding "convergence rates per (RU)" are 0.576 and 0.652 for the V-, respectively for the W-cycle. This explains why practitioners often prefer the "efficient" $V(2,1)$-cycle in spite of its "worse" asymptotic behavior.*

Within the multigrid literature there is no unique definition both for work and efficiency. Thus it is quite natural to measure the numerical work in terms of one relaxation sweep over the finest grid. To avoid the counting of operations per relaxation the average computing time for the relaxation method used is determined. Such timing allows—for a fixed machine—to "normalize" the investigated relaxation schemes with respect to one selected relaxation scheme and with respect to the underlying application problem. Such a normalization is applied to compare the efficiency of several relaxation schemes for the diffusion model problem in Section 4.3.3.

Remark 2.18. *Exploiting characteristic features of multigrid components may help to save additional computational work. Especially the defect calculation and its transfer can be modified.*
Both the full and the half weighting are approximately as expensive as one relaxation step is. But for five-point discretization schemes in combination with coloured relaxation methods the computational effort to compute weighted restrictions may be reduced considerably. For the Poisson equation discretized by the usual five-point stencil after red-black relaxation the residuals at all "black" points are zero. The full weighting operator in this case simplifies to

$$\frac{1}{16}\begin{bmatrix} 1 & 0 & 1 \\ 0 & 4 & 0 \\ 1 & 0 & 1 \end{bmatrix}.$$

Similarly, if the half weighting operator is used, the simplified restriction coincides with the very cheap half injection operator. CGS-ZEBRA relaxation produces by the second partial sweep, treating all odd lines, zero residuals for all equations concerning points on just these lines. It is not necessary to evaluate the defect there, and the restriction operator reduces to

$$\frac{1}{16}\begin{bmatrix} 0 & 2 & 0 \\ 0 & 4 & 0 \\ 0 & 2 & 0 \end{bmatrix}.$$

So far only model problems including Dirichlet boundary conditions have been considered. Because it is not intended to discuss theoretical aspects of "Neumann" problems (existence, uniqueness, continuous and discrete regularity condition,...) the remainder of this section is restricted to some details which are important with respect to the application of multigrid. The main difference to the Dirichlet case is the necessary discretization of the boundary conditions. For rectangular domains there are some standard approaches to get first or second order approximations.
A first one uses one sided difference approximations of the normal derivative at the boundary, excluding the corners of the rectangle. This method gives only a first order approximation. The equations originating from the boundary conditions can be used to eliminate those unknowns

corresponding to grid points at the boundary. This is expressed by a modified stencil for the Laplacian near the boundary. For instance, near the left boundary, position (a) in the left picture of Figure 2.23, the stencil changes from

$$
\frac{1}{h^2}\begin{bmatrix} & 1 & \\ 1 & -4 & 1 \\ & 1 & \end{bmatrix} \quad \text{to} \quad \frac{1}{h^2}\begin{bmatrix} & 1 & \\ 0 & -3 & 1 \\ & 1 & \end{bmatrix},
$$

(and similarly at other boundary positions). Near the lower left corner, position (b) in the left picture of Figure 2.23, two boundary equations can be used to eliminate two corresponding unknowns, which leads to

$$
\frac{1}{h^2}\begin{bmatrix} & 1 & \\ 0 & -2 & 1 \\ & 0 & \end{bmatrix}
$$

and similar at the other corners.

The approximation order is improved if a symmetric difference formula with step size $h/2$ is used to discretize the normal derivative. In order to do this, the grid is shifted by $h/2$ in both directions. Additional grid points outside of the original domain are introduced to discretize the boundary condition (see the staggered grid in the right picture of Figure 2.23 where these points are marked by ∘). Again the discrete boundary conditions may help to eliminate exactly those unknowns which correspond to the introduced "ghost points". For such grids no coarse grid point coincides with fine grid nodes. This influences both restriction and interpolation operators.

The third approach to treat Neumann boundary conditions is of second order, too, but a grid is used where the grid lines coincide with the borders of the rectangle (fitting grid). The further technique, discretizing the differential operator even on $\partial\Omega$, introducing ghost points outside the domain (distance h) and approximating the normal derivative by symmetric central differences with second order accuracy but step size $2h$ corresponds to the above

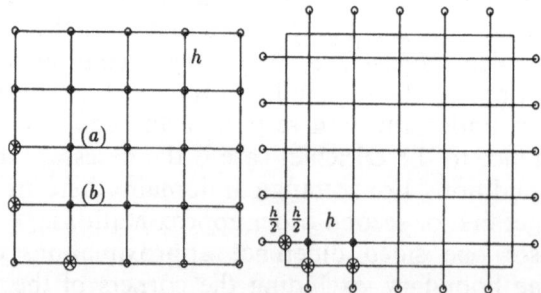

Fig. 2.23 Standard discretization of Neumann boundary conditions

procedure. Again, the discrete boundary condition helps to eliminate the "ghost" unknowns.

Due to the "symmetry" of the (homogeneous) Neumann boundary conditions the eliminated boundary conditions influence the difference operator. At the lower border of $\partial\Omega$, away from corners, Δ_h is modified from

$$\frac{1}{h^2}\begin{bmatrix} & 1 & \\ 1 & -4 & 1 \\ & 1 & \end{bmatrix} \quad \text{to} \quad \frac{1}{h^2}\begin{bmatrix} & 2 & \\ 1 & -4 & 1 \\ & 0 & \end{bmatrix}$$

and at the upper right corner, for instance to

$$\frac{1}{h^2}\begin{bmatrix} & 0 & \\ 2 & -4 & 0 \\ & 2 & \end{bmatrix}.$$

In Section 4.2.2 the same process is explained for an operator of the form $\nabla(f(u)\nabla u)$ in more detail.

Remark 2.19. *In* [5] *a detailed analysis of multigrid methods applied to the Helmholtz equation with Neumann boundary conditions is performed. The staggered grid approach as well as the fitting grid approach is discussed. The boundary conditions are either eliminated or they are treated separately.*

A modification of the full weighting operator according to the changes of the difference operator seems natural. Near the boundary symmetry arguments suggest a mirror imaging that transforms, for instance, the full weighting stencil near the lower boundary from

$$\frac{1}{16}\begin{bmatrix} 1 & 2 & 1 \\ 2 & 4 & 2 \\ 1 & 2 & 1 \end{bmatrix} \quad \text{to} \quad \frac{1}{16}\begin{bmatrix} 2 & 4 & 2 \\ 2 & 4 & 2 \\ 0 & 0 & 0 \end{bmatrix}.$$

This modification works analogously at the other boundary segments. The imaging applies twice at corners and leads, for instance at the upper right corner, to

$$\frac{1}{16}\begin{bmatrix} 0 & 0 & 0 \\ 4 & 4 & 0 \\ 4 & 4 & 0 \end{bmatrix}.$$

The heuristic consideration is formally confirmed by the same idea as mentioned in Remark 2.15, but applied now to the intersection of nine point surroundings of boundary nodes and the rectangular domain Ω. A further modification near curved boundaries is explained in Section 4.2.3.

2.5 The Full Approximation Scheme

The CS formulation (Algorithm 2.6) is based upon the principle of corrections, especially exploiting the idea to improve an approximate solution w_h
of the problem $L_h u_h = f_h$ by the interpolated coarse grid correction $I_H^h \tilde{e}_H$
which is (in the two-level method) the exact solution of the coarse grid correction equation

$$L_H \tilde{e}_H = I_h^H r_H (= I_h^H (f_h - L_h w_h)) \tag{2.69}$$

which itself is the coarse grid approximation of the defect equation $L_h \tilde{e}_h = r_h$.
Within the multigrid context this idea provides an approximate discrete
solution on the finest grid and corrections on all coarser grids. By different
reasons it may be desirable to have complete (full) approximations to the
fine grid solution everywhere instead of only corrections. To derive the
defect equation $L_h \tilde{e}_h = r_h$ the linearity of L_h is explicitly used. This manifests
a natural drawback of the correction scheme—it is directly applicable to
linear problems only.
Using $u_h = w_h + \tilde{e}_h$ the fine grid problem $L_h u_h = f_h$ is trivially replaced by

$$L_h(w_h + \tilde{e}_h) - L_h w_h = f_h - L_h w_h = r_h \tag{2.70}$$

which is a valid identity even for nonlinear operators L_h. If the operator is a
linear one, the above equation is equivalent to the defect equation. The
approximation of (2.70) on the coarse grid is

$$L_H(\hat{I}_h^H w_h + \tilde{e}_H) - L_H \hat{I}_h^H w_h = I_h^H(f_h - L_h w_h) = I_h^H r_h, \tag{2.71}$$

or after rearranging and introducing the new coarse grid variable
$u_H := \hat{I}_h^H w_h + \tilde{e}_H$

$$L_H u_H = L_H \hat{I}_h^H w_h + I_h^H r_h. \tag{2.72}$$

$\hat{I}_h^H$ is some restriction operator which maps fine grid functions to coarse
grid functions and is not necessarily the same as the residual restriction I_h^H.
The standard choice for $\hat{I}_h^H$ is the simple injection.
With an actual approximation w_H to the solution of the above coarse grid
problem, the new approximate coarse grid correction is $\tilde{e}_H := w_H - \hat{I}_h^H w_h$. It
is this quantity which has to be transferred back to the fine grid to obtain
the improved fine grid approximation

$$\tilde{w}_h := w_h + I_H^h \tilde{e}_H = w_h + I_H^h(w_H - \hat{I}_h^H w_h). \tag{2.73}$$

Similar to the correction scheme (Algorithm 2.6) a multigrid algorithm is
recursively defined by a two-grid scheme where (2.72) and (2.73) are incorporated.

Algorithm 2.7. *Full approximation scheme*

(1) *Pre-smoothing* $\bar{w}_h^{(n)} := RELAX^{v_1}(w_h^{(n-1)}, L_h, f_h)$
(2) *Computation of residuals* $r_h^{(n)} := f_h - L_h \bar{w}_h^{(n)}$

(3) *Restriction of residuals* $r_H^{(n)} := I_h^H r_h^{(n)}$
(4) *Exact solution of the*
    coarse grid problem $L_H u_H^{(n)} = L_H \hat{I}_h^H \bar{w}_h^{(n)} + r_H^{(n)}$
(5) *Transfer of the correction* $\tilde{e}_H^{(n)} := u_H^{(n)} - \hat{I}_h^H \bar{w}_h^{(n)}$
                                   $\tilde{e}_h^{(n)} := I_H^h \tilde{e}_H^{(n)}$
(6) *Correction* $\tilde{w}_h^{(n)} := \bar{w}_h^{(n)} + \tilde{e}_h^{(n)}$
(7) *Post-smoothing* $w_h^{(n)} = RELAX^{\nu_2}(\tilde{w}_h^{(n)}, L_h, f_h)$

$$(2.74)$$

The new coarse grid variable u_H approximates $\hat{I}_h^H u_h$, the full fine grid approximation represented on the coarse grid, because in case of convergence $(r_h \to 0)$ there is $u_H = \hat{I}_h^H u_h$. This justifies the name "full approximation scheme (FAS)" for this approach. Algorithm 5.5 in Section 5.3.1 shows a procedure how to implement the FAS-algorithm in practice on a sequence of grids.

Remark 2.20.

1. *The derivation of the FAS equations does not exploit the linearity of the problem. The equivalence of CS and FAS for linear problems is easily verified.*
2. *The step from CS to FAS is straight forward by introducing the new coarse grid variable u_H, composed of the restricted recent fine grid approximation and some coarse grid correction term. Therefore it is not surprising that Algorithm 2.7 differs only slightly from Algorithm 2.6. At first the restricted fine grid approximation $\hat{I}_h^H \bar{w}_h^{(n)}$ has to be computed to formulate the modified coarse grid problem and the computation of "full approximations" requires the determination of the correction $\tilde{e}_H^{(n)} = u_H^{(n)} - \hat{I}_h^H \bar{w}_h^{(n)}$ to transfer it to the finer grid.*
3. *The transfer of only corrections within the FAS (although full approximations are available) is highly important: only error terms are smoothed by relaxation, only smooth fine grid errors are approximated by coarse grid errors and only smooth H-grid quantities should be interpolated back to finer grids.*
   If a full solution $u_H^{(n)}$ were transferred back instead of the correction $u_H^{(n)} - \hat{I}_h^H \bar{w}_h^{(n)}$ this would introduce large interpolation errors. The smooth error guarantees small interpolation errors while the interpolation of the full solution not only enlarges the interpolation error but may also cause additional problems, for instance oscillations.
4. *It is essential to use the same restriction $\hat{I}_h^H$ in (4) and (5) of Algorithm 2.7.*

An interesting feature turns out by rewriting the FAS-coarse grid equation (2.72) to

$$L_H u_H = f_H + \tau_H^h \quad \text{with}$$
$$f_H := I_h^H f_h \text{ interpreted as a "special discretization" of } f \text{ and}$$
$$\tau_H^h := L_H \hat{I}_h^H w_h - I_h^H L_h w_h \text{ the fine to coarse defect correction.}$$

This has to be interpreted in the following sense: instead of the "original" coarse grid problem $L_H u_H = f_H$ a problem with modified right hand side is solved in order to make the coarse grid solution coincide with the fine grid solution restricted to the coarse grid.

The quantity τ_H^h—also called "relative local truncation (discretization) error of the H-grid with respect to the h-grid" is a cheaply determined by-product of the FAS-processing and establishes the basis for algorithmic variants. τ-estimation and τ-extrapolation techniques are key-words in this context. Of growing importance are algorithms which make use of the τ_H^h-quantity to determine regions where to refine grids adaptively—for instance the multilevel adaptive technique (Section 2.7). Comparing τ_H^h with the local discretization error

$$\tau_h := L_h I^h u - I^h L u$$

which is introduced by the discretization of the problem (interpreted as a transition from the continuous to the h-discrete formulation) shows the connection: τ_h is the "relative" local truncation error of the h-grid with respect to continuity, while τ_H^h is the corresponding quantity with respect to the $h - H$-transition.

For nonlinear problems the correction scheme is no longer applicable due to the missing superposition principle. Of course, the nonlinear problem can be linearized by some (outer) linearization process like the Newton's method. The CS may then serve to solve the corresponding linear problem within the outer iteration.

If $L(u) = f$ now defines a nonlinear problem and if $L_h(u_h) = f_h$ is its discrete nonlinear counterpart, an iterative linearization method to solve the discrete problem is characterized by a sequence of linear problems

$$L_h(u_h^{(k)}) + L_h^{(k)} e_h^{(k+1)} = f_h$$
$$u_h^{(k+1)} = u_h^{(k)} + e_h^{(k+1)}, \quad k = 0, 1, 2, \ldots$$

(to emphasize the nonlinearity the argument is put into brackets). $L_h^{(k)}$ is an approximation of $L_h'(u_h^{(k)})$, and particularly for the Newton's method $L_h^{(k)}$ is the Jacobian $L_h'(u_h^{(k)})$ which depends on the actual approximation $u_h^{(k)}$.

Example 2.18. *Let* $L(u) := \Delta u + g(u) = f$ *be the nonlinear equation to be solved on* Ω *with Dirichlet boundary conditions. The discretization of the partial differential operator establishes*

$$L_h(u_h) := \Delta_h u_h + g_h(u_h) = f_h.$$

The derivative of this operator with respect to u_h *is*

$$L_h'(u_h) := \Delta_h + g_h'(u_h)$$

with $g_h'(u_h) = \dfrac{\partial g_h}{\partial u_h}$. *Writing the (global) Newton's method for the discrete non-*

linear equation in terms of $u_h^{(k)}$ and $u_h^{(k+1)}$

$$L_h(u_h^{(k)}) + L_h'(u_h^{(k)})(u_h^{(k+1)} - u_h^{(k)}) = f_h$$

and inserting the expressions for $L_h(u_h^{(k)})$ and $L_h'(u_h^{(k)})$ ends in the sequence of linear problems

$$\Delta_h u_h^{(k+1)} + g_h'(u_h^{(k)})u_h^{(k+1)} = f_h - g_h(u_h^{(k)}) + g_h'(u_h^{(k)})u_h^{(k)}$$

for all $k = 0, 1, 2, \ldots$ with an initial guess $u_h^{(0)}$.

The computation of the Jacobian for each iteration step and the required extra storage for the coefficients of the linearized equations are the main drawbacks of this approach. To avoid this, the Jacobian is sometimes frozen to $L_h'(u_h^{(0)})$ in hope to have a good approximation to $L_h'(u_h^{(k)})$ for all $k > 0$. Such "approximate Newton's" methods usually converge only with first order while the Newton's method has convergence order greater or equal to two. Generally, the conflict of different convergence properties of inner and outer algorithm has to be resolved.

One strategy is to perform only one multigrid cycle within one linearization step. The relation between linearization work and multigrid solution work is not well balanced and as another disadvantage the second order convergence of the Newton's method may be destroyed. A more sophisticated strategy uses the order of the linearization method to adapt the number of cycles. If, for instance, the Newton's method converges quadratically, the number of multigrid cycles should be roughly doubled per step in order to resolve the conflict of different convergence properties.

The indirect approach (linearization and linear problem solver) is no longer required if the "direct" multigrid approach, using FAS and local nonlinear relaxation, is chosen. The FAS characterizing equations show that no global linearization is necessary. Due to the multigrid philosophy the intergrid transfer operators have to deal with smooth quantities. To achieve this, the linear smoothing now has to be replaced by nonlinear variants of relaxation schemes [109, 132]. The nonlinear relaxation method within FAS demands for each unknown, associated to a certain grid point, the solution of a nonlinear equation (point relaxation assumed). In principle, the same linearization methods can be used which are applied for global linearization approaches. Newton's method and simplified versions are commonly used.

Example 2.19. *The nonlinear equation to solve at a certain point (x_i, y_j) after some five-point discretization of $L(u) = f$ may be given by*

$$L_{i,j}(u_{i-1,j}, u_{i,j-1}, u_{i,j}, u_{i,j+1}, u_{i+1,j}) = f_{i,j}.$$

Assigning this equation to $u_{i,j}$ the Newton's method to compute a correction

$\tilde{e}_{i,j}^{(k+1)}$ *and a new approximation* $u_{i,j}^{(k+1)} = u_{i,j}^{(k)} + \tilde{e}_{i,j}^{(k+1)}$ *is defined by*

$$L'_{i,j}(u_{i-1,j}^{(k)}, u_{i,j-1}^{(k)}, u_{i,j}^{(k)}, u_{i,j+1}^{(k)}, u_{i+1,j}^{(k)}) \tilde{e}_{i,j}^{(k+1)}$$
$$= f_{i,j} - L_{i,j}(u_{i-1,j}^{(k)}, u_{i,j-1}^{(k)}, u_{i,j}^{(k)}, u_{i,j+1}^{(k)}, u_{i+1,j}^{(k)})$$

with

$$L'_{i,j}(\ldots, u_{i,j}^{(k)}, \ldots) = \frac{\partial L_{i,j}(\ldots, u_{i,j}, \ldots)}{\partial u_{i,j}}\bigg|_{(u_{i-1,j}^{(k)}, u_{i,j-1}^{(k)}, u_{i,j}^{(k)}, u_{i,j+1}^{(k)}, u_{i+1,j}^{(k)})}$$

The partial differential equation of Example 2.18 defines $L_{i,j}$ *to be*

$$\frac{1}{h^2}(u_{i-1,j} + u_{i,j-1} - 4u_{i,j} + u_{i,j+1} + u_{i+1,j}) + g_{i,j}(u_{i,j}).$$

The derivative $L'_{i,j}(\ldots u_{i,j}^{(k)} \ldots)$ *is* $g'_{i,j}(u_{i,j}^{(k)}) - \frac{4}{h^2}$ *and the Newton's method determines* $\tilde{e}_{i,j}^{(k+1)}$ *by*

$$\left(g'_{i,j}(u_{i,j}^{(k)}) - \frac{4}{h^2}\right)\tilde{e}_{i,j}^{(k+1)} = f_{i,j} - \left(\frac{1}{h^2}\begin{bmatrix} & 1 & \\ 1 & -4 & 1 \\ & 1 & \end{bmatrix} u_{i,j}^{(k)} + g_{i,j}(u_{i,j}^{(k)})\right).$$

For scalar equations and pointwise relaxation the "Jacobian" reduces to just a single value. If several coefficients of the equation depend on the actual unknown $u_{i,j}$ the computation of the derivative becomes more complicated and the decision has to be made whether this value is actualized within each iteration step or not. Additionally, the question arises when to actualize the solution dependent coefficients. This is discussed in Chapter 4 for a nonlinear diffusion operator of the form $\nabla(f(u)\nabla u)$. It is not a trivial decision to prescribe the accuracy upto which the local nonlinear equation has to be solved. For many applications only one Newton step per relaxation turned out to be sufficient. This is not surprising if a good initial guess is available. On the other hand, depending on the type of nonlinearity it may become imperative to get a very accurate approximation to the solution. Then more than one linearization steps become necessary to obtain good smoothing properties. This has been observed for the semiconductor device equations (Section 2.12).

Multigrid cycles based on FAS proceed analogously to CS-cycles. The components are chosen similar to the linear case. The following protocol of a $V(2,1)$-cycle reports about the solution process of a nonlinear Helmholtz-type equation which arises from the implicit time discretization of the nonlinear diffusion model problem (Chapter 4). Measuring the L_2-norms of the defect on different levels shows the same efficient behavior as demonstrated earlier for a CS-based algorithm applied to a linear Poisson problem.

```
GRID     5 L2-DEF-OLD 0.69316D+04 the defect at the beginning of
GRID     5 L2-DEF-NEW 0.54700D+04 the first cycle and after
GRID     5 L2-DEF-NEW 0.44515D+04 two pre-smoothing steps;
GRID     4 L2-DEF-NEW 0.30915D+04 the defect on the next coarser
GRID     4 L2-DEF-NEW 0.25796D+04 level after residual restriction
GRID     4 L2-DEF-NEW 0.12979D+04 and two pre-smoothing steps;
                           . . .
                             .
                           . . .
GRID     4 L2-DEF-NEW 0.92021D+03
GRID     4 L2-DEF-NEW 0.93554D+02
GRID     5 L2-DEF-NEW 0.31573D+04
GRID     5 L2-DEF-NEW 0.22341D+03
GRID     5 L2-DEF-RED 0.32230D-01 reduction of the defect by one cycle
```

Conclusion 2.4. *The main reasons to prefer the direct multigrid approach (FAS) for nonlinear problems are:*

1. *There is no need for a global linearization. A local one which avoids the computation of large Jacobian matrices and the storing of coefficients is sufficient.*
2. *From the applications point of view it is more promising to linearize as late as possible, so transporting characteristica of the problem into the solution process as far as possible.*
3. *The structure of the algorithm and its programming is similar to that of linear problems. A CS program is easily modified to an FAS program. The only difference is the calculation of both the coarse grid correction, step (5) in Algorithm 2.7, and the right hand side of the coarse grid equation, step (4) in Algorithm 2.7.*
4. *In many cases the components can be chosen with the aid of the same tools as in the linear case and oriented at similar linearized problems.*

Of course, a good initial guess is necessary to guarantee the convergence—but this condition is implicitly posed for all strategies due to the local convergence of Newton-like linearization methods.

2.6 Full Multigrid

Both the CS and the FAS have been applied iteratively to solve a discrete problem. For linear problems an often used initial guess is the zero grid function. But this choice is not always recommended—for nonlinear applications this approach may be impossible. Then one is faced with the problem to provide a reasonable initial guess. This should be obtained cheaply, that means the start value should be computed with an amount of work which is small compared to the further work to get a good approximation to the solution.

To solve the fine grid problem $L_h u_h = f_h$ by multigrid iteration, a sequence of grids characterized by decreasing meshsizes $h_1 > h_2 > \cdots > h_{M-1} > h_M$ is defined. The cycling starts on G_M for the problem $L_M u_M = f_M$ and proceeds to coarser grids. If, for instance the FAS-approach is used, the corresponding sequence of problems to be solved is

$$L_M u_M = f_M$$
$$\vdots$$
$$L_{l-1} u_{l-1} = f_{l-1} + \tau_{l-1}^l \quad \text{with} \quad f_{l-1} = I_l^{l-1} f_l \qquad (2.75)$$
$$\vdots$$
$$L_1 u_1 = f_1 + \tau_1^2 \quad \text{with} \quad f_1 = I_2^1 f_2$$

where for all $1 \le l \le M - 1$ the equations corresponding to two consecutive levels are mainly coupled by the "relative local discretization error". This formulation opens the way to calculate a coarse grid solution which is of fine grid accuracy in the intersection of the two levels.

If the continuous problem is discretized separately but with the same order of approximation on each level, the resulting hierarchy of problems differs from (2.75):

$$L_M u_M = f_M$$
$$\vdots$$
$$L_{l-1} u_{l-1} = f_{l-1} \qquad (2.76)$$
$$\vdots$$
$$L_1 u_1 = f_1$$

There is no coupling information between all these discrete problems. Consequently an accuracy statement like the one for the FAS-case should not be expected. The only direct link is given with respect to the continuous problem: the solution u_1 is an approximation to u upto discretization order, u_2 is an approximation to u upto $O(h_2^p), \ldots$, and u_M is an approximation to u upto discretization order $O(h_M^p)$, too. For a proper discretization the solution u_{l-1} is close to u_l, at least upto an order $O(h_{l-1}^p)$. If u_{l-1} is a solution with this property, it may serve on the next finer grid as an initial guess which is already very close to u_l, if the transfer is done properly. It is natural to hope that this initial guess can be improved to $O(h_l^p)$ accuracy by a small amount of computational work.

In principle, this idea—nested iteration—to find a reasonable initial guess for the fine grid iteration by computing approximations on coarser grids which are successively interpolated to finer grids can be used in combination with any method to solve discrete problems. If nested iteration and the multi-

$G_h = G_4$

$G_{2h} = G_3$

$G_{4h} = G_2$

$G_{8h} = G_1$

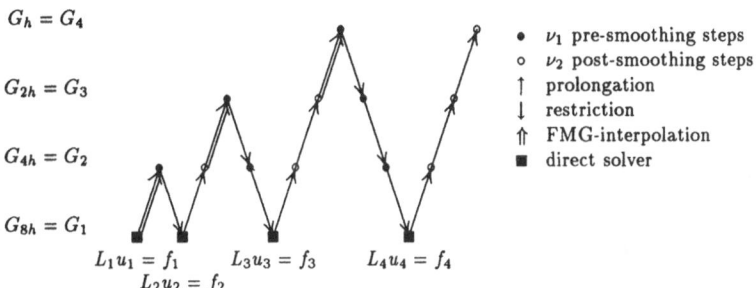

- • ν_1 pre-smoothing steps
- ○ ν_2 post-smoothing steps
- ↑ prolongation
- ↓ restriction
- ⇑ FMG-interpolation
- ■ direct solver

$$L_1 u_1 = f_1 \qquad L_3 u_3 = f_3 \qquad L_4 u_4 = f_4$$
$$L_2 u_2 = f_2$$

Fig. 2.24 The structure of FMG—nested iteration

grid iteration per level are plugged together, the final algorithm is called the "full multigrid (FMG)" method. The h-independent multigrid convergence essentially contributes to the final properties of FMG.

The standard FMG-procedure—successfully applied to many linear and nonlinear problems—solves the coarsest-grid problem exactly and continues with interpolation and cycling for all finer grids until the finest grid is reached, performing only one cycle on each level. The structure of such an FMG-process is shown in Figure 2.24.

Algorithm 2.8. *Full multigrid*
On a given hierarchy of grids defined by the meshsizes $h_1 > h_2 > \cdots > h_{M-1} > h_M$
perform

1. on the coarsest grid G_1 solve $L_1 u_1 = f_1$
2. for $l = 2, \ldots, M$
   (a) interpolate the initial guess $u_l^{(0)} = \mathbb{I}_{l-1}^l u_{l-1}$
   (b) start with this value to perform on level l n_l-multigrid cycles using the grids $G_1, G_2, \ldots, G_l$ to calculate the new approximation u_l.

The interpolation operator $\mathbb{I}_{l-1}^l$ or $\mathbb{I}_H^h$ is also called "FMG-interpolation" to express, in general, the difference from the correction transfer operator I_H^h. For FMG the different grids serve different purposes: a certain grid, considered as actually finest one, is used to compute an initial guess for the problem on the next finer grid. For this computation the coarser grids provide corrections within the multigrid cycling.

Instead of the V-cycle any other cycle type can be chosen. Fixed cycling schemes are possible as well as accomodative strategies. Of course, the number of multigrid cycles per level may vary and it is not absolutely necessary to start the FMG-procedure on the coarsest grid of the grid hierarchy used. A corresponding piece of FORTRAN-code shows these features. It is part of the program to solve the diffusion model problem of Chapter 4. Compare with Algorithm 5.5, too.

```
C
C===> FMG PROCEDURE
C===> the first level LIN is not necessary the coarsest one
C
      DO 370 L=LIN,M
      IF (L.EQ.LIN) GO TO 375
C
C===> FMG-interpolation only from level L-1 to L if L > LIN
C
      CALL FMGI4(L-1,L,.....)
  375 NCYC = 1
C===> determine the number of cycles on the
C===>           initial grid (where FMG starts)      NCLIN
C===>           grid levels LIN < L < M              NCL
C===>           finest grid of the FMG procedure     NCM
C===> REMARK:   the standard FMG cycle is NCLIN=NCL=NCM=1 and LIN=1
C===>           but: solve exactly on level LIN=1
      IF (L.EQ.LIN .AND. LIN.GT.1)      NCYC = NCLIN
      IF (L.GT.LIN .AND. L.LT.M)        NCYC = NCL
      IF (L.EQ.M)                       NCYC = NCM
C
C===> perform the determined and fixed number of cycles
C===> with fixed cycling structure using the FAS-formulation
C
      DO 350 ICYC=1,NCYC
      CALL FASFIX(L,...)
  350 CONTINUE
  370 CONTINUE
```

The FMG-algorithm as previously discussed combines the main ideas which have been known and applied independently from each other: nested iteration, coarse grid correction and the smoothing property of relaxation methods.

The second and third idea lead to multigrid cycling with the convergence behavior which is independent from h. The questions to be posed now are: Does it pay out to combine all three ideas? Does FMG offer further properties which are, perhaps, even more attractive? In this section, considering only model problems, the answer is "yes" in any case. But it is shown in Chapters 4 and 5 that the attractivity carries over to more complex situations.

At first one should remember the following: the aim is to have a solution u of $Lu = f$; but it is only possible to compute an approximation w_h. The final total error $\hat{e}_h = u - w_h$ is composed of the global error $u - u_h$, introduced by the discretization process, and of the algebraic error $u_h - w_h$, caused by the numerical solution of the discrete problem. It is appropriate to require a maximal error norm of magnitude ε:

$$\|\hat{e}_h\| \le \|u - u_h\| + \|u_h - w_h\| < \varepsilon$$

This condition is satisfied if each of the two contributing terms is less than $\varepsilon/2$. So it is sufficient to solve the discrete problem $L_h u_h = f_h$ on G_h (where h is chosen to guarantee $\|u - u_h\| < \varepsilon/2$) with the goal of $\|u_h - w_h\| < \varepsilon/2$ to be at the level of truncation error.

The most interesting theoretical result on FMG states that if the FMG-interpolation is chosen "properly" and if the multigrid convergence factor is "sufficiently" small, the algebraic error $u_h - w_h$ after one standard FMG-iteration is below the level of truncation

$$\|u_h - w_h\| \le (C + O(h^2))\|u - u_h\|$$

without any further cycling on the finest grid, where C depends on the norm of both the multigrid iteration operator and the FMG-interpolation operator and on n_l, [12, 49, 132]. Then

$$\|u - w_h\| \le (1 + C + O(h^2))\|u - u_h\|$$

follows immediately. The detailed discussion of C in [132] leads to the advice to use cycles with a multigrid convergence factor of approximately 0.1 ("sufficiently small"). This constraint is not severe because for most standard applications and even in more difficult applications the observed convergence rates are smaller than this value. To be sure that the interpolation is "chosen properly" one should follow the rule to have an interpolation order which is larger or equal to the discretization order. For second order discretization the cubic polynomial interpolation is a well established choice. For the diffusion model problem of Chapter 4 a variety of further interpolation methods satisfying additional conditions is proposed (Section 4.6).

The nice property to solve a problem upto the level of truncation has to be paid for with additional work comparing a standard FMG-step and a single multigrid cycle. The total work for a standard FMG-iteration envolving M levels of discretization is approximately

$$W_M^{FMG} \approx \sum_{l=2}^{l=M} (n_l W_l + W_{l-1}^{FMG-INT})$$

where $W_{l-1}^{FMG-INT}$ is the work to interpolate the G_{l-1}-solution u_{l-1} to the initial guess $u_l^{(0)} = \mathbb{I}_{l-1}^l u_{l-1}$. For a d-dimensional problem and standard coarsening the above estimate is simplified further if $W_{M-1}^{FMG-INT}$ is assumed to be equivalent to one RU and if W_M, the work corresponding to one cycle, is expressed in terms of relaxation units, too:

$$W_M^{FMG} \approx \frac{2^d}{2^d - 1}\left(\frac{2^d}{2^d - \gamma} n_M(\nu + 1) + 1\right) RU \qquad (2.77)$$

This formula then reduces for standard FMG ($n_M = 1$) in two dimensions ($d = 2$) and V-cycles ($\gamma = 1$) to

$$W_M^{FMG} \approx \frac{4}{3}\left(\frac{4}{3}(v + 1) + 1\right)RU.$$

Thus it is not more expensive than 6 to 9 relaxation units to solve upto the level of truncation if $V(1, 1)$- or $V(2, 1)$-cycles are used.

Because it only makes sense to compute solutions to the level of the truncation error, a natural question is to ask what method is more promising to do this—pure multigrid cycling or FMG-iteration. The finest grid of a d-dimensional model problem may consist of $\mathcal{N}$ gridpoints. The cycle with fixed v_1 and v_2 is assumed to have a certain convergence rate ρ, independent on the actual meshsize $h = 1/(\mathcal{N}^{1/d})$. If p-th order discretization is used, the reduction to the level of truncation $O(h^p)$ requires κ iterations. The total error reduction then is

$$\rho^\kappa = O(h^p) = O(\mathcal{N}^{-p/d}).$$

In other words, there are $\kappa = O(\log \mathcal{N})$ cycles necessary to reach the goal. Because the work per cycle is proportional to $\mathcal{N}$, the numerical cost of converging to the level of truncation by multigrid cycling is $O(\mathcal{N} \log \mathcal{N})$.

The reasoning with respect to FMG starts from the knowledge that the coarse grid problem with $H = 2h$ has already been solved by the FMG-procedure upto the level of truncation with respect to this grid

$$\|u_{2h} - w_{2h}\| = O((2h)^p) = O(2^p h^p)$$

or $\dfrac{1}{2^p}\|u_{2h} - w_{2h}\| = O(h^p).$

The aim is to get

$$\|u_h - w_h\| = O(h^p) = \frac{1}{2^p}\|u_{2h} - w_{2h}\|.$$

Therefore the algebraic error on grid G_h has to be reduced only by a factor $1/2^p$, starting from $w_h^{(0)} = \mathbb{I}_{2h}^h w_{2h}$ and assuming the proper choice of $\mathbb{I}_{2h}^h$. κ cycles have to be performed to yield $\rho^\kappa = 1/2^p$, that means, there are $\kappa = O(1)$ cycles necessary to reduce the h-grid error by the required factor. With similar arguments as above the computational cost to solve upto the level of truncation error with FMG is determined to be $O(\mathcal{N})$.

These considerations show why it is important to combine nested iteration with multigrid cycling to get the "optimality", the $O(\mathcal{N})$-property of FMG. If nested iteration is combined with "some" singlegrid method, the h-dependent convergence rate leads to $\rho(h)^\kappa = 1/2^p$ and therefore κ depends on h, respectively $\mathcal{N}$. The work to solve as accurate as FMG then is $O(f(\mathcal{N})\mathcal{N})$ with some function $f(\mathcal{N})$.

Example 2.20. *To demonstrate the potential of FMG in practice, this method is used to solve the problem of the Examples 2.3 and 2.10. One single FMG-step reduces errors as well as defects to values which are obtained by pure multigrid iteration only after seven to ten cycles. This fact is impressively demonstrated in Figure 2.25 where* $\log \| u - w_h \|_\infty$ *is plotted. The values after one FMG-step are marked by* $\bullet_h$ *(similar for 2h and 4h) and placed at positions corresponding to the numerical work which is invested. The horizontal and vertical lines are only drawn to support the easy determination of the number of cycles which is necessary to yield the same accuracy. Obviously, this number depends on the meshsize, that means on* $\mathcal{N}$, *thus pointing to the* $O(\mathcal{N} \log \mathcal{N})$*-property of multigrid iteration to reach the level of truncation.*

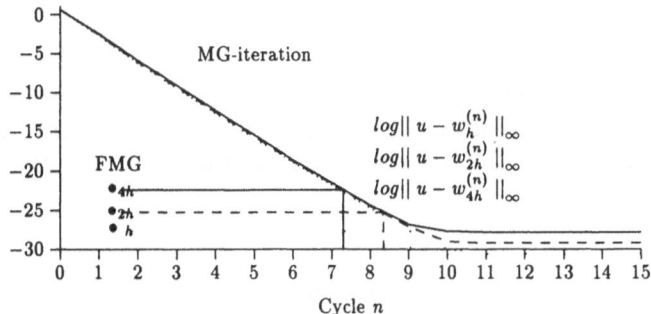

Fig. 2.25 Multigrid and FMG convergence, $h = \frac{1}{256}$

Conclusion 2.5. *There are two main features which authorize to understand FMG as an "optimal approximate direct solver" for* $L_h u_h = f_h$:

1. *an approximation* w_h *to the exact discrete solution* u_h *is computed upto an algebraic error* $\| u_h - w_h \|$ *which approximately has the size of the discretization error* $\| u - u_h \|$.
2. *the required work to do this is proportional to the number of gridpoints of* G_h.

FMG is especially important in connection with nonlinear problems because this approach provides excellent initial guesses which are necessary due to the only local convergence of the commonly used linearization schemes.

2.7 Multigrid Methods for Refined Grids

Many, perhaps most of the technically important applications request a special accuracy of the solution all over the computational domain. The goal to reduce the error everywhere below a given border leads in general to small meshsizes and therefore to a large number of equations on uniform grids. To avoid this, it is necessary to concentrate highly resolving meshes to the critical parts of the domain. As an example one may think at shocks, singularities, boundary layers or non-smooth boundaries which demand for the increased resolution. Following this condition a mesh like the one of Figure 2.12-(d) is not the best choice, although easy to handle. A more promising approach is the generation of refinements as indicated by Figure 2.26-(b). Such a (nearly) optimal generation of refinement areas usually assumes a well-known behavior of the discretization error. But such an information is not available in general and the *a priori* selection of refinement areas is extremely difficult or even impossible. This is the reason for adaptive strategies which determine the critical regions as the solution process progresses.

There are essentially two approaches: the first one uses physically based information of the problem to refine grids, while the second one, minimizing the local discretization error, is numerically motivated and is, from the numerical point of view, more reliable. Additionally, the principle is extendable to systems of equations, to higher order equations and to problems in more dimensions and it does not depend explicitly on the underlying physical problem.

2.7.1 Multilevel Adaptive Technique, MLAT

Within the MG approach "local refinements" are realized usually on a composite mesh which is the union of a sequence of grids with decreasing meshsizes. We call G_h a "refinement" or "patch" if $G_h \subset \mathscr{G}_h \cap (\Omega \cup \partial\Omega)$, where $\mathscr{G}_h$ is the infinite mesh of stepsize h. Grids that cover the complete computational domain are called "global" grids.

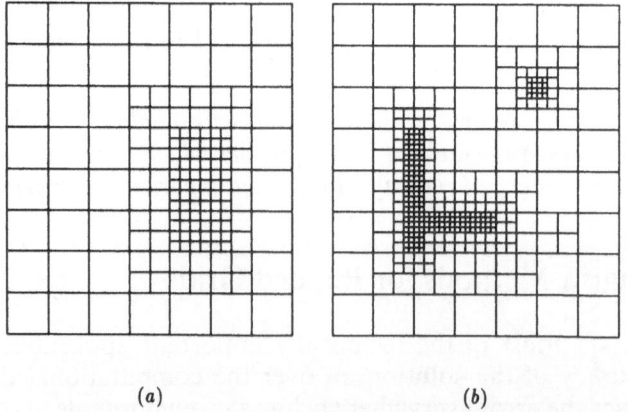

(a)   (b)

Fig. 2.26 Refined grids composed of rectangular patches

To explain the principle how to imbed local refinements into the multigrid algorithm a restriction to one simply connected refinement and some additional technical conventions suffice. Thus, the patches are assumed to be nested rectangular grids all having the same orientation. Additionally, the boundary lines of a fine grid coincide with lines of the next coarser one, and the meshsizes are decreased by a factor of $\frac{1}{2}$, from one grid to the next finer one. A composite mesh will, under these conditions, look like that of Figure 2.26-(a).

Dropping the technical convention of "nested rectangles" enables refined meshes which can be handled in the same way and the treatment of interior boundaries still remains relatively inexpensive. Additionally, several disjoint refinements of arbitrary shape within the domain may be necessary (see Figure 2.26-(b)). In any case, the standard discretization is performed on uniform subgrids. This offers a variety of advantages (see also Chapter 3). The principle of MG on a composite mesh is presented again by the corresponding two grid scheme. The set G_h° of "regular points" consists of all those points of G_h where the operator L_h (including the discrete boundary operator) is defined. Within the grid sequence $\{G_h\}_{h \in \mathcal{H}}$, indexed by $\mathcal{H} = \{h_l | h_l = h_1/2^{l-1}, l = 1, \ldots, M\}$, there should exist at least one global grid and at least one refinement $G_h \neq \phi$.

To solve the discrete problem on the composite grid $G_{\mathcal{H}} := \bigcup_{h \in \mathcal{H}} G_h$ the discretization with the finest possible meshsize has to be used. In this context, the coarse grid points of two consecutive grids G_h and G_H have different functions:

1. Those regular points of G_H which are also regular points of G_h $(G_h^\circ \cap G_H^\circ, \circ$ in Figure 2.27), are used to calculate corrections to the approximate solution on G_h°.

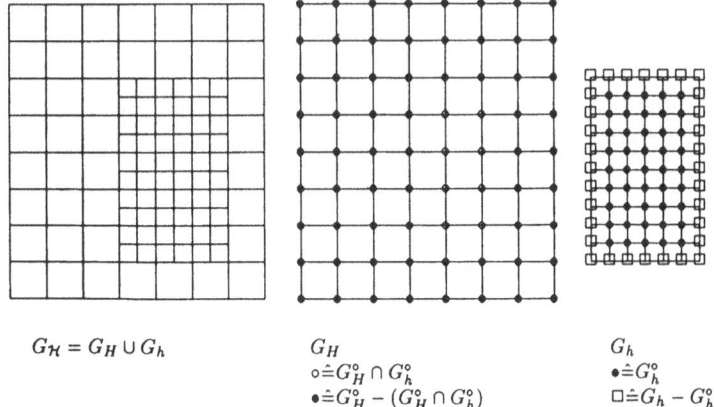

$$G_{\mathcal{H}} = G_H \cup G_h \qquad\qquad G_H \qquad\qquad\qquad G_h$$
$$\circ \doteq G_H^\circ \cap G_h^\circ \qquad\qquad \bullet \doteq G_h^\circ$$
$$\bullet \doteq G_H^\circ - (G_H^\circ \cap G_h^\circ) \qquad \square \doteq G_h - G_h^\circ$$

Fig. 2.27 Example of a composite mesh

2. For those regular points of G_H which are not regular points of G_h $(G_H^\circ - (G_H^\circ \cap G_h^\circ)$, • in Figure 2.27), G_H is the finest discretization level. A solution to the discrete problem has to be computed there.
3. To evaluate the discrete operator L_h at all regular points of G_h it is necessary to determine approximations to the solution at all "interior boundary points" of G_h $(G_h - G_h^\circ$, □ in Figure 2.27) because the solution has to fit together.

Satisfying these aims—computing an approximate solution as well as corrections to an approximate solution on the same level of discretization— becomes possible by the use of a slightly modified FAS within the MLAT (Multilevel Adaptive Technique) approach due to Brandt [12].
With an approximation u_H to the coarse grid problem and with $w_h^{(n)}$ as first approximation to the solution u_h of the problem (2.78) let $RELXLV^\nu(w_h, L_h, f_h, u_H)$ denote the result of ν relaxations for

$$L_h u_h = f_h \text{ on } G_h^\circ \text{ and}$$
$$u_h = \hat{I}_H^h u_H \text{ on } G_h - G_h^\circ. \tag{2.78}$$

Defining
$$F_H := \begin{cases} L_H \hat{I}_h^H u_h + r_H & \text{on} \quad G_H^\circ \cap G_h^\circ \\ f_H & \text{on} \quad G_H^\circ - (G_H^\circ \cap G_h^\circ) \end{cases} \tag{2.79}$$

simplifies the formulation of the FAS two grid scheme for composite grids. F_H due to (2.79) formally represents the fact that the FAS fine to coarse correction τ_H^h is omitted where it is not defined, or equivalently, that in those points where no corresponding difference equation with respect to a smaller meshsize exists, the original problem (with right hand side f_H) is solved. If, for a given point a corresponding grid point within a finer subgrid does exist, then the modified (corrected) coarse grid problem (with right hand side equal to $I_h^H f_h + \tau_H^h$) is considered.

Algorithm 2.9. MLAT

(1) *Pre-smoothing* $\qquad\qquad\qquad \bar{w}_h^{(n)} := RELXLV^{\nu_1}(w_h^{(n-1)}, L_h, f_h, u_H^{(n-1)})$
(2) *Computation of residuals* $\qquad r_h^{(n)} := f_h - L_h \bar{w}_h^{(n)}$
(3) *Restriction of residuals* $\qquad\;\; r_H^{(n)} := I_h^H r_h^{(n)}$
(4) *Exact solution of the* $\qquad\qquad L_H u_H^{(n)} = F_H \text{ from (2.79)}$
    coarse grid problem $\qquad\qquad\qquad\qquad\qquad\qquad\qquad\qquad\qquad$ (2.80)
(5) *Transfer of the correction* $\quad\;\; \tilde{e}_H^{(n)} := u_H^{(n)} - \hat{I}_h^H \bar{w}_h^{(n)}$
    $\qquad\qquad\qquad\qquad\qquad\qquad\qquad\;\; \tilde{e}_h^{(n)} := I_H^h \tilde{e}_H^{(n)}$
(6) *Correction* $\qquad\qquad\qquad\qquad\;\; \tilde{w}_h^{(n)} := \bar{w}_h^{(n)} + \tilde{e}_h^{(n)}$
(7) *Post-smoothing* $\qquad\qquad\qquad\; w_h^{(n)} = RELXLV^{\nu_2}(\tilde{w}_h^{(n)}, L_h, f_h, u_H^{(m)})$

For G_h a refinement, the interior boundary $G_h - G_h^\circ$ has to be treated carefully. The determination of function values on interior boundaries makes high order interpolation $\hat{I}_H^h$ necessary in order to maintain the order of consistency. The aforementioned conventions simplify these interpolations because the position of interior fine grid boundaries which coincide with coarse grid lines simplifies the determination of fine grid (interior) boundary values. Usually, the (bi)-cubic polynomial interpolation which is used as FMG-interpolation, too, is of sufficiently high order.

Another technique like "interpolation by relaxation of the interior boundaries" which is applicable without these conventions is presented in [118]. Both approaches have similar properties concerning accuracy and convergence. Further, this approach is a cheap one and uses the actually best available information—the difference equation.

To prolongate corrections the standard bilinear interpolation is sufficient, but it has to be pointed out explicitly that near interior boundaries the already updated interior boundary values have to be taken into account, too.

Remark 2.21. *The interpolation by relaxation of the interior boundary is based upon the use of rotated difference stencils. In Figure 2.28 this method is explained as well for the situation where the interior boundary of G_h coincides with H-grid lines (a) as for the situation where fine grid boundary lines are no coarse grid lines (b).*

In the first case, for points of the fine grid (interior) boundary (marked by • in picture (a) of Figure 2.28) which are also coarse grid points, the corresponding fine grid values are transferred directly. Then the values for additional points (marked by ∗) are computed by means of a rotated stencil with meshsize $\sqrt{2}h$, in case of the Laplacian

$$\frac{1}{2h^2}\begin{bmatrix} 1 & & 1 \\ & -4 & \\ 1 & & 1 \end{bmatrix}.$$

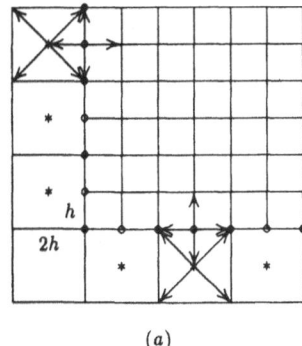
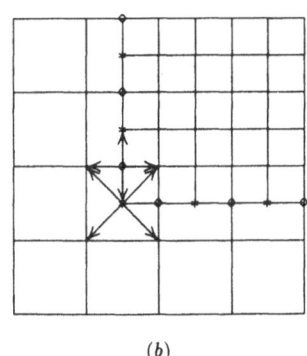

(a) (b)

Fig. 2.28 Interpolation by "boundary relaxation"

The result helps to calculate the grid function in those fine grid (interior) boundary points which do not belong to G_H ($\circ$ in Figure 2.28-(a)).
The procedure is similar and Figure 2.28-(b) has to be interpreted analogously if the interior boundary does not coincide with coarse grid lines. The rotated stencil is used to calculate values in $*$-points while the non-modified stencil with meshwidth h is applied for $\circ$-points.

2.7.2 An Easy to Determine Refinement Criterion

An analytic representation of the discretization error is impossible for many problems. Even good estimations are not available in general. But at least such estimations are necessary to guarantee that large errors are recognized and to avoid a refinement where it is not necessary at all. Fortunately, even for nonlinear cases, the FAS-scheme (2.80) offers a numerically cheap possibility to estimate the local discretization error with the aid of two consecutive grids.

Rewriting the FAS coarse grid equation (2.72) with $f_H = I_h^H f_h$ as

$$L_H u_H = L_H \hat{I}_h^H u_h + I_h^H (f_h - L_h u_h) = \tau_H^h + f_H$$

introduces the "relative local discretization error of G_H and G_h", $\tau_H^h := L_H \hat{I}_h^H u_h - I_h^H L_h u_h$. This quantity stands for the error which is caused by the substitution of u_h restricted to G_H into the H-grid equations. The local discretization error τ_H on G_H may be considered to be composed of the discretization error τ_h with respect to G_h and the relative local discretization error τ_H^h on G_H with respect to G_h:

$$\tau_H = \tau_h + \tau_H^h \quad \text{or better} \quad \tau_H = I_h^H \tau_h + \tau_H^h.$$

Assuming the existence of asymptotic expansions for the discretization error it is easy to show that $\tau_H^h = \tau_H - I_h^H \tau_h = \dfrac{H^q - h^q}{H^q} \tau_H$. Thus the relative local discretization error τ_H^h equals the discretization errors of interest; up to high order terms and a factor depending on the mesh size ratio and the order of consistency. For meshsizes $H = 2h$ and second order discretization this relation is $\tau_H = \frac{4}{3}\tau_H^h$. τ_H^h is an inherent information of the FAS scheme and is cheaply analyzed on the coarse grid G_H in order to determine where to refine the h-grid by a $h/2$-patch (see Section 4.9.3).

Having found a condition for the decision "where" to refine a grid the next question is "when" to adapt the grid. As there are no exact solutions but only approximations the aforementioned quantities change within the iteration process. This raises the question: when is the local discretization error approximated well enough by τ_H^h for the grid adaptation to be performed? In other words: when is the approximate solution well enough to ensure a good approximation of the local discretization error by τ_H^h? This in fact occurs only after some iterations as it is demonstrated in Section 4.9.4 for the diffusion model problem of Chapter 4.

The above described analysis of τ_H^h-information does not consider the relation of accuracy and invested work to achieve it. If the FMG-method is used in combination with refined subgrids the total work to solve upto the level of truncation is no longer proportional to the number of grid points because the number of grid points belonging to coarser levels may be considerably larger than that on the finer grids. The $O(\mathcal{N})$-property is regained if the FMG-algorithm is modified as described in the following section.

2.7.3 λ-FMG

In [10] adaptive discretization techniques are proposed which offer criteria to adapt both the meshsize and the order of approximation. The idea is to consider the two discretization parameters to be dependent from the position $(x, y) \in \Omega$. For technical simplicity in most real applications the approximation order is chosen once and never changed. It is more appropriate to vary the meshsize. This automatically leads to locally refined meshes. The discretization error τ then depends both on $h(x, y)$ and on the chosen $p: \tau = \tau(h(x, y), p)$.

For the following standard coarsening is assumed. To determine locally that $h(x, y)$ which minimizes the error due to a fixed amount of work (of course, also depending on p), the local work per unit volume and the weighted local discretization error $\tau(h(x, y), p)$ are integrated over the entire domain Ω to

$$\mathcal{W} = \int_\Omega \frac{w(p)}{h(x, y)^2} \, dx \, dy \quad \text{and} \quad \mathcal{E} = \int_\Omega G(x, y) \tau(h(x, y), p) \, dx \, dy.$$

The work per grid point w usually increases monotonously with p: with a fourth order approximation it is certainly larger than the solution work per grid node with a second order formula.

The optimization problem with respect to the meshsize then looks for an $h(x, y)$ which minimizes the error

$$\mathcal{E} = \int_\Omega G(x, y) \tau(h(x, y), p) \, dx \, dy,$$

investing a fixed amount of work C such that

$$\mathcal{W} = \int_\Omega \frac{w(p)}{h(x, y)^2} \, dx \, dy = C.$$

The corresponding Euler-Lagrange equation requires

$$\frac{\partial L}{\partial h} = 0 \quad \text{with} \quad L := G(x, y)\tau(h(x, y), p) + \lambda \frac{w(p)}{h(x, y)^2},$$

that means

$$\lambda = \frac{G(x, y)\dfrac{\partial \tau}{\partial h}(h(x, y), p)h(x, y)^3}{2w(p)}. \tag{2.81}$$

λ is the Lagrange multiplyer, a constant, and is viewed as a global control parameter in the following sense: provided that the functions $\tau(h(x, y), p)$ and $w(p)$ are known, as soon as an optimal pair $(\lambda^{opt}, h^{opt}(x, y))$ is determined, it satisfies (2.81), the local condition for the optimal grid size at (x, y). Selecting a $\lambda_1 > \lambda^{opt}$ and computing $h(x, y)$ due to (2.81) leads to grids where a larger error can be expected, while a smaller λ_1 improves the accuracy, provided the properly determined meshsize h is used. The optimal λ^{opt} expresses the change of optimal accuracy with respect to the invested work:

$$\lambda^{opt} = -\frac{d\mathscr{E}_{min}}{d\mathscr{W}}.$$

Usually, the optimal global information is not present. But local approximations to it are available. Within the solution process

$$\tau^h_{2h} = \tau_{2h} - \tau_h = \tau(2h, p) - \tau(h, p) \text{ approximates } h\frac{\partial \tau}{\partial h}. \text{ Inserting this into (2.81)}$$

establishes an approximate local optimization condition for h

$$\lambda = \frac{G\tau^h_{2h}h^2}{2w}.$$

The local relation of the error reduction due to the change of invested numerical work is

$$Q = -\frac{\Delta\mathscr{E}_{2h}}{\Delta\mathscr{W}_{2h}} = \frac{4G\tau^h_{2h}h^2}{3w} \approx -\frac{d\mathscr{E}_{min}}{d\mathscr{W}} = \lambda^{opt}, \tag{2.82}$$

a quantity which is easily computed on the $2h$-grid when $G(x, y)$ and $w(p)$ are given. As constant factors do not influence the optimization problem, the really used control quantity is

$$Q = G\tau^h_{2h}h^2.$$

If this "local" value Q is considerably larger than the optimal global control value λ^{opt}, a further refinement is recommended at the considered position. Similarly, a coarsening is allowed if the ratio is smaller than λ^{opt}. Because the h-dependency of Q is of fourth order, "considerably larger" stands for "16 times larger" to introduce a new $h/2$ level.

The refinement process is not restricted to introduce only a single new level. Depending on the actual size of Q on the evaluated grid, several new levels are possible. If $Q((x, y), h) > 16\lambda^{opt}$ then, due to the $O(h^4)$-estimation, $Q((x, y), h/2) > \lambda^{opt}$ and a refinement is required at the actual grid position. For a $Q((x, y), h) > 256\lambda^{opt}$ a further refinement with meshsize $h/4$ is justified, and so on.

The idea of λ-FMG is to use a sequence of control parameters $\lambda_0 > \lambda_1 > \cdots > \lambda_n = \lambda^{opt}$ with $\lambda_{i-1}/\lambda_i = 16$. For each of these λ_i the local optimization criterion determines a hierarchy of corresponding (dependent on λ_i) composite meshes

$$G_{\mathcal{H}(\lambda_0)}, \ldots, \; G_{\mathcal{H}(\lambda_i)} = \bigcup_{j=1}^{j=M_i} G_{ij}, \; \ldots, G_{\mathcal{H}(\lambda_n)}.$$

λ_0 is chosen large enough to get a rough approximation to the solution cheaply, for instance using only two (coarse) global grids. During the continuation process $\lambda_0 \to \lambda_n = \lambda^{opt}$ the solution of the λ_{i-1}-step is used as an initial approximation for the subsequent λ_i-step solution on the actually new generated grid sequence $G_{\mathcal{H}(\lambda_i)}$. The old solution is transferred directly where it is possible to do, or by higher order interpolation (FMG-interpolation). Such an interpolation is necessary if a new refinement area covers regions which were not yet refined (see also Sections 4.6 and 4.8). Figure 2.29 explains this approach. Instead of the sketched V-cycle any other cycle can be used.

At the end of the λ_0-step, the refinement criterion ($Q > 16\lambda_0$), for instance, forces the creation of two additional refinement levels. The transfer of the old solution from $G_{\mathcal{H}(\lambda_{i-1})}$ to $G_{\mathcal{H}(\lambda_i)}$ is indicated by $\to$. The start-approximations on actually created meshes are computed by higher-order (FMG-) interpo-

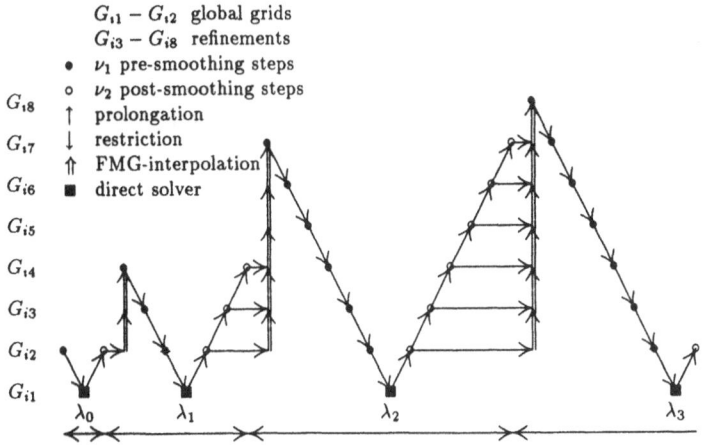

Fig. 2.29 The structure of λ-FMG

lation (⇑). Later on, the continuation process from λ_1 to λ_2 in Figure 2.29 requires the new levels G_{25} to G_{27}. Again, the initial guesses for the actual problem are obtained by a direct transfer on the global grids and on the non-vanishing intersections of corresponding grid levels ($\rightarrow$) while values for grid points in formerly not existing refinement areas are interpolated as before (⇑). Completely new levels are supplied with grid functions by higher order interpolation, anyhow.

This λ-FMG creates several new levels all at once. A modification due to [118] is more "adapted" to the development of the solution and very promising for hard problems. Marching from the λ_{i-1}-step to the λ_i-step, only one new level is created. After one cycle using this new level, the actual information is used to decide whether another refinement is necessary within the current λ_i-step. By this way the best available information is used to generate a finer grid patch. Figure 2.30 shows how the λ_2-step of Figure 2.29 looks like with the "adapted" λ-FMG.

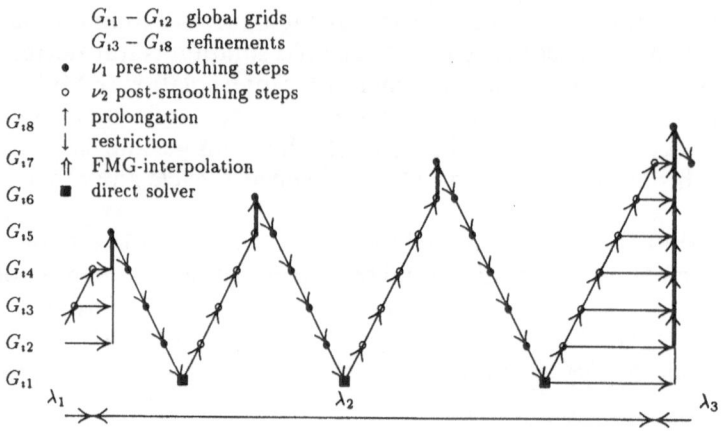

Fig. 2.30 The λ_2-step of the previous Figure with the "adapted" λ-FMG

Both of the mentioned λ strategies have been applied to Poisson-problems with singular behavior of the solution. The following results, due to [118] for the Poisson equation on $(-1,1)^2$ with cut from the origin to the point $(1,0)$, show the power of the method comparing it with data coming from an algorithm using exclusively global grids.

	exclusively global grids	refined grids
finest global meshsize	$\frac{1}{128}$	$\frac{1}{4}$
number of global grids	7	2
number of refined grids	0	5
number of gridpoints	64.897	315
$\|e_h\|_\infty = \|u_h - u\|_\infty$	1.7D-2	1.8D-2

Remark 2.22. *One "RU" has been considered so far as the numerical work corresponding to one relaxation over the finest global grid. As long as exclusively global grids are used, the overall work per cycle does not depend on the meshsize, respectively on the number of levels used. This was derived using the argument that coarse grid work is neglectable compared to fine grid work. As soon as refinements are introduced, the number of points on a refined grid may be smaller than that of the finest global grid and the assumption is no longer valid. The proposed λ-FMG regains the well-known efficiency and makes the cost of multigrid on refined meshes again proportional to the number of grid points belonging to the composite mesh.*

2.7.4 The Fast Adaptive Composite Grid Method, FAC

A different approach to combine multigrid and locally refined grids is FAC, the "Fast Adaptive Composite grid method [96]" with the variant AFAC "Asynchronous Fast Adaptive Composite grid method [96]". Comparing the descriptions of MLAT and FAC one may get the impression of very different approaches. In [97] the main differences are summarized. But, at least for model problems, the two approaches are quite similar.

For the discussion of the main ideas, the Poisson equation with Dirichlet boundary conditions on the unit square is used. Generalizations to other boundary conditions, nonlinear equations, to systems of equations and

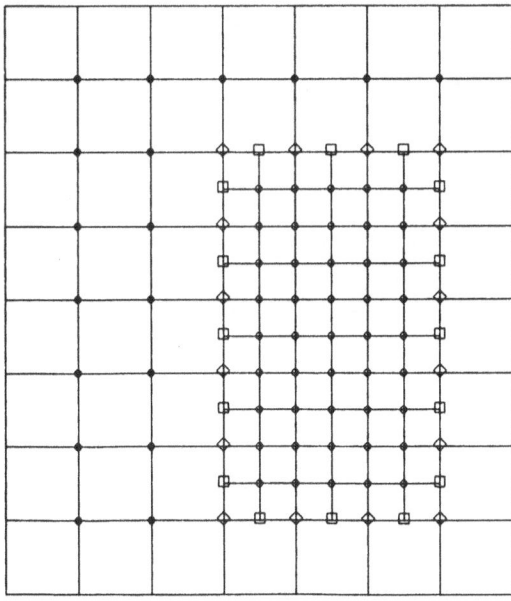

Fig. 2.31 Composite grid for the model problem

multilevel multigrid are similar as in standard multigrid on global grids. The analogon to λ-FMG in MLAT is the FACNI (FAC Nested Iteration). A complete presentation of FAC and AFAC is given in [96].

The partial differential equation to be solved is discretized on a composite grid, which is for the subsequently discussed model problem the grid of Figure 2.31. It is the union of two nested grids, a global square grid G_H of meshsize H and a local grid G_h of meshsize $h = \frac{1}{2}H$.

The standard discretization approach in the context of FAC is to use *finite volume elements* (FVE) which combine ideas both from finite elements and finite volumes. Here it is considered as a finite element approximation v expressed in terms of its nodal values. For the nodal values a matrix equation is derived by the integration over "control volumes".

In case of the model problem a triangulation as indicated in Figure 2.32 is defined. The approximation v is expressed in the form

$$v = \sum_k u_k \Phi_k$$

where the summation runs over all interior nodes of Figure 2.32. u_k is the value of v at node k and Φ_k is the so-called "hat" function associated with the node k. The u_k are determined by the requirement

$$\int_V -\Delta u = \int_V f \, dV$$

for all control volumes V (indicated by the dashed lines in Figure 2.32). The final discretization is characterized by:

- For points marked by • in Figure 2.31 the standard five-point star with

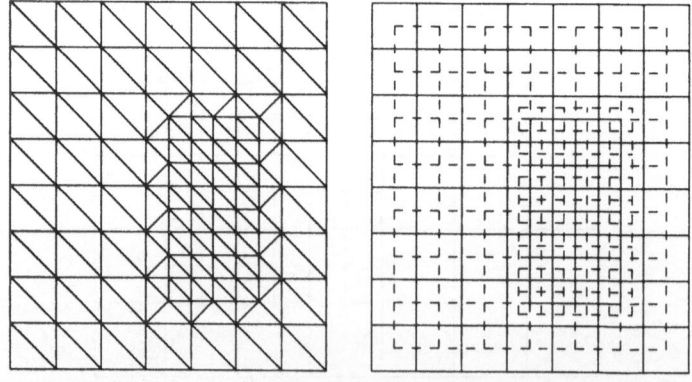

Fig. 2.32 Triangulation and control volumes for the model problem

meshsize H

$$\frac{1}{H^2}\begin{bmatrix} & -1 & \\ -1 & 4 & -1 \\ & -1 & \end{bmatrix}$$

is applied.
- At points marked by ∘ in Figure 2.31 the five-point star with meshsize h is obtained.
- The so-called composite grid star is derived for points marked by ◇ in Figure 2.31. At the upper surface of the patch this star is, for instance, given by

$$\frac{4}{3H^2}\begin{bmatrix} & & -1 & & \\ -0.5 & & 4 & & -0.5 \\ & -0.5 & -1 & -0.5 & \end{bmatrix}.$$

As indicated by the notation of the star and by Figure 2.33, the discretization involves points both of the coarse grid G_H and of the fine grid G_h. Similar stencils are given on the other boundary segments of G_h.
- At points marked by □ in Figure 2.31 the value is obtained by linear interpolation from two neighboring points (marked by ◇).

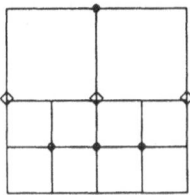

Fig. 2.33 Points involved in the composite grid star (model problem)

From now the special composite grid $G_{\mathcal{H}}$ is the union of the global regular grid G_H and a local regular grid G_h with $h = \frac{1}{2}H$ (Figure 2.31). Quantities defined on them are marked by the corresponding formal discretization quantity. The discrete operators L_H and L_h are derived by FVE, too. For the model problem L_H and L_h are described by the normal five-point stars with meshsize H and h.

In addition some grid transfer operators for the composite grid are necessary. This are $I_{\mathcal{H}}^H$, $I_H^{\mathcal{H}}$, $I_{\mathcal{H}}^h$ and $I_h^{\mathcal{H}}$ (in the usual notation; e.g. $I_{\mathcal{H}}^H$ is the restriction from $G_{\mathcal{H}}$ to G_H). Outside $\bar{G}_h$ these operators are defined in a natural way, e.g. $I_{\mathcal{H}}^H$ is the identity there.

Now one FAC-cycle is given by (following the notation in [96] the cycle is described in the form of a V(0,1)-Cycle):

Algorithm 2.10. *FAC-cycle*

Step 1 (solution on G_H)

$$f_H := I_{\mathcal{H}}^H (f_{\mathcal{H}} - L_{\mathcal{H}} u_{\mathcal{H}}^{(n)})$$
$$L_H \tilde{e}_H = f_H$$
$$u_{\mathcal{H}}^{(n+0.5)} = u_{\mathcal{H}}^{(n)} + I_H^{\mathcal{H}} \tilde{e}_H$$

Step 2 (solution on G_h)

$$f_h := I_{\mathcal{H}}^h (f_{\mathcal{H}} - L_{\mathcal{H}} u_{\mathcal{H}}^{(n+0.5)})$$
$$L_h \tilde{e}_h = f_h$$
$$u_{\mathcal{H}}^{(n+1)} = u_{\mathcal{H}}^{(n+0.5)} + I_h^{\mathcal{H}} \tilde{e}_h$$

In Step 2 homogeneous Dirichlet boundary conditions are posed on ∂G_h.

Remark 2.23. *Due to [96], p. 90: "FAC should not be interpreted as a method that requires exact solvers, even though we have defined it this way. In fact, while FAC allows for direct methods to resolve the subgrid equations, its predominant use in practise has been with iterative methods. Our main reasons … are … to pave the way for AFAC."*

AFAC uses the same discretization approach as FAC does. The main idea of AFAC is to decouple Step 1 and Step 2 of the FAC-Cycle. The straight forward decoupling generally does not converge. The reason is that smooth errors on G_h are overcorrected. This overcorrection is avoided by introducing Step 3 of the following AFAC-algorithm. Let G_F be the restriction of G_H to the refinement region, $I_H^{\mathcal{H}}$ and $\bar{I}_{\mathcal{H}}^H$ the intergrid transfer operators restricted to the refinement region.

Algorithm 2.11. *AFAC-cycle*

Step 1 (solution on G_H)

$$f_H := I_{\mathcal{H}}^H (f_{\mathcal{H}} - L_{\mathcal{H}} u_{\mathcal{H}}^{(n)})$$
$$L_H \tilde{e}_H = f_H$$

Step 2 (solution on G_h)

$$f_h := I_{\mathcal{H}}^h (f_{\mathcal{H}} - L_{\mathcal{H}} u_{\mathcal{H}}^{(n)})$$
$$L_h \tilde{e}_h = f_h$$

Step 3 (solution on G_F)

$$f_F := \bar{I}_{\mathcal{H}}^H (f_{\mathcal{H}} - L_{\mathcal{H}} u_{\mathcal{H}}^{(n)})$$
$$L_H \tilde{e}_F = f_F$$

Step 4 (summation of the corrections)

$$u_{\mathcal{H}}^{(n+1)} = u_{\mathcal{H}}^{(n)} + I_H^{\mathcal{H}} \tilde{e}_H + I_h^{\mathcal{H}} \tilde{e}_h - \bar{I}_H^{\mathcal{H}} \tilde{e}_F$$

The boundary conditions in Step 2 and Step 3 are corresponding to the boundary conditions in Step 2 of Algorithm 2.10.

Remark 2.24 *Since G_F is both local and coarse, Step 3 is much cheaper than Step 2. Moreover if multigrid is to be used as the subgrid solver, the additional work of Step 3 can be reduced further. The cost of AFAC in terms of the total number of arithmetic operations is therefore comparable to that of FAC.*

Because FAC and AFAC use the same discretization, the most obvious difference between the methods should be observed when considering their numerical performance.

The disadvantage of AFAC is that, due to the decoupling of the problems on the coarse and fine grid, the convergence rates of FAC and AFAC are related by $\rho_{AFAC} = \sqrt{\rho_{FAC}}$. Under special assumptions this is proved in [96]. The relation is also confirmed by numerical results. $\rho_{AFAC} = \sqrt{\rho_{FAC}}$ implies that two cycles of AFAC are roughly equivalent to one cycle of FAC. Because the cost of AFAC in terms of the total number of arithmetic operations is comparable to that of FAC, AFAC is useless on a sequential computer. The advantage of AFAC is that AFAC allows a more efficient parallelization than FAC. The parallelization of AFAC is described in [96], the appropriate parallelization of both approaches in [76].

It is worth to compare MLAT and FAC—the used discretizations and multigrid cycles. In FAC the discretization is given explicitly. The discussion with respect to MLAT assumes to be "at convergence", that means the equations are considered to have zero-residuals. Then the discretization properties of the two methods can be summarized as follows:

- Both methods can be interpreted in the way, that the unknowns are the values at all points of $G_{\mathcal{H}}$.
- At the interior points of G_h and of $G_H \backslash G_h$ (marked by • and ∘ in Figure 2.31) both approaches use the discretization for regular grids with meshsizes h and H, respectively. In the case of the model problem these discretizations are given by the standard five-point star.
- Both methods interpolate values at those points of the interface which are not coarse grid points (marked by ▫ in Figure 2.31). The order of the interpolation, which is used typically, is different. MLAT uses cubic interpolation and FAC linear interpolation.
- MLAT and FAC use different discretization formula at points on the interface belonging to G_H (marked by a ◇ in Figure 2.31). MLAT uses the normal discretization at points of the mesh G_H. FAC uses the composite grid star involving points of both G_H and G_h.

The following list summarizes the deviations of a FAC cycle from the standard MLAT cycle. The first item is the most important one while the others are more formally.

- At coarse grid points on the interface, the right hand side of the FAC coarse grid problem involves also the last approximate solution at some fine grid points. This is a consequence of the composite grid star.
- For linear equations FAC is described as a correction scheme. A corresponding reformulation of MLAT is possible, but unusual.
- In the description of FAC the equations on each level are solved exactly (compare Remark 2.24).

The numerical results in [96] and in [118] show, that the convergence properties on sequential machines are similar. The appropriate parallelization strategies of MLAT and FAC are similar, too.

Conclusion 2.6. *At least for standard situations the MLAT and the FAC approaches are very similar. Further investigations are necessary to allow a more detailed comparison of the numerical behavior of MLAT and FAC with respect to the actual error, to the convergence rates per multigrid cycle and with respect to parallelization aspects.*

2.8 Parabolic Initial Boundary Value Problems

Because it is not intended to discuss the total variety of numerical methods for time-dependent problems, the class of problems is chosen in such a way that the application of multigrid methods is a reasonable and attractive approach. There is only a sparse experience with multigrid applied to general parabolic equations. But for evolution problems which are characterized by the model problem "heat equation"

$$Lu := \frac{\partial u}{\partial t} - Au = f \text{ on } \Omega \times (0, T]$$

$$Bu = g \text{ on } \partial\Omega \text{ for all } t \in (0, T] \tag{2.83}$$

$$u = u^0 \text{ on } \bar{\Omega} = \Omega \cup \partial\Omega \quad \text{and} \quad t = 0$$

the results are very promising. In (2.83) A stands for a second order linear or nonlinear elliptic differential operator containing exclusively spatial derivatives, while B denotes the boundary operator. The wanted solution $u(x, y, t)$ is defined on $\bar{\Omega} \times [0, T]$. $[0, T]$ is the time interval of interest and $\Omega \subset \mathbb{R}^2$ is the space domain. The right hand side $f(x, y, t)$ is a given function, the boundary condition $g(x, y, t)$ is defined on $\partial\Omega$ for all $t \in (0, T]$, while the initial condition $u(x, y, 0) = u^0(x, y)$ is valid on $\bar{\Omega}$ for $t = 0$.
In [127] and [128] a representative set of both discretization schemes and solution methods is investigated considering a nonlinear heat transfer problem whith $Au := \nabla(a(u)\nabla u)$ where $a(u)$ allows a transformation into $Au := a(u)\Delta u$ (after renaming). They conclude that implicit schemes in combi-

nation with a time stepping procedure and fast elliptic solution methods lead to efficient algorithms. Because many of the evolution problems of process simulation are similar to the cited heat transfer problem it is justified to expect efficient algorithms by similar approaches.

Nevertheless, to apply multigrid on evolution problems one should keep in mind problem characteristica and investigate whether there are reasons which make extremely small time-steps imperative. If such restrictions are valid then the use of explicit time discretization schemes (for instance the explicit Euler method (ExE)) or implicit discretization combined with classical solution methods is recommended (for instance ADI or SOR).

If large time-steps are allowed from the physical problem background an implicit discretization technique may be chosen—for instance in combination with a time stepping procedure. Well-known and to some degree easily analyzed versions are the implicit (backward) Euler method (BE) and the Crank-Nicolson (CN) scheme. Of course, accuracy requirements and numerical stability conditions have to be considered, too.

To get a more compact representation, the boundary conditions and the initial condition are not always mentioned. Although the operator A and its discrete counterpart may be nonlinear, a global linearization is not considered because the representation assumes the application of FAS, where the local linearization is hidden within the nonlinear relaxation method.

Due to their importance and exploiting the well-known and compact homotopy-representation of the ExE, CN and BE schemes, only these methods are discussed further. With a $\beta \in [0, 1]$ the time discrete representation of (2.83) is formally given by

$$\frac{u(t + \Delta t) - u(t)}{\Delta t} = \beta A u(t + \Delta t) + (1 - \beta) A u(t) + f$$

where β selects the method: $\beta = \begin{cases} 1 & \text{Backward-Euler (BE)} \\ \frac{1}{2} & \text{Crank-Nicolson (CN)} \\ 0 & \text{Explicit Euler (ExE)} \end{cases}$

Introducing $t^{n+1} = t^n + \Delta t^{n+1}$ and $u^n = u(t^n)$ the above formal representation can be rewritten to

$$\frac{u^{n+1} - u^n}{\Delta t^{n+1}} = \beta A u^{n+1} + (1 - \beta) A u^n + f^n.$$

Collecting terms of the same time level leads to

$$\left(-\beta A + \frac{Id}{\Delta t^{n+1}} \right) u^{n+1} = \left((1 - \beta) A + \frac{Id}{\Delta t^{n+1}} \right) u^n + f^n$$

$$\text{or } \cdot L_{\Delta t^{n+1}} u^{n+1} = F_{\Delta t^{n+1}}^n \quad \text{where}$$

$$L_{\Delta t^{n+1}} := -\beta A + \frac{Id}{\Delta t^{n+1}} \quad \text{and}$$

$$F^n_{\Delta t^{n+1}} := \left((1-\beta)A + \frac{Id}{\Delta t^{n+1}} \right) u^n + f^n. \tag{2.84}$$

Example 2.21. *Setting $A := \Delta$ and $f \equiv 0$ the time discrete equations are given by*

$$\frac{u^{n+1}}{\Delta t^{n+1}} = \Delta u^n + \frac{u^n}{\Delta t^{n+1}} \quad for \quad \beta = 0, (ExE)$$

$$\left(-\frac{1}{2}\Delta + \frac{Id}{\Delta t^{n+1}} \right) u^{n+1} = \left(\frac{1}{2}\Delta + \frac{Id}{\Delta t^{n+1}} \right) u^n \quad for \quad \beta = \frac{1}{2}, (CN)$$

$$\left(-\Delta + \frac{Id}{\Delta t^{n+1}} \right) u^{n+1} = \frac{u^n}{\Delta t^{n+1}} \quad for \quad \beta = 1, (BE).$$

In case of implicit methods, $L_{\Delta t}$ obviously is a "Helmholtz-like" operator. For such elliptic problems within each time level fast elliptic solver and especially multigrid algorithms can be expected to be efficient. Nevertheless, one should keep in mind the above mentioned restrictions for the application of multigrid methods to parabolic problems.

The formal description of the completely discretized problem is obtained from (2.84) by replacing continuous quantities by their discrete counterparts

$$\left(-\beta A_h + \frac{Id_h}{\Delta t^{n+1}} \right) u_h^{n+1} = \left((1-\beta)A_h + \frac{Id_h}{\Delta t^{n+1}} \right) u_h^n + f_h^n$$

$$\text{or} \quad L_{h,\Delta t^{n+1}} u_h^{n+1} = F^n_{h,\Delta t^{n+1}} \quad \text{with}$$

$$L_{h,\Delta t^{n+1}} := -\beta A_h + \frac{Id_h}{\Delta t^{n+1}} \quad \text{and}$$

$$F^n_{h,\Delta t^{n+1}} := \left((1-\beta)A_h + \frac{Id_h}{\Delta t^{n+1}} \right) u_h^n + f_h^n. \tag{2.85}$$

This shows again, that a sequence of—now discrete—"Helmholtz-like" problems has to be solved if an implicit time discretization is used.

Example 2.22. *If $A = \Delta$ is discretized by the standard second order finite difference approach and if $L_{h,\Delta t^{n+1}} = -\beta A_h + \frac{Id_h}{\Delta t^{n+1}}$ is written in stencil notation, the difference operator is, for inner points of the domain, represented* by

$$\frac{1}{h^2}\begin{bmatrix} & -\beta & \\ -\beta & 4\beta + \dfrac{h^2}{\Delta t^{n+1}} & -\beta \\ & -\beta & \end{bmatrix}$$

For fixed h and with $\beta \neq 0$ it is obvious: if Δt^{n+1} is small, the term $\dfrac{h^2}{\Delta t^{n+1}}$ increases the diagonal term. In other words, it improves the diagonal dominance. This explains the above-mentioned possibility to apply classical solution methods if either coarse meshsizes or small Δt are used, because they are very efficient in such situations. On the other hand, multigrid convergence does not dramatically depend on the degree of the diagonal dominance. Therefore it can be expected to get efficient methods even if $\dfrac{h^2}{\Delta t^{n+1}}$ becomes small. The convergence rate should not be worse than that of the corresponding stationary problem. So multigrid as a method to solve the discrete problem within each time step may still work when the convergence of classical methods may have been detoriated.

The underlying problem is simplified by assuming f and g to be independent from t. For only space-dependent operators (here $A = \Delta$) and Dirichlet boundary conditions ($B = Id$) this allows a splitting (separation)

$$u(x, y, t) = v(x, y) + w(x, y, t),$$

where $v(x, y)$ denotes the solution of the corresponding inhomogeneous stationary boundary value problem

$$- Av = f \text{ on } \Omega$$
$$v = g \text{ on } \partial\Omega$$

showing that $w(x, y, t)$ is the solution of an homogeneous initial boundary value problem:

$$\frac{\partial w}{\partial t} - Aw = 0 \text{ on } \Omega \times (0, T]$$

$$w = 0 \text{ on } \partial\Omega \text{ for all } t \in (0, T] \tag{2.86}$$
$$w = w^0(x, y) := u^0(x, y) - v(x, y) \text{ on } \bar{\Omega} \text{ at } t = 0.$$

For $\Omega = (0, 1)^2$ the solution $w(x, y, t)$ possesses a representation

$$w(x, y, t) = \sum_{l,m=1}^{\infty} c_{l,m} e^{-(l^2 + m^2)\pi^2 t} \sin l\pi x \cdot \sin m\pi y \tag{2.87}$$

The $\varphi_{l,m}(x, y) = \sin l\pi x \cdot \sin m\pi y$ are the eigenfunctions of the Laplacian and

$\lambda_{l,m} = -(l^2 + m^2)\pi^2$ are the corresponding eigenvalues. $c_{l,m}$ are the Fourier-coefficients of w_0 with respect to the above set of eigenfunctions. Generalizing the notation of the elliptic case, where $\varphi_{l,m}$ is called a "frequency" or a "mode", it is now the product $e^{-(l^2+m^2)\pi^2 t}\varphi_{l,m}$ which one should have in mind if these words are used.

To get some deeper insight into the structure of the solution of evolution problems, the approach of discretizing with respect to time and then with respect to the space variables is reversed now. The space discrete problem derived from (2.86) then can be considered as a system of ordinary differential equations (A_h includes the boundary conditions):

$$\dot{w}_h - A_h w_h = 0 \text{ for all } (x, y) \in G_h, t \in (0, T] \tag{2.88}$$
$$w_h(0) = w_{0,h} \text{ on } \bar{\Omega}_h$$

$\dot{w}_h$, w_h and $w_h(0)$ are vectors with $(N-1)^2$ components if $h = 1/N$. A_h has to be interpreted as a $(N-1)^2 \times (N-1)^2$ matrix. The transposed of the vector $\dot{w}_h(t)$ is, for instance,

$$\left(\ldots, \frac{\partial w}{\partial t}(x, y, t), \frac{\partial w}{\partial t}(x, y+h, t), \ldots, \frac{\partial w}{\partial t}(x+h, y, t), \ldots \right),$$

and

$$w_h(0)^T = (\ldots, w(x, y, 0), w(x, y+h, 0), \ldots, w(x+h, y, 0), \ldots).$$

The component of w_h corresponding to the node $(x, y) \in G_h$ is given by

$$w_h(x, y, t) = \sum_{l,m=1}^{l,m=N-1} c_{l,m}^h e^{-\lambda_{l,m}^h t} \sin l\pi x \cdot \sin m\pi y \text{ with} \tag{2.89}$$

$$\lambda_{l,m}^h = \frac{1}{h^2}[4 - 2\cos l\pi h - 2\cos m\pi h].$$

The coefficients $c_{l,m}^h$ again depend on the initial condition. Some properties of the $\lambda_{l,m}^h$ are:

1. $\lambda_{l,m}^h > 0$ for all l, m, that means $e^{-\lambda_{l,m}^h t} \to 0$ for $t \to \infty$.
2. $\lambda_{l,m}^h = (l^2 + m^2)\pi^2 + O(h^2)$.
   Again, the discrete eigenvalues are shifted compared to the eigenvalues of the underlying continuous problem.
3. $\lambda_{1,1}^h$ and $\lambda_{N-1,N-1}^h$ are the absolutely smallest and largest eigenvalues, respectively.

$$\lambda_{1,1}^h = \frac{4}{h^2}(1 - \cos \pi h) \approx 2\pi^2 + O\left(\frac{1}{N^2}\right).$$

$$\lambda_{N-1,N-1}^h = \frac{4}{h^2}(1 + \cos \pi h) \approx 8N^2 - 2\pi^2 + O\left(\frac{1}{N^2}\right).$$

$$\frac{\lambda_{N-1,N-1}^h}{\lambda_{1,1}^h} = O(N^2).$$

Especially the last point shows that the eigenvalues are of significantly different order of magnitude and that they differ the more the smaller h becomes. This is the "stiffness" of the system (2.88). From the theory of systems of ordinary differential equations it is known that implicit methods are well-suited for such stiff problems. This is a more theoretically based reasoning to use implicit time discretization methods leading to sequences of (discrete) elliptic problems which can be solved efficiently by fast elliptic solvers — including multigrid algorithms, of course.

Continuing with the time discretization ends in the completely discretized problem (compare with (2.85))

$$\left(-\beta A_h + \frac{Id_h}{\Delta t^{n+1}}\right) u_h^{n+1} = \left((1-\beta)A_h + \frac{Id_h}{\Delta t^{n+1}}\right) u_h^n. \qquad (2.90)$$

Inserting a general eigenfunction $\Phi_{l,m}^h(x,y,t)$, for instance $\exp[-\lambda_{l,m}^h t]\sin l\pi x \cdot \sin m\pi y$ in case of $A := \Delta$, the relation

$$\left(-\beta \Lambda_{l,m}^h + \frac{1}{\Delta t^{n+1}}\right)\Phi_{l,m}^{h,n+1} = \left((1-\beta)\Lambda_{l,m}^h + \frac{1}{\Delta t^{n+1}}\right)\Phi_{l,m}^{h,n} \qquad (2.91)$$

or in divided form
$$\Phi_{l,m}^{h,n+1} = \frac{1 + \Delta t^{n+1}(1-\beta)\Lambda_{l,m}^h}{1 - \Delta t^{n+1}\beta\Lambda_{l,m}^h}\Phi_{l,m}^{h,n} \qquad (2.92)$$

holds with corresponding general eigenvalues $\Lambda_{l,m}^h$. The concrete relation for $A_h = \Delta_h$ is

$$\Phi_{l,m}^{h,n+1} = g(\Delta t^{n+1}\lambda_{l,m}^h)\Phi_{l,m}^{h,n} \quad \text{with } g(z) = \begin{cases} 1-z & \beta = 0, (\text{ExE}) \\[2mm] \dfrac{1-\frac{1}{2}z}{1+\frac{1}{2}z} & \beta = \frac{1}{2}, (\text{CN}) \\[2mm] \dfrac{1}{1+z} & \beta = 1, (\text{BE}) \end{cases}$$

Then the discrete solution at $t^{n+1} = (n+1)\Delta t$ is the sum

$$w_{h,\Delta t}(x,y,t^{n+1}) = \sum_{l,m=1}^{l,m=N-1} c_{l,m}(g(\Delta t \lambda_{l,m}^h))^{n+1}\sin l\pi x \cdot \sin m\pi y \qquad (2.93)$$

with the above g. The representation (2.89) of w_h and the first point of the above properties show, that different components of the solution decrease differently — due to the behavior of the eigenvalues — as t becomes large. It is this quantitative behavior which the solution $w_{h,\Delta t}$ of the completely discretized problem has to follow. This requirement then automatically leads to the theoretically and practically important question of stability. The desired behavior (decrease as the continuous solution decreases) requires at least

$$|g(\Delta t \lambda_{l,m}^h)| < 1 \text{ for all } l, m.$$

The discussion of this condition shows that time-steps used by the explicit Euler's method have to satisfy the condition $\Delta t \le h^2/4$. On the other hand, it is easily verified that Δt can be chosen arbitrarily for CN and BE, the step sizes are not restricted by a "stability condition".

It is interesting to see how CN behaves for large $\Delta t \lambda_{l,m}^h$. If l, m are large, that means $2 - \cos l\pi h - \cos m\pi h$ is close to 4, and Δt is large compared to h^2, then $g(\Delta t \lambda_{l,m}^h)$ is approximately minus one. The changing sign in $\Phi_{l,m}^{h,n+1} \approx -\Phi_{l,m}^{h,n}$ stands for the "oscillating behavior" of the Crank-Nicolson scheme. These oscillations may occur for Δt large compared to h^2 and as long as the high-frequent modes, components with l, m close to $N-1$, dominate. This is true, especially for small t. So the initial time steps have to be chosen very careful to avoid this phenomenon.

Another feature of the above time discretization schemes concerns their accuracy. The truncation error with respect to time and in combination with a second order space discretization is

$$\tau_{\Delta t}(\tau_{h,\Delta t}) = \begin{cases} O(\Delta t) & (O(h^2 + \Delta t)) & \text{for ExE} \\ O(\Delta t^2) & (O(h^2 + \Delta t^2)) & \text{for CN} \\ O(\Delta t) & (O(h^2 + \Delta t)) & \text{for BE} \end{cases}$$

The explicit Euler method has to satisfy a condition like $\Delta t \le h^2/4$. Then the scheme is of second order. But small h require small Δt. This makes the explicit method only useful in case of low accuracy requirements. The fully implicit method (BE) is characterized by its excellent stability properties (strongly A-stable) which make the scheme well suited for time stepping procedures even for nonlinear problems. Nevertheless, the time-step size has to be chosen due to $\Delta t = O(h^2)$ to guarantee second order consistence and convergence. With respect to the second order discretization in time and space, the CN-method requires only $O(h)$ time-step sizes. But the already mentioned oscillations during an initial phase require either small time-steps or more stable methods. It has to be pointed out, that only in case of linear problems the statement "CN has to satisfy no stability condition" is valid. For nonlinear operators A this is no longer true, because the eigenvalues will, in general, depend on $\Phi_{l,m}^{h,n+1}$ and $\Phi_{l,m}^{h,n}$ as the equations (2.90)–(2.92) show (compare Section 4.2.6).

What rules have to be observed in practice with this indirect multigrid approach for the considered class of evolution problems? Everything within a time-step is more or less standard. But before applying an "elliptic" multigrid solver to the "t^{n+1}-Helmholtz-type" problem the discrete equations have to be set up with the aid of the "old" results of the previous (t^n) time-step (see Figure 2.34). It is assumed to have approximations to the solution on each level. One way to construct the discrete problem, is to calculate F_h^{n+1} from u_h^n due to (2.85) and to take the old solution as an initial guess $u_h^{n+1(0)}$ and to restrict these grid functions to the coarser levels (solid arrows). On the other hand, it is possible to transfer the grid function to the respective grid level directly (dashed arrows). The latter choice is more expensive with respect to the right hand side, because the coarse grid values have to be determined by computations which are more expensive than a restriction from the finer grid. The use of the "old" solutions as an initial guess for the "new" solution is realizable with neglectable computational costs.

For cycles without post relaxation ($v_2 = 0$) and with straight injection both ways round will set up a very similar initial situation for t^{n+1}. If post relaxation improves the approximations after the correction step, the "direct transfer" is not the best choice, because it neglects on coarser grids the improvements of fine grid results by both correction and relaxation. To take into account the whole information of the previous time-step and looking for a minimum amount of work a proceeding as indicated by the solid arrows in Figure 2.34 is recommended.

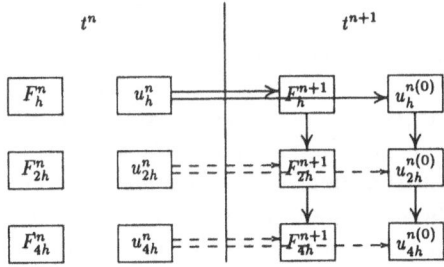

Fig. 2.34 Setting up the discrete problem for t^{n+1}

Additionally one has to decide between FMG and pure multigrid iteration per time-step. On the other hand, the knowledge about the structure of the solution provides a good reason to prefer FMG: For time-dependent problems it is necessary to treat those components first, which dominate in time direction. This is easily done first on coarse grids before the more oscillating components are subsequently handled on fine grids.

Another approach, the "direct" or "parabolic" multigrid method has to be mentioned, although upto now the only practical results have been reported for the model equation with $A = \Delta$ [14, 25, 40, 73]. Nevertheless, it provides many valuable ideas for practical use.

The main idea is to use fine scales to supply that information which allows the computation of coarse grid quantities with fine grid accuracy (often called "the dual point of view for multigrid"). Therefore the relative local discretization error τ_H^h is used to correct the coarse grid equations. To get this information, the fine grid has to be used. If a very slow change of the fine grid quantity is known, its actual state may be frozen and updated only after some calculations on coarser grids. Such a frozen-τ-technique activates fine grids seldom and only at times when the fine grid information needs an update. This may be applied to the τ_H^h-term with respect to the space discretization as well as to the complete defect correction term $\tau_{H,\Delta T}^{h,\Delta t}$ which is used for frozen-τ algorithms in [40].

High-frequent components decrease fast and low-frequent modes dominate the larger t becomes. The difference between the stationary solution and the solution of the full problem becomes more and more smooth in time. It is legal to assume a similar behavior for the local discretization error—and approximations to it, namely τ_H^h. For large t the variation of τ_H^h is therefore dominated by the small changes of low frequencies. This observation substantiates the idea to apply the frozen-τ-technique to evolution problems.

In general, τ_H^h has to be known at each time t^n to modify the coarse grid equations in such a way that the fine grid accuracy on the coarse grid is achieved. Due to the above reasoning, the τ_H^h may be frozen for a certain time period—that means for some time-steps—to correct the coarse grid equations but still maintaining fine grid accuracy. As a benefit of working on coarse grids, larger time-steps are allowed there. Only after some marching in time direction on coarse grids using the last determined τ_H^h the fine grid is revisited to get new τ_H^h information. The principle is shown in Figure 2.35. ΔT denotes the enlarged time-step size on the coarser grid, while t_h^H and t_H^h indicate the time when to switch from the h-grid to the H-grid and vice versa. An important and automatically arising question concerns the choice of t_h^H and t_H^h. From the above consideration it is adequate to work on fine grids as long as the relative changes of τ_H^h with respect to time are large:

$$\frac{\|\tau_H^h(t^{n-1}) - \tau_H^h(t^n)\|}{\|\tau_H^h(t^{n-1})\|} \quad \text{``large''}$$

In other words, as soon as this quantity becomes smaller than an ε, the marching in time will continue on coarser grids, improving the corresponding coarse grid problems with the aid of the last determined $\tau_H^h(t^n)$.

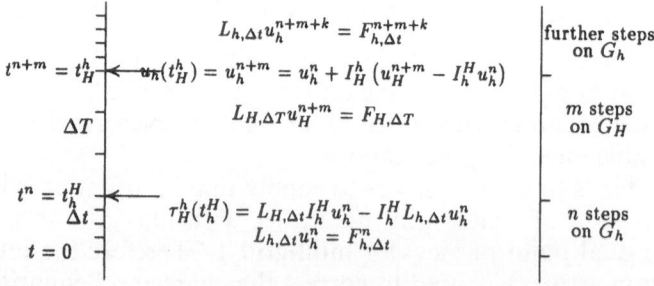

Fig. 2.35 The principle of parabolic multigrid

To return back to the fine grid as well fixed strategies (fixed number of time steps on the coarse level) as adaptive strategies are applied. In [73] a criterion due to A. Brandt is applied, which simply says: switch back to the finest grid and update τ_H^h as soon as the error forced by the frozen $\tau_H^h(t^n)$ reaches the magnitude of the discretization error.

Remark 2.25. *Recent research on parabolic multigrid for the heat equation* [14] *allows some further hints to construct efficient algorithms.*

1. *The design goal concerning computational work is summarized by the following: the work per time-step should be proportional to the increment in time* $\| u_h^{n+1} - u_h^n \|$. *That means, if the solution varies slowly, the invested work to compute it should be small. Considering the above criterion for a switching from fine to coarse grids, the consequence will be a rare use of fine grids, especially when the solution approaches the steady state.*
2. *In FMG algorithms higher order interpolations are necessary to keep interpolation errors small, particularly for high-frequent modes. For parabolic problems the high-frequent components disappear soon, and only smooth quantities have to be interpolated. Then lower order methods may be sufficient.*
3. *The use of red-black relaxation is not unconditionally recommended due to its capability to "convert" high frequencies into dominating low-frequent ones. To avoid the conversion of frequencies, W-cycles without post-relaxation* $(v_2 = 0)$ *are favoured.*

Although derived for a typical model problem, these recommendations are of interest. Especially the second one seems to be confirmed in practice, by the results of Section 4.6, where the bilinear FMG-interpolation works surprisingly good.

2.9 Systems of Partial Differential Equations

So far, the presented multigrid strategies offered no features which restrict their use to scalar equations. Similar to the scalar case, the main questions to discuss for systems are the structure of the grids—directly connected to the problem of stable discretization—the transfer of grid functions between different levels and, of course, the proper choice of relaxation schemes.

The first question automatically runs into the discussion of stable discretization schemes on staggered or non-staggered approaches. "Staggered" in this context not only denotes a certain (shifted and cell-centered) grid construction but also stands for a discretization technique where different components of the solution vector live on different places of the grid structure (see Figure 2.12-(b)).

Example 2.23. *Without specifying characteristica like the domains, the boundary values and special features of the system (for instance the nonlinearity) some typical and widely investigated model problems (1.–3.) and two examples (4.–5.) which are of immediate interest for process and device simulation are given below.*

1. *The Cauchy-Riemann equations*

$$\frac{\partial u}{\partial x} + \frac{\partial v}{\partial y} = F_1$$

$$\frac{\partial u}{\partial y} - \frac{\partial v}{\partial x} = F_2$$

and in matrix notation:

$$\begin{pmatrix} \dfrac{\partial}{\partial x} & \dfrac{\partial}{\partial y} \\ \dfrac{\partial}{\partial y} & -\dfrac{\partial}{\partial x} \end{pmatrix} \begin{pmatrix} u \\ v \end{pmatrix} = \begin{pmatrix} F_1 \\ F_2 \end{pmatrix}$$

2. *The Stokes equations in matrix notation:*

$$\begin{pmatrix} -\Delta & 0 & \dfrac{\partial}{\partial x} \\ 0 & -\Delta & \dfrac{\partial}{\partial y} \\ \dfrac{\partial}{\partial x} & \dfrac{\partial}{\partial y} & 0 \end{pmatrix} \begin{pmatrix} u \\ v \\ p \end{pmatrix} = \begin{pmatrix} F_1 \\ F_2 \\ F_3 \end{pmatrix}$$

3. *The incompressible steady state Navier-Stokes equations with primitive variables in matrix notation:*

$$\begin{pmatrix} \mathcal{Q} & 0 & \dfrac{\partial}{\partial x} \\ 0 & \mathcal{Q} & \dfrac{\partial}{\partial y} \\ \dfrac{\partial}{\partial x} & \dfrac{\partial}{\partial y} & 0 \end{pmatrix} \begin{pmatrix} u \\ v \\ p \end{pmatrix} = \begin{pmatrix} F_1 \\ F_2 \\ F_3 \end{pmatrix}$$

with $\mathcal{Q} := -\Delta + Re\left(u\dfrac{\partial}{\partial x} + v\dfrac{\partial}{\partial y} \right).$

4. *The unscaled stationary semiconductor device equations with "natural variables"*

$$\nabla(\varepsilon\nabla\psi) - q(n - p - N) = 0$$
$$\nabla(D_n\nabla n - \mu_n n\nabla\psi) - R(\psi, n, p) = 0$$
$$-\nabla(D_p\nabla p + \mu_p p\nabla\psi) + R(\psi, n, p) = 0$$

5. *A system of diffusion equations coupled by their diffusion coefficients* D_i *which depend on the diffusing species* $u_1, \ldots, u_{N_{eq}}$:

$$\nabla(D_i(u_1, \ldots, u_{N_{eq}})\nabla u_i) - \frac{\partial u_i}{\partial t} = 0 \quad for \quad i = 1, \ldots, N_{eq}$$

The matrix notation is well suited to describe linear or quasilinear systems of partial differential equations. A system of N_{eq} partial differential equations defined on a two-dimensional domain Ω is then given by an operator matrix

$$\left(L_{lm} \right)_{N_{eq} \times N_{eq}} \begin{pmatrix} u_1 \\ \vdots \\ u_{N_{eq}} \end{pmatrix} = \begin{pmatrix} F_1 \\ \vdots \\ F_{N_{eq}} \end{pmatrix}, \tag{2.94}$$

where the boundary values are not explicitly mentioned. The discrete counterpart on G_h is denoted by

$$\left(L_{lm}^h \right)_{N_{eq} \times N_{eq}} \begin{pmatrix} u_1^h \\ \vdots \\ u_{N_{eq}}^h \end{pmatrix} = \begin{pmatrix} F_1^h \\ \vdots \\ F_{N_{eq}}^h \end{pmatrix} \tag{2.95}$$

For complex nonlinear systems it is sometimes more appropriate to write down all equations separately (compare System 4 in Example 2.23).

The "model problems" for systems of partial differential equations are the Cauchy-Riemann equations and the Stokes equations. Staggered grid approaches have been widely used for both these model systems and the more complex Navier-Stokes equations (in their various variants).

If a staggered approach is preferred, the multigrid extension to systems requires a modified multigrid realization. For instance, any restriction will be an averaging of fine grid values. Additionally, the relative position of fine, respectively coarse grid points complicates the interpolation of values near the boundaries considerably.

Flow problems are posed on general regions. A widely used technique is to use boundary fitted grids. On such grids the lines usually are no longer orthogonal and the advantages of staggered grids disappear. On the other hand, even for simple flow problems the standard central differencing of the Navier-Stokes equations is known to become unstable if the unknowns reside on non-staggered regular meshes. Fortunately, non-staggered discretizations may be stabilized in a way which can be used on boundary fitted grids, too (artificial pressure term, flux difference splitting [26, 80, 82]). As a consequence, even for situations where staggered approaches have been dominating for a long time, non-staggered approaches become more and more attractive.

Conclusion 2.7. *As main benefit of this observation and generalizing from the above "model systems" the statement becomes true, that the extension of multigrid methods to systems is straight forward if the components of the vector solution to the system live at same grid locations. Thus, for the following non-staggered approaches are assumed.*

Similar to the scalar case, the proper choice of an adequate smoothing scheme remains. Being the basic process in multigrid, the selected relaxation method has to possess a good smoothing property which additionally has to be independent from the meshsize to get h-independent multigrid convergence.

For systems the question whether there are relaxation schemes with h-independent smoothing properties, at all, becomes even more important than in the scalar case. There is no general rule to construct efficient and robust smoothing schemes. But under certain conditions, the so-called measure of h-ellipticity has to be greater than zero, there exists a point relaxation scheme whose smoothing factor is bounded away from one [12, 79]. A similar statement can be given for block relaxation schemes. Unfortunately, the corresponding proof is not a very constructive one. So the concrete choice of the scheme to be used still depends on the problem and on the experience of the user. Specific tools, based upon Fourier analysis, may help to decide for a concrete scheme.

A very general class of smoothing procedures which show a satisfying applicability even for systems (especially if there is no unique correspondence between equation and grid point) is the class of distributive relaxations [12, 25, 82]. They are characterized by the change of more than one unknown within a small surrounding of the specific point whose equation is actually relaxed. One has to take care that the residuals in the neighbouring points are changed only a little. Distributive relaxations turned out to be very efficient smoothing procedures for staggered approaches to solve the Stokes and Navier-Stokes equations. A particular example is the Kaczmarz relaxation.

Another general class are the so-called "box-relaxations". Their feature in common is to satisfy the discrete equations for all unknowns within a discretization cell (box), that means: box-relaxations treat all equations for all unknowns which may reside on different points within a discretization cell simultaneously.

Both types of relaxation are capable for use within staggered and non-staggered formulations.

In case of non-staggered grids there are some straight-forward extensions of schemes which are applied to scalar equations successfully. If there is a correspondence between each of the equations of the system and one of the solution components and if the discrete problem possesses a "nearly diagonal" matrix representation, the discrete equations may be scanned in a

Gauss-Seidel type manner for all points and for each of the equations, one after another.

This global ordering "for each equation scan all points of G_h lexicographically within each equation" applied to (2.95) produces an $N_{eq}N^2 \times N_{eq}N^2$-matrix, composed of $N^2 \times N^2$-submatrices due to the operators L^h_{lm}:

$$\begin{pmatrix} (L^h_{11})_{N^2 \times N^2} & \cdots & (L^h_{1N_{eq}})_{N^2 \times N^2} \\ \vdots & \ddots & \vdots \\ (L^h_{N_{eq}1})_{N^2 \times N^2} & \cdots & (L^h_{N_{eq}N_{eq}})_{N^2 \times N^2} \end{pmatrix}$$

Interchanging the ordering to "for each point scan all the equations for the unknowns living at this particular grid point" produces a more local splitting into a $N^2 \times N^2$ block matrix consisting of $N_{eq} \times N_{eq}$ submatrices. From the iterative scheme's point of view the convergence of a Gauss-Seidel type scheme with these orderings does not change, because the spectral radius of Gauss-Seidel relaxations is independent from the ordering of equations. From the smoothing factor's point of view the ordering is of importance. So one has to investigate for each application whether the more "local" approach or the "global" splitting is advantageous with respect to the smoothing property.

The "local" splitting naturally leads to another relaxation scheme, if the sets of local equations are not only scanned in a Gauss-Seidel manner, but if they are relaxed together. That means, the local subset of discrete equations for all discrete unknowns living at a certain grid point is relaxed simultaneously. These relaxation schemes are called "collective" schemes.

Example 2.24. *The "local" ordering of equations requires the solution of a set of discrete equations one after another at a certain grid position (x_i, y_j).*
The "collective" variants combine this set to a local subsystem of equations, which is solved for the variables placed at (x_i, y_j).
If $N_{eq} = 3$, the "local" subsystem to solve for the unknowns $u_{i,j}, v_{i,j}$ and $w_{i,j}$ is

$$f(u_{i,j}, v_{i,j}, w_{i,j}) = 0$$
$$g(u_{i,j}, v_{i,j}, w_{i,j}) = 0$$
$$h(u_{i,j}, v_{i,j}, w_{i,j}) = 0.$$

Just for simplicity and to focus on the unknowns corresponding to the point (x_i, y_j) the usually envolved unknowns $u_{i-1,j}, \ldots, w_{i+1,j}$ are not included in this representation. If f, g and h are nonlinear in $u_{i,j}, v_{i,j}$, and $w_{i,j}$, a Newton's method can be used locally to solve the nonlinear system. Only a "local" and therefore small Jacobian has to be computed.

The extension to collective "block"-relaxation schemes is straight forward. It is only the size of the systems and, in case of nonlinearity the size of the Jacobian, which becomes larger. Additionally, all schemes allow a modified ordering of points or combining points to blocks whose equations are solved simultaneously (collective line or column relaxation with different ordering of blocks, too). Of course, the underlying iterative scheme may be selected from the large set of available methods (Gauss-Seidel, Jacobi, weighted or not, SOR, ILU,...).

The collective relaxation methods are very attractive within the FAS-framework. The discrete problem remains close to the underlying physical problem in the sense that a linearization or decoupling becomes necessary only at a late time within the solution process.

Remark 2.26. *To get a feeling about the system to solve and to get an impression which smoothing scheme might be an appropriate choice, the principle of the "principal determinant operator" is introduced by* [12]:

The smoothing factor for a given PDE system L can be as good as the smoothing obtainable for the factors of the subprincipal part (terms) of det L. The subprincipal terms of a matrix operator L are—roughly—those terms contributing to the highest order derivative in the determinant of the—eventually linearized—operator L. Assuming L to have subprincipal terms only, det L can be factorized into simpler factors.

If det $L = l_1 l_2$, *where each* l_i *is a (scalar) differential operator, then one can factorize the* $N_{eq} \times N_{eq}$ *operator L into* $L = L_1 L_2$, *where* L_i *are* $N_{eq} \times N_{eq}$ *matrix operators, such that* det $L_i = l_i$.

With this information, the factorized system $L_1 L_2 u = f$ *is split into two systems* $L_2 u = v$ *and* $L_1 v = f$. *The accepted smoothing rate for* $Lu = f$ *should not be worse than the worst of those for the two auxiliary systems. And these may be determined using the above principle.*

For the Cauchy-Riemann system the principal determinant operator is just the Laplacian. Following the above idea, one should not be satisfied with a relaxation scheme for the discrete Cauchy-Riemann equations which smoothes worse than schemes for the Laplacian.

Similar to the scalar case systems of difference operators with constant coefficients possess eigenfrequencies, too. The formal description in terms of matrix operators uses the fact, that for each single difference operator $L_{lm}^h u_h(x, y) = \sum_{\substack{\kappa \in \mathbb{I} \times \mathbb{I} \\ \kappa = (\kappa_1, \kappa_2)}} a_\kappa^{lm} u_h(x + \kappa_1 h, y + \kappa_2 h)$ the relation

$$L_{lm}^h e^{i\theta \cdot (x/h)} = \left(\sum_{\substack{\kappa \in \mathbb{I} \times \mathbb{I} \\ \kappa = (\kappa_1, \kappa_2)}} a_\kappa^{lm} e^{i\theta_1 \kappa_1} e^{i\theta_1 \kappa_2} \right) e^{i\theta \cdot (x/h)} = \lambda_{lm}^h(\theta) e^{i\theta \cdot (x/h)}$$

holds. The matrix $\Lambda_h(\theta) := (\lambda_{lm}^h(\theta))_{N_{eq} \times N_{eq}}$ is "the symbol" of $L_h = (L_{lm}^h)$ and $\Lambda^h(\theta)$ is the amplification matrix of L_h for components of the form $\mathbf{a} e^{i\theta \cdot (x/h)}$ with $\mathbf{a} \in \mathbb{C}^{N_{eq}}$: $L_h(\mathbf{a} e^{i\theta \cdot (x/h)}) = (\Lambda_h(\theta) \mathbf{a}) e^{i\theta \cdot (x/h)}$. This representation allows a natural extension of smoothing analysis and other more sophisticated theoretical tools from the scalar case to systems of partial differential equations.

2.10 Tools to Estimate Multigrid Convergence

By experience, there are situations where a naive multigrid approach applied to real life problems will run into severe difficulties. One decision may then be to return to "more robust classical" approaches. The natural question is to ask "why" the special algorithm does not work. What kind of characteristic difficulties are observed and how can they be resolved within a multigrid environment?

Based on Fourier analysis there exist advices for the analysis of algorithms and their improvement. This mathematically based tool provides asymptotic two grid convergence information and smoothing factors. Although it seems to be somewhat theoretically oriented, the Fourier analysis is absolutely necessary to understand the interplay of smoothing and coarse grid correction, the correct treatment of high and low frequencies and the influence of different components on the final algorithm. The following section summarizes the already given information for model problems and combines it systematically with additional information to get a feeling for more complex problems.

The questions concerning multigrid convergence, that means the convergence of an algorithm where "many" levels are involved are answered by two-grid tools. Is the restriction to only two grids severe? Is it a restriction at all? The answer simply is no! The recursive definition of the multigrid operator and the proof of the h-independent multigrid convergence essentially depends on the convergence of the corresponding two-grid iteration. In more detail: the two-grid convergence is sufficient for multigrid convergence because the other technical conditions required for the proof are usually satisfied [47, 132].

For a convergent two-grid method with $\rho(M_l^{l-1}) \leq \rho* < 1$ the multigrid iteration operator $M_h(h = h_M)$ allows an estimation $\| M_h \| \leq \eta_M$ where η_M is recursively defined by ($\rho*$ and C independent from l)

$$\eta_1 = \rho* \quad and \quad \eta_l = \rho* + C\eta_{l-1}^{\gamma} \quad for \quad l = 2, \ldots, M.$$

Assuming $\rho*$ sufficiently small and choosing $\gamma = 2$ (W-cycle) then the corresponding multigrid method has similar convergence properties.

For other choices of γ the h-independent convergence of multigrid methods can be proven under additional conditions [9, 47, 132], too. In any case, the two-grid convergence plays a key role. This finally justifies the two-grid analysis.

Conclusion 2.8. *There are two theoretically founded tools. The first one takes into account all the components of a given algorithm—called two-level analysis. It does not only provide asymptotic two-grid convergence rates but also gives advices for an adequate composition of methods. The second one—the* *smoothing analysis—concentrates on the task to find relaxation schemes which smooth out high-frequent modes efficiently. Additionally, the smoothing rates allow a careful estimation of the multigrid behavior to be expected—if the limitations are kept in mind. The presented ideas and techniques are extendable to higher dimensions [134] and to systems of equations [12,79].*

2.10.1 Two-Grid Analysis—Global

The two-grid analysis will be used here mainly as a tool to get insight into the algorithm and to describe the dependence of two-grid convergence on different components. This finally leads to rules for an approximately optimal choice of components for a given problem. In principle, two different approaches have to be distinguished. The "model problem" or "global" analysis and the "local mode analysis" which are related to each other in a way which is explained in detail in [131] and [132].

The model problem analysis provides exact and realistic predictions for complete ("global") model problems with constant coefficients on bounded rectangular domains and takes the boundary conditions (Dirichlet, Neumann or periodic) into account. The restriction to selected problems is no disadvantage because the experience shows that the obtained results are valid for more general classes of problems (variable coefficients, general domains, general boundary conditions,...) where the considered model equations are representatives of.

The task of any two-grid method is to determine the spectral radius of the two-grid iteration operator

$$\rho(M_h^H) = \rho(\mathscr{S}_h^{\nu_2} M_h^H (CGC) \mathscr{S}_h^{\nu_1}) = \rho(\mathscr{S}_h^{\nu_2} (Id_h - I_H^h L_H^{-1} I_h^H L_h) \mathscr{S}_h^{\nu_1})$$

If it is possible to find an appropriate basis within the space of grid functions on G_h consisting of eigenfunctions to M_h^H, the computation of $\rho(M_h^H)$ may be complicated but straight forward. The h-dependency is eliminated by taking the supremum over all admissable h, defining $\rho^* := \sup_{h \le h_0} \{\rho(M_h^H)\}$,

the asymptotic two grid convergence rate. The hope is to find $\rho(M_h) \approx \rho^* \ll 1$ for the multigrid iteration operator.

The previous sections supplied with a lot of information which is valuable to understand both the global model problem analysis and the local mode analysis. To demonstrate the principle, the already known facts for the Poisson equation with Dirichlet boundary conditions on the unit square are collected and rewritten in a more formal manner. Although not necessary, the simplifying standard coarsening ($H = 2h$) is assumed. A basis of discrete eigenfunctions is known (see (2.49) in Section 2.1.3). The iteration operator M_h^H can be represented with respect to exactly this eigenbasis by a block diagonal matrix consisting of block matrices of dimension equal to or smaller than four. The submatrices correspond to the subspaces

$$E_{h,(l,m)} := \langle \varphi_{l,m}^h, \varphi_{N-l,N-m}^h, \varphi_{N-l,m}^h, \varphi_{l,N-m}^h \rangle \quad \text{for} \quad l,m \le N/2. \quad (2.96)$$

The following representation uses $\xi = \sin^2 \dfrac{l\pi h}{2}$ and $\eta = \sin^2 \dfrac{m\pi h}{2}$

1. Due to the aliasing phenomenon some h-grid basisfunctions coincide with $2h$-grid basisfunctions on G_{2h} (see Sections 2.1.4 and 2.2.1). On G_{2h} the generating functions of $E_{h,(l,m)}$ and $\varphi^{2h}_{l,m}$ coincide:

$$\varphi^h_{l,m} = \varphi^h_{N-l,N-m} = \varphi^h_{N-l,m} = \varphi^h_{l,N-m} = \varphi^{2h}_{l,m} \quad \text{for} \quad l,m < N/2.$$

That means, the spaces $E_{h,(l,m)}$ are spanned by those modes which coincide on the H-grid. This gives the name "H-harmonics" to these spaces.

2. For Jacobi smoothing schemes the eigenfunctions of $\mathbf{P}_J(\omega)$ and Δ_h coincide (see Section 2.1.4). It is obvious that $\mathbf{P}_J(\omega): E_{h,(l,m)} \to E_{h,(l,m)}$ and the matrix representation restricted to the subspace $E_{h,(l,m)}$ is

$$\mathbf{P}_J(\omega)_{(l,m)} \triangleq \begin{cases} \begin{pmatrix} 1 - \omega(\xi + \eta) & & & \\ & 1 - \omega(2 - \xi - \eta) & & \\ & & 1 - \omega(1 - \xi + \eta) & \\ & & & 1 - \omega(1 + \xi - \eta) \end{pmatrix}_{4 \times 4} & \text{if } l,m < N/2 \\[2em] \begin{pmatrix} 1 - \omega(\xi + \eta) & \\ & 1 - \omega(2 - \xi - \eta) \end{pmatrix}_{2 \times 2} & \text{if either } l \text{ or } m = N/2 \\[1em] (1 - \omega(\xi + \eta))_{1 \times 1} & \text{if } l = m = N/2 \end{cases}$$

3. The restriction operators map low-frequent h-grid modes into the set of H-grid frequencies with the same wave numbers (see for example (2.60) in Section 2.3 for full weighting).
$I^{2h}_h: E_{h,(l,m)} \to \langle \varphi^{2h}_{l,m} \rangle$ for $l,m < N/2$ and
$I^{2h}_h \varphi^h_{l,m} = 0$ if l or m is equal to $N/2$.
For full weighting (2.60) the concrete specification of the above mapping is

$$F^{2h}_h \begin{Bmatrix} \varphi^h_{l,m} \\ \varphi^h_{N-l,N-m} \\ \varphi^h_{N-l,m} \\ \varphi^h_{l,N-m} \end{Bmatrix} = \begin{Bmatrix} (1-\xi)(1-\eta) \\ \xi\eta \\ \xi(1-\eta) \\ (1-\xi)\eta \end{Bmatrix} \varphi^{2h}_{l,m}$$

4. A prolongated $2h$-grid eigenfunction is a linear combination of fine grid eigenfunctions (see (2.61) and (2.62) for bilinear interpolation).
$I^h_{2h}: \langle \varphi^{2h}_{l,m} \rangle \to E_{h,(l,m)}$ for $l,m < N/2$.
The rule for bilinear interpolation and $l,m < N/2$ is

$$I^h_{2h}\varphi^{2h}_{l,m} = (1-\xi)(1-\eta)\varphi^h_{l,m} + \xi\eta\varphi^h_{N-l,N-m} + \xi(1-\eta)\varphi^h_{N-l,m} \\ + (1-\xi)\eta\varphi^h_{l,N-m}.$$

All these facts show that the problem to calculate the spectral radius of M^{2h}_h is reduced to the determination of spectral radii of—in the worst case—

Table 2.7. *Two-grid convergence factors,*
weighted Jacobi

v	$\omega = 0.5$	$\omega = 0.8$
1	0.750	0.600
2	0.563	0.360
3	0.422	0.216
4	0.316	0.137

4×4-dimensional matrices $M_{h,(l,m)}^{2h}$ representing the two-grid iteration operator restricted to the spaces $E_{h,(l,m)}$. These values depend on h, v and, for the above weighted Jacobi relaxation, on ω, too:

$$\rho = \rho(h, v, \omega) = \rho(M_h^{2h}) = \max_{l,m \leq N/2} \rho(M_{h,(l,m)}^{2h})$$

The supremum over all "small" meshsizes, ρ^*, still depends on v and ω. The calculated values ρ^* for several v and $\omega = 0.5$, respectively $\omega = 0.8$, are given in Table 2.7 for an algorithm with full weighting and bilinear interpolation. The analysis of ρ^* with respect to ω shows that the weighted Jacobi relaxation with $\omega = 0.8$ yields optimal values for the considered problem.

The presented ideas are extendable to more general problems $L_h u_h = f_h$ with constant coefficients on rectangular grids. Assuming Dirichlet boundary conditions (similar for other types of boundary conditions) L_h possesses eigenfunctions which form a basis of the space of grid functions on G_h:

$$\varphi_{l,m}^h(x, y) = \sin l\pi x \cdot \sin m\pi y$$

For the more general formalism frequencies $\theta = (\theta_1, \theta_2)$ replace the wave numbers (l, m):

$$\varphi_\theta^h(x, y) = \sin \frac{\theta_1 x}{h} \sin \frac{\theta_2 y}{h}$$

with $\theta \in T_h := \{(l\pi/N, m\pi/N) | h = 1/N; 0 < l, m < N\}$. If L_h is given by a compact nine-point stencil $[s_\kappa]$ the φ_θ^h form a basis of eigenfunctions with corresponding eigenvalues $\lambda^h(\theta) = \displaystyle\sum_{\substack{\kappa \in \mathbb{I} \times \mathbb{I} \\ \kappa = (\kappa_1, \kappa_2)}} s_\kappa \cos(\theta_1 \kappa_1 + \theta_2 \kappa_2)$.

The idea to call eigenfunctions "low-frequent" when they can be represented on the coarse grid G_H shows, that being "low-frequent" essentially depends on the structure of the coarse grid. The index set T_h is split into the union of disjoint subsets $T_h = T_h^{\text{high}} \cup T_h^{\text{low}}$ of corresponding high and low frequencies. If $\varphi_\theta^H := \varphi_\theta^h|_{G_H}$ then the $\varphi_\theta^H, \theta \in T_h^{\text{low}}$, establish a basis for the G_H-grid functions. The "aliasing effect" is expressed by the condition that for all $\tilde{\theta} \in T_h^{\text{high}}$ and $(x, y) \in G_H$ the function $\varphi_{\tilde\theta}^h$ either coincides with a $\pm \varphi_\theta^H$ for a certain $\theta \in T_h^{\text{low}}$ or degenerates to the zero grid function on G_H. T_h^H is defined to be that superset of T_h^{low} which additionally contains the degenerate cases. Finally, within the space of grid functions on G_h the H-harmonics are defined to be the lower-dimensional subspaces $E_{h,\theta}^H$ which are the span of those basisfunctions which coincide on G_H with some φ_θ^H.

Example 2.25. *Due to the above rules the spaces $E_{h,\theta}^H$ and the "low-frequent" modes essentially depend on the coarse grid, that means on the coarsening:*

standard coarsening:

$$H = 2h; \; T_h^{\text{low}} = \{\theta \in T_h | 0 \le \theta_1, \theta_2 < \pi/2\}; \; T_h^H := \overline{T_h^{\text{low}}};$$
$$E_{h,\theta}^H = \langle \varphi_{(\theta_1,\theta_2)}^h, \varphi_{(\theta_1-\pi,\theta_2-\pi)}^h, \varphi_{(\theta_1-\pi,\theta_2)}^h, \varphi_{(\theta_1,\theta_2-\pi)}^h \rangle;$$
if $\theta \in T_h^{\text{low}}$ then dim $E_{h,\theta}^H = 4$, otherwise dim $E_{h,\theta}^H < 4$;

x-coarsening, semi-coarsening:

$$H = (2h, h); \; T_h^{\text{low}} = \{\theta \in T_h | 0 \le \theta_1 < \pi/2\}; \; T_h^H := \{\theta \in T_h | 0 \le \theta_1 \le \pi/2\};$$
$$E_{h,\theta}^H = \langle \varphi_{(\theta_1,\theta_2)}^h, \varphi_{(\theta_1-\pi,\theta_2)}^h \rangle$$

If the components of the coarse-grid correction operator satisfy

$$I_h^H : E_{h,\theta}^H \to \langle \varphi_\theta^H \rangle \qquad \text{for } \theta \in T_h^H$$
$$I_H^h : \langle \varphi_\theta^H \rangle \to E_{h,\theta}^H \qquad \text{for } \theta \in T_h^{\text{low}}$$
$$L_h : \langle \varphi_\theta^H \rangle \to \langle \varphi_\theta^H \rangle \qquad \text{for } \theta \in T_h^{\text{low}}$$

and if L_H^{-1} exists, then $M_h^H(CGC): E_{h,\theta}^H \to E_{h,\theta}^H$ for all $\theta \in T_h^H$ leaves the H-harmonics invariant and is therefore equivalent to a block matrix, where the dimension of each block is equal to the dimension of the corresponding space $E_{h,\theta}^H$. Because many important smoothing operators allow a decomposition with respect to the H-harmonics, the two-grid operator M_h^H is represented by a block matrix, too.

The framework of the two-grid analysis permits the definition of a general measure for the smoothing property of a given relaxation scheme. Assuming the coarse grid operator to be "ideal", that means high-frequent modes on G_h are unchanged and low-frequent ones are eliminated (denoting this ideal operator by P_h^H), its matrix representation with respect to the H-harmonics is a block diagonal matrix consisting of blocks with diagonal entries equal to one only at positions which correspond to high-frequent basis modes of $E_{h,\theta}^H$ (all other entries are zero). The quantity

$$\rho(\mathscr{S}_h^{\nu_2} P_h^H \mathscr{S}_h^{\nu_1}) = \rho(P_h^H \mathscr{S}_h^\nu) = \max_{\theta \in T_h^H} \{\rho((P_h^H)_{(E_{h,\theta})}(\mathscr{S}_h)_{(E_{h,\theta})}^\nu)\}$$

thus describes the total smoothing of the two-grid method with "ideal" coarse grid operator and ν relaxation sweeps. The "smoothing factor" per relaxation step then is defined by

$$\mu^*(\nu) = \sup_{h \le h_0} \{\sqrt[\nu]{\rho(P_h^H \mathscr{S}_h^\nu)}\}.$$

Table 2.8. *Influence of the coarsening strategy on the smoothing property*

$\mu^*(v), v = 1$ coarsening	relaxation			
	red-black	x-ZEBRA	y-ZEBRA	AD-ZEBRA
	$-\Delta u = f$			
standard	0.250	0.250	0.250	0.048
	$-\varepsilon\dfrac{\partial^2 u}{\partial x^2} - \dfrac{\partial^2 u}{\partial y^2} = f, \varepsilon = 0.1$			
standard	0.826	0.826	0.125	0.102
y-SEMI	0.125	0.125	0.008	0.002

A comparison between $(\mu^*(v))^v$ and $\rho^*(v)$ shows good agreement, especially for small v (see for instance Table 2.9). For the Poisson model problem red-black relaxation is one of the most efficient smoothing methods, especially if the number of operations per point is taken into account. If there are no further difficulties to expect the red-black relaxation in combination with standard coarsening may be considered as the most recommended first attempt for Poisson-like problems.

As the values in Table 2.8 show, the smoothing quality becomes worse if the anisotropic model equation with standard five-point discretization and (still moderate) $\varepsilon = 0.1$ is considered. The smoothing property of red-black relaxation with standard coarsening approximately vanishes. Such a behavior especially occurs for point relaxations or, more general, for relaxations which do not consider the direction of strong coupling (see Section 2.4 and Remark 2.17). To overcome these difficulties the use of y-ZEBRA or alternating (AD) ZEBRA relaxation methods may help—even in combination with standard coarsening. Such a decision is of technical importance, because standard coarsening is easy to realize. If there are good reasons to use point relaxation methods like red-black or Gauss-Seidel relaxation then a more sophisticated coarsening strategy, for instance y-SEMI coarsening should be used as the values in Table 2.8 prove. Then red-black relaxation remains attractive. Nevertheless, alternating (AD) ZEBRA is the better smoother—but the more expensive one, too.

2.10.2 Two-Grid Analysis—Local

The "local mode analysis (local Fourier analysis)" in principle aims to the same objectives as the previously described "global" model problem analysis. "Local" in this context stands for neglecting geometry, boundary conditions and non-constant coefficients (including a "frozen" non-linearity). In detail, a given operator with variable coefficients is evaluated locally at a certain position $(x_0, y_0) \in G_h$ and replaced by an operator on the infinite grid $\mathscr{G}_h$ with constant coefficients, where the globally used coefficients are exactly those gained by the local evaluation. If the coefficients vary within a wide numerical range a lot of "local mode analysis problems" has to be solved to get the desired representative overview on the expected behavior (worst/best case estimation). If the observed behavior of a concrete algorithm strongly differs from the prediction of local mode analysis, a detailed investigation of boundary conditions, singularities, geometries, ..., is recommended. Experience shows that local changes near these local sources of trouble may improve the algorithm considerably.

The close connection between local mode and global analysis becomes transparent if problems with periodic boundary conditions and symmetric operators are investigated on regular rectangular grids. They require the choice of basis functions $\varphi_\theta^h(x) = e^{i\theta \cdot (x/h)}$ instead of the sine and cosine functions for problems with Dirichlet or Neumann boundary conditions.

To avoid additional notation, all grid operators defined now on the infinite grids as well as the (now infinite-dimensional) spaces of grid functions or the (continuous) index sets for frequencies are denoted as before. Standard coarsening is assumed. Then the "frozen" operator $L_h \equiv [s_\kappa]$ is defined on $\mathscr{G}_h$ and operates on the space

$$E_h := \{\varphi_\theta^h | \varphi_\theta^h(x) = e^{i\theta \cdot (x/h)} = e^{i\theta_1 (x/h)} e^{i\theta_2 (y/h)},$$
$$x = (x, y) \in \mathscr{G}_h, \theta = (\theta_1, \theta_2) \in (-\pi, \pi]^2\}.$$

It is sufficient to consider only $\theta \in T := (-\pi, \pi]^2$ because of the periodicity of the φ_θ^h on $\mathscr{G}_h$. All $\varphi_\theta^h \in E_h$ are eigenfunctions of L_h satisfying

$$L_h \varphi_\theta^h(x) = \left(\sum_{\substack{\kappa \in \mathbb{I} \times \mathbb{I} \\ \kappa = (\kappa_1, \kappa_2)}} s_\kappa e^{i\theta_1 \kappa_1} e^{i\theta_2 \kappa_2} \right) \varphi_\theta^h(x) = \lambda_\theta^h \varphi_\theta^h(x). \tag{2.97}$$

Because of standard coarsening the introduction of low and high frequencies is not surprising and similar to the already known definitions. E_h is split into the spaces E_h^{low} and E_h^{high}. φ_θ^h is low-frequent for $\theta \in (-\pi/2, \pi/2]^2$ and high-frequent otherwise. Again, there is an identification of E_h^{low} and

$$E_{2h} = \langle \varphi_\theta^{2h} | \varphi_\theta^{2h}(x) := \varphi_\theta^h(x) \text{ for } x \in \mathscr{G}_{2h} \text{ and } \theta \in (-\pi/2, \pi/2]^2 \rangle.$$

Taking into account that $\varphi_\theta^h = \varphi_{\hat\theta}^h$ on $\mathscr{G}_h$ if and only if $\theta = \hat\theta \mod \pi$, the spaces $E_{h,\theta}$ of 2h-harmonics are for all $\theta \in (-\pi/2, \pi/2]^2$ generated by all $\varphi_{\hat\theta}^h$ with $\hat\theta \in (-\pi, \pi]^2$ and $\hat\theta = \theta \mod \pi$. The idea behind this is once more the existence of E_h basis functions which coincide on the 2h-grid ("aliasing"). With $\alpha = (\alpha_1, \alpha_2) \in \{(0,0)(1,1)(1,0)(0,1)\}$ and $\theta_\alpha := (\theta_1 - \alpha_1 \text{ sign } \theta_1 \pi, \theta_2 - \alpha_2 \text{ sign } \theta_2 \pi)$ it is easy to verify—compare (2.96)—that

$$E_{h,\theta} = \langle \varphi_{\theta_\alpha}^h | \alpha \in \{(0,0)(1,1)(1,0)(0,1)\} \rangle.$$

For standard coarsening the range of high and low frequencies (identifying

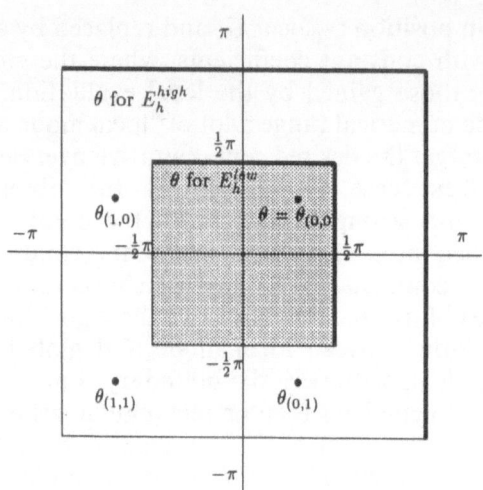

Fig. 2.36 Frequencies defining E_h^{low} and E_h^{high} for standard coarsening

E_h^{high}, respectively E_h^{low} and E_{2h}) together with a frequency $\theta = (\theta_1, \theta_2) = \theta_{(0,0)}$ and those frequencies θ_α (marked by •) which generate $E_{h,\theta}$ are shown in Figure 2.36.

With this notation and operators I_h^H, I_H^h and L_{2h} which are represented by the stencils $[t_\kappa]$, $[T_\kappa]$ and $[S_\kappa]$, respectively, for all $\theta \in (-\pi, \pi]^2$ the relations

$$L_{2h}\varphi_\theta^{2h}(\mathbf{x}) = \Lambda_\theta^h \varphi_\theta^{2h}(\mathbf{x}) \qquad \text{with} \qquad \Lambda_\theta^h = \sum_{\kappa \in \mathbb{I} \times \mathbb{I}} s_\kappa e^{i2\theta \cdot \kappa}$$

$$I_h^{2h}\varphi_\theta^{2h}(\mathbf{x}) = q(\theta_\alpha)\varphi_\theta^{2h}(\mathbf{x}) \qquad \text{with} \qquad q(\theta_\alpha) = \sum_{\kappa \in \mathbb{I} \times \mathbb{I}} t_\kappa e^{i\theta_\alpha \cdot \kappa} \qquad (2.98)$$

$$I_{2h}^h\varphi_\theta^{2h}(\mathbf{x}) = \sum_\alpha p_\alpha(\theta)\varphi_{\theta_\alpha}^h(\mathbf{x}) \quad \text{with} \quad p_\alpha(\theta) = \frac{1}{4} \sum_{\kappa \in \mathbb{I} \times \mathbb{I}} T_{-\kappa} e^{i\theta_\alpha \cdot \kappa}$$

hold [132].

This makes the coarse grid correction operator well defined and leaves the spaces $E_{h,\theta}$ (neglecting frequencies with $\Lambda_\theta^h = 0$ or $\lambda_\theta^h = 0$) invariant. The further procedure to compute $\mu^*(\nu)$ and $\rho^*(\nu)$ is straight forward, that means computing the spectral radii of submatrices with dimensions smaller than or equal to four.

The mentioned Fourier representation of many common relaxation methods is an important feature of both the "global" and the "local" mode analysis. To obtain it, the index set $\mathbb{I} \times \mathbb{I}$ used to describe the difference operator is subdivided into three disjoint subsets which denote points of the infinite grid

$\mathscr{G}_h$ with the following use:

$\mathbb{I} \times \mathbb{I}^c$: new values are actually calculated simultaneously
$\mathbb{I} \times \mathbb{I}^o$: old values are used within the current relaxation step
$\mathbb{I} \times \mathbb{I}^n$: new values of the current step are available and used

Example 2.26. *Jacobi type relaxations are characterized by* $\mathbb{I} \times \mathbb{I}^n = \varnothing$. *Point relaxation methods require* $\mathbb{I} \times \mathbb{I}^c = \{(0,0)\}$. *Block relaxation methods use an* $\mathbb{I} \times \mathbb{I}^c$ *which is a proper superset of* $\{(0,0)\}$ *within* $\mathbb{I} \times \mathbb{I}$.

According to the subsets of indices both the operator and the eigenvalues are splitted, too:

$$L_h = L_h^n + L_h^c + L_h^o \quad \text{and} \quad \lambda_\theta^h = \lambda_\theta^{hn} + \lambda_\theta^{hc} + \lambda_\theta^{ho},$$

where

$$\lambda_\theta^{hn} = \sum_{\kappa \in \mathbb{I} \times \mathbb{I}^n} s_\kappa e^{i\theta \cdot \kappa}, \quad \lambda_\theta^{hc} = \sum_{\kappa \in \mathbb{I} \times \mathbb{I}^c} s_\kappa e^{i\theta \cdot \kappa}, \quad \lambda_\theta^{ho} = \sum_{\kappa \in \mathbb{I} \times \mathbb{I}^o} s_\kappa e^{i\theta \cdot \kappa} \quad (2.99)$$

If $\tilde{e}_h$ and $\bar{\tilde{e}}_h := \mathscr{S}_h \tilde{e}_h$ denote the error before, respectively after relaxation, this relaxation (for simplicity without parameter ω) can be written as

$$(L_h^c + L_h^n)\bar{\tilde{e}}_h + L_h^o \tilde{e}_h = 0 \text{ on } \hat{\mathscr{G}}_h$$
$$\bar{\tilde{e}}_h = \tilde{e}_h \text{ on } \mathscr{G}_h - \hat{\mathscr{G}}_h. \quad (2.100)$$

The subset $\hat{\mathscr{G}}_h \subset \mathscr{G}_h$ is introduced to describe partial relaxation sweeps. If $\hat{\mathscr{G}}_h$ is equal to $\mathscr{G}_h$ (2.100) reduces to

$$(L_h^c + L_h^n)\bar{\tilde{e}}_h + L_h^o \tilde{e}_h = 0, \quad \text{or} \quad \mathscr{S}_h \tilde{e}_h := -(L_h^c + L_h^n)^{-1} L_h^o \tilde{e}_h. \quad (2.101)$$

The application to φ_θ^h then leads to

$$\mathscr{S}_h \varphi_\theta^h = \frac{-\lambda_\theta^{ho}}{\lambda_\theta^{hc} + \lambda_\theta^{hn}} \varphi_\theta^h. \quad (2.102)$$

Example 2.27. *For Gauss-Seidel relaxation with lexicographic ordering of points one has* $\hat{\mathscr{G}}_h = \mathscr{G}_h$, $\mathbb{I} \times \mathbb{I}^c = \{(0,0)\}$ *and*
$\mathbb{I} \times \mathbb{I}^n = \{\kappa \in \mathbb{I} \times \mathbb{I} | \kappa_1 < 0 \text{ or } (\kappa_1 = 0 \text{ and } \kappa_2 < 0)\}$.
Applied to the five-point stencil for the discretized Poisson equation the splitting of both the operator and the eigenvalues due to (2.99) and (2.100) is

$$L_h^c = \frac{1}{h^2}[4], \quad L_h^n = \frac{1}{h^2}\begin{bmatrix} & 0 & \\ -1 & 0 & 0 \\ & -1 & \end{bmatrix}, \quad L_h^o = \frac{1}{h^2}\begin{bmatrix} & -1 & \\ 0 & 0 & -1 \\ & 0 & \end{bmatrix}$$

$$\lambda_\theta^{hc} = \frac{4}{h^2}, \quad \lambda_\theta^{hn} = -\frac{1}{h^2}(e^{-i\theta_1} + e^{-i\theta_2}), \quad \lambda_\theta^{ho} = -\frac{1}{h^2}(e^{i\theta_1} + e^{i\theta_2}).$$

Because $\hat{\mathcal{G}}_h$ is equal to $\mathcal{G}_h$ the smoothing operator is represented by (2.101) and the application to a frequency φ_θ^h yields

$$\mathcal{S}_h \varphi_\theta^h = -(L_h^c + L_h^n)^{-1} L_h^o \varphi_\theta^h = \frac{-\lambda_\theta^{ho}}{\lambda_\theta^{hc} + \lambda_\theta^{hn}} \varphi_\theta^h = \frac{e^{i\theta_1} + e^{i\theta_2}}{4 - (e^{-i\theta_1} + e^{-i\theta_2})} \varphi_\theta^h.$$

The splitting for point Gauss-Seidel relaxation with lexicographic ordering applied to the five-point stencil of an anisotropic Helmholtz operator corresponds to

$$L_h^c = \frac{1}{h^2}[2(1+\varepsilon) + ch^2], \quad L_h^n = \frac{1}{h^2}\begin{bmatrix} & 0 & \\ -\varepsilon & 0 & 0 \\ & -1 & \end{bmatrix},$$

$$L_h^o = \frac{1}{h^2}\begin{bmatrix} & -1 & \\ 0 & 0 & -\varepsilon \\ & 0 & \end{bmatrix}$$

$$\lambda_\theta^{hc} = \frac{1}{h^2}[2(1+\varepsilon) + ch^2], \quad \lambda_\theta^{hn} = -\frac{1}{h^2}(\varepsilon e^{-i\theta_1} + e^{-i\theta_2}),$$

$$\lambda_\theta^{ho} = -\frac{1}{h^2}(\varepsilon e^{i\theta_1} + e^{i\theta_2}),$$

and

$$\mathcal{S}_h \varphi_\theta^h = \frac{(\varepsilon e^{i\theta_1} + e^{i\theta_2})}{2(1+\varepsilon) + ch^2 - (\varepsilon e^{-i\theta_1} + e^{-i\theta_2})} \varphi_\theta^h.$$

If instead of the lexicographic point Gauss-Seidel relaxation the lexicographic y-line relaxation is chosen, the above operator splitting and the working on frequencies changes to

$$L_h^c = \frac{1}{h^2}\begin{bmatrix} & -1 & \\ 0 & 2(1+\varepsilon) + ch^2 & 0 \\ & -1 & \end{bmatrix}, \quad L_h^n = \frac{1}{h^2}\begin{bmatrix} & 0 & \\ -\varepsilon & 0 & 0 \\ & 0 & \end{bmatrix},$$

$$L_h^o = \frac{1}{h^2}\begin{bmatrix} & 0 & \\ 0 & 0 & -\varepsilon \\ & 0 & \end{bmatrix},$$

respectively, $\mathcal{S}_h \varphi_\theta^h = \dfrac{\varepsilon e^{i\theta_1}}{2(1+\varepsilon) + ch^2 - (\varepsilon e^{-i\theta_1} + e^{-i\theta_2} + e^{i\theta_2})} \varphi_\theta^h.$

To conclude the "two-grid analysis", some results for both the Poisson equation and for the anisotropic model problem with Dirichlet boundary conditions on the unit square are summarized in the following tables and shortly discussed.

Table 2.9. *Smoothing rate and two-grid convergence factor (Poisson equation)*

v	red-black/FW		red-black/HW		GS-lex/FW	
$-\Delta u = f$	$(\mu^*)^v$	$\rho^*(v)$	$(\mu^*)^v$	$\rho^*(v)$	$(\mu^*)^v$	$\rho^*(v)$
1	0.250	0.250	0.250	0.500	0.500	0.400
2	0.063	0.074	0.063	0.125	0.250	0.193
3	0.034	0.053	0.034	0.034	0.125	0.119
4	0.025	0.041	0.025	0.025	0.063	0.084

The values in Table 2.9 demonstrate that the use of red-black relaxation and HW for restriction provide cheap and fast converging methods. If the number of operations is considered, the choice of $v = 2$ and $v = 3$ are recommended for FW and HW, respectively. The special situation of the Poisson equation should not divert from the fact that FW in most practical situations is the more robust restriction operator. Obviously, red-black relaxation is for the considered model problem more efficient than the Gauss-Seidel relaxation with lexicographically ordered points. Although the GS-lex values are only obtained by local mode analysis, the same effect is observed for concrete algorithms on finite grids, too. This table also shows that $(\mu^*(v))^v$ approximates the two-grid convergence factor well, especially for small values of v.

As pointed out before, point relaxation methods become bad for the anisotropic model equation. The snap-shot of Table 2.8 with $\varepsilon = 0.1$ is extended by Table 2.10 where the varying ε demonstrates the dramatic deterioration of red-black smoothing. Because of $\varepsilon \ll 1$ the choice of y-ZEBRA relaxation already yields both good smoothing and two-grid convergence. The best results with this respect are obtained by means of the alternating (AD) ZEBRA relaxation. This type of relaxation is very attractive if the anisotropy changes direction and magnitude within a given domain.

Further details concerning results of the two-grid analysis (both the global and the local one) are collected in [131] and [132].

Table 2.10. *Smoothing rate and two-grid convergence factor (anisotropic model equation)*

$-\varepsilon\dfrac{\partial^2 u}{\partial x^2} - \dfrac{\partial^2 u}{\partial y^2} = f$	red-black/FW		y-ZEBRA/FW		AD-ZEBRA/FW	
	$(\mu^*(3))^3$	$\rho^*(3)$	$(\mu^*(2))^2$	$\rho^*(2)$	$(\mu^*(2))^2$	$\rho^*(2)$
$\varepsilon = 1$	0.034	0.053	0.063	0.063	0.014	0.009
$\varepsilon = 0.5$	0.088	0.088	0.053	0.028	0.020	0.013
$\varepsilon = 0.1$	0.564	0.564	0.053	0.047	0.041	0.038
$\varepsilon = 0.01$	0.942	0.942	0.053	0.052	0.051	0.051

2.10.3 Smoothing Analysis

At first it has to be pointed out that even the very basic computation of smoothing factors, often called one-grid analysis, in principle has to be interpreted as a two-grid analysis. The simple reason is the distinction of "high-" and "low-" frequent modes, whose definition strongly depends on the coarse grid structure relative to the fine grid. The real two-grid character of smoothing analysis becomes more apparently as soon as the basic assumption is mentioned: the smoothing analysis considers a specific two-grid method assuming an idealized coarse grid correction operator. "Ideal" denotes a coarse grid operator which annihilates the low-frequent modes and produces no high-frequent components. The smoothing factor in this sense is an extension of that quantity which is often characterized by the worst damping of high-frequent modes. Consequently, the damping of high-frequent modes may not only serve to analyze the smoothing properties of a given scheme but also gives a first impression of the multigrid convergence. The previous sections supplied the theoretical fundamentals for a convenient presentation of the smoothing analysis. The assumption of the "local" mode analysis, that means thinking in terms of infinite grid operators and basis functions $\varphi_\theta^h(\mathbf{x}) = e^{i\theta \cdot (\mathbf{x}/h)}$, is still valid, and the procedure to determine a smoothing factor becomes straight forward. The following motivation in terms of a compact five-point stencil $\begin{bmatrix} & \mathscr{C} & \\ \mathscr{A} & \mathscr{X} & \mathscr{D} \\ & \mathscr{B} & \end{bmatrix}_h$ for the linear problem $L_h u_h = f_h$ is sufficient, because the extension to more complex stencils is natural. The exact solution u_h to the above problem satisfies the difference formula

$$\mathscr{A}u_{i-1,j} + \mathscr{B}u_{i,j-1} + \mathscr{X}u_{i,j} + \mathscr{C}u_{i,j+1} + \mathscr{D}u_{i+1,j} = f_{i,j} \qquad (2.103)$$

for all $(x_i, y_j) \in \mathscr{G}_h$. If w_h and $\tilde{w}_h$ denote an approximation to u_h before, respectively after one point Gauss-Seidel relaxation sweep with lexicographical scanning of grid points then

$$\mathscr{A}\tilde{w}_{i-1,j} + \mathscr{B}\tilde{w}_{i,j} + \mathscr{X}\tilde{w}_{i,j} + \mathscr{C}w_{i,j+1} + \mathscr{D}w_{i+1,j} = f_{i,j} \qquad (2.104)$$

is satisfied for all $(x_i, y_j) \in \mathscr{G}_h$. With corresponding algebraic errors $\tilde{e}_{i,j} := u_{i,j} - w_{i,j}$ before, respectively $\tilde{e}_{i,j} := u_{i,j} - \tilde{w}_{i,j}$ after relaxation, taking the difference of the equations (2.103) and (2.104) leads to

$$\mathscr{A}\overline{\tilde{e}_{i-1,j}} + \mathscr{B}\overline{\tilde{e}_{i,j-1}} + \mathscr{X}\overline{\tilde{e}_{i,j}} + \mathscr{C}\tilde{e}_{i,j+1} + \mathscr{D}\tilde{e}_{i+1,j} = 0. \qquad (2.105)$$

Assuming expansions $\tilde{e}_h(\mathbf{x}) = \sum_{\theta \in T} A_\theta \varphi_\theta^h(\mathbf{x})$, respectively $\overline{\tilde{e}}_h(\mathbf{x}) = \sum_{\theta \in T} \tilde{A}_\theta \varphi_\theta^h(\mathbf{x})$, and inserting the corresponding components into (2.105) shows that the amplitudes A_θ and $\tilde{A}_\theta$ are related by

$$\tilde{A}_\theta(\mathscr{A}e^{-i\theta_1} + \mathscr{B}e^{-i\theta_2} + \mathscr{L}) + A_\theta(\mathscr{C}e^{i\theta_2} + \mathscr{D}e^{i\theta_1}) = 0, \text{ or}$$

$$\tilde{A}_\theta = \frac{-(\mathscr{C}e^{i\theta_2} + \mathscr{D}e^{i\theta_1})}{(\mathscr{A}e^{-i\theta_1} + \mathscr{B}e^{-i\theta_2} + \mathscr{L})} A_\theta$$

for all $\theta \in T$. That means, the amplitude A_θ is damped by PGS-lex for the above stencil to $\tilde{A}_\theta$. The natural consequence is to define the damping factor for a frequency $\theta \in T$ to be

$$\mu(h, \theta) := \frac{\| \tilde{A}_\theta \|}{\| A_\theta \|}. \tag{2.106}$$

The smoothing factor $\tilde{\mu}(h)$ of PGS-lex for the above L_h then is the worst (the largest) factor by which high-frequent modes are damped:

$$\tilde{\mu}(h)_{PGS\text{-}lex} := \max_{\theta \in T^{high}} \mu(h, \theta) = \max_{\theta \in T^{high}} \frac{\| -(\mathscr{C}e^{i\theta_2} + \mathscr{D}e^{i\theta_1}) \|}{\| (\mathscr{A}e^{-i\theta_1} + \mathscr{B}e^{-i\theta_2} + \mathscr{L}) \|} \tag{2.107}$$

In a similar way for the lexicographical y-line (column) Gauss-Seidel relaxation (CGS-lex) the smoothing factor

$$\tilde{\mu}(h)_{CGS\text{-}lex} := \max_{\theta \in T^{high}} \frac{\| -\mathscr{D}e^{i\theta_1} \|}{\| \mathscr{A}e^{i\theta_1} + \mathscr{B}e^{i\theta_2} + \mathscr{L} + \mathscr{C}e^{i\theta_2} \|} \tag{2.108}$$

is derived. Fortunately, the dependency from the meshsize h is small and for many relaxation schemes h-independent upper bounds $\mu^* := \max_{h \in \mathscr{H}} \tilde{\mu}(h)$ exist. For the standard five-point approximation of the Laplacian these values are $\mu^*_{PGS\text{-}lex, \Delta_h} = 0.5$ and $\mu^*_{CGS\text{-}lex, \Delta_h} = 1/\sqrt{5}$. Comparing the above presentation with the decomposition of smoothing operators as introduced in Section 2.10.2 shows that exactly the idea of breaking the smoothing operator into parts is used again. If the coefficients of the five-point stencil are set to $\mathscr{A} = \mathscr{B} = \mathscr{C} = \mathscr{D} = -1/h^2$ and $\mathscr{L} = 4/h^2$, the equation (2.107) reduces to

$$\frac{e^{i\theta_1} + e^{i\theta_2}}{4 - (e^{-i\theta_1} + e^{-i\theta_2})}$$

as it was derived in Example 2.27 for the discrete Laplacian.
If $\mathscr{A} = \mathscr{D} = -\varepsilon/h^2$, $\mathscr{B} = \mathscr{C} = -1/h^2$ and $\mathscr{L} = 1/h^2\ (2(1+\varepsilon) + ch^2)$ (2.107) and (2.108) reduce to the factors as presented in Example 2.27, again.
Expressing the relaxation operator $\mathscr{S}_h$ by the decomposition L_h^c, L_h^n and L_h^o of the difference operator and applying it to error terms (2.101) and (2.102) shows that the amplification factor for the θ-frequency is composed of the

corresponding symbols λ_θ^{ho} and $\lambda_\theta^{hc} + \lambda_\theta^{hn}$:

$$\mu(h, \theta) = \frac{|-\lambda_\theta^{ho}|}{|\lambda_\theta^{hc} + \lambda_\theta^{hn}|}$$

$\tilde{\mu}(h)$ and μ^* are defined as before.

Example 2.28. *Splitting the standard five-point stencil for the Laplacian due to Example 2.26 for the Jacobi relaxation* $(\mathbb{I} \times \mathbb{I}^n = \emptyset,\ \mathbb{I} \times \mathbb{I}^c = \{(0,0)\})$ *allows the description of the weighted Jacobi relaxation (Algorithm 2.3) by*

$$L_h^c \tilde{\bar{e}}_h = ((1 - \omega)L_h^c - \omega L_h^o)\tilde{e}_h.$$

The corresponding amplification factor for a component with $\theta \in T$ *is*

$$\mu(h, \theta) = \frac{\left| (1 - \omega)\dfrac{4}{h^2} + \dfrac{\omega}{h^2}(e^{-i\theta_1} + e^{-i\theta_2} + e^{i\theta_1} + e^{i\theta_2}) \right|}{\dfrac{4}{h^2}}$$

$$= 1 - \omega(\sin^2 \tfrac{1}{2}\theta_1 + \sin^2 \tfrac{1}{2}\theta_2).$$

This value does not depend on h and the discussion for $\theta \in T^{\text{high}}$ *shows* $\mu^*(\omega) = \max\{|1 - \tfrac{1}{2}\omega|, |1 - 2\omega|\}$. *Consequently the weighted Jacobi relaxation only shows a smoothing property for* $0 < \omega < 1$. $\omega = 0.8$ *provides the optimal damping of high-frequent modes:* $\mu^*(0.8) = 0.6$ *for* $-\Delta_h$. *This finally justifies the choice of* $\omega = 0.8$ *for the experiments of Section 2.1.4 and additionally shows the poor smoothing of the weighted Jacobi scheme compared to PGS-lex, CGS-lex or coloured schemes, even for* ω_{opt}.

μ^* not only allows a careful estimation of the asymptotic two-level-convergence factor—and by this the multigrid convergence—but also gives advices what relaxation or coarsening strategy is recommended.
The PGS-lex for the anisotropic model problem in Example 2.27 provided amplification factors

$$\mu(h, \theta) = \frac{(\varepsilon e^{i\theta_1} + e^{i\theta_2})}{2(1 + \varepsilon) + ch^2 - (\varepsilon e^{-i\theta_1} + e^{-i\theta_2})}.$$

With $c = 0$ and $\theta = (\pi, 0)$ $\mu((\pi, 0))$ becomes $\dfrac{1 - \varepsilon}{1 + 3\varepsilon}$. Therefore μ^* tends towards one for $\varepsilon \to 0$ and for $\varepsilon \to \infty$. This explains the bad smoothing of PGS in combination with standard coarsening and ε not close to one (see Table 2.10).
The first already mentioned alternative for the same problem with $\varepsilon \ll 1$ is block relaxation in direction of strong coupling. y-line relaxation leads to

the amplification factor

$$\frac{\varepsilon e^{i\theta_1}}{2(1 + \varepsilon) - (\varepsilon e^{-i\theta_1} + e^{-i\theta_2} + e^{i\theta_2})}$$

which finally shows

$$\mu^* = \begin{cases} 1/\sqrt{5} & \text{for} \quad \varepsilon \leq 1 \\ \\ \rightarrow 1 & \text{for} \quad \varepsilon \rightarrow \infty \end{cases}.$$

The second approach is the combination of point relaxation and y-SEMI coarsening. A first benefit, the decreased anisotropy of the coarse grid operator has already been shown. But the essential characteristic with respect to smoothing analysis is the set of high-frequent modes which differs from that with respect to standard coarsening. μ^* is then determined with respect to all θ with $|\theta_2| > 0$.

A direct transformation of the above technique for the lexicographic relaxation schemes to coloured schemes is impossible. The coloured treatment of points within partial relaxation sweeps acts similar to a coarsening, that means identifying or coupling frequencies, within a current relaxation step. Nevertheless, the above calculation of amplification and smoothing factors has been generalized to chequered (Gauss-Seidel) schemes in [78] by a careful consideration of the coupled frequencies. [132] use the matrix representation of partial relaxation steps with respect to the invariant spaces (harmonics). The complete relaxation then is described by the corresponding matrix product and the smoothing factor is determined as the maximum spectral radius of these submatrices as before.

2.11 Multigrid on Parallel Computers

The importance of numerical simulations for a large variety of different application classes has increased dramatically during the last two decades. Two main reasons are of equal significance for this development. On the algorithmic side, new developments such as multigrid have been utilized for increasingly complicated large applications. Concerning hardware and system software, the impressive increase in supercomputer performance has given further impact to the success of numerical simulations.

For many applications, however, even today's supercomputers are not powerful enough, even if the most efficient numerical methods were used. It is generally agreed that further accelerations in the supercomputer range will principally be achieved by an increasing degree of parallelism since possible improvements of single processors seem to be much more limited.

There is no doubt that parallelism will be a decisive criterion for future algorithmic developments, which are significant for scientific and industrial applications. Fast and efficient numerical algorithms, however, are of similar importance. Methods lacking either parallelism or numerical efficiency will not be suitable for the challenging problems of the future. The parallelization of multigrid methods, which are sequentially optimal for many applications, is thus of particular importance.

Though the future belongs to parallel computers, it is not yet clear, which particular architecture will prevail. Today, there are many different parallel machines with different architectures ranging from SIMD (Single Instruction Multiple Data) machines to MIMD (Multiple Instruction Multiple Data) computers with shared or distributed memory, combined with a corresponding variety of programming models. In general, the parallelization for SIMD machines is based on the vector model, e.g., on the use of vector or array constructs as contained in Fortran 90. The system then takes care of the work distribution to the processors. On machines with shared (global) memory, parallelization tools such as automatically parallelizing compilers are available, too. They utilize the existence of the shared memory, to which all processors are connected. Typical machines such as the Alliant support in particular vectorization and parallelization on the loop level, provided that no data dependencies occur. A knowledge of the whole application is not required.

The situation is different for MIMD machines with distributed (local) memory. The programming model available on most of these machines is based on explicit communication among processes by message passing. The concept of a virtual shared memory on such machines does not yet provide sufficiently efficient solutions for typical parallel applications. Nevertheless, this class of machines is of particular importance because the number of processors employed is significantly larger than in shared memory computers, where hardware and efficiency aspects restrict the number of processors. Since future parallel computers will be composed of even more processors than they are today, MIMD machines with distributed memory represent the current challenge for parallel computing. We will thus focus on such computers and the corresponding programming model in the rest of this section.

According to the overall nature of this book, only some basic aspects of parallel multigrid can be discussed, here. Further information can be obtained from many recent publications (see, e.g., [11, 17, 36, 37, 45, 54, 56, 81, 85]). A survey, which contains more detailed discussions on the parallelization of multigrid algorithms and on a classification of parallel machines, is [95]. Aspects of advanced applications such as the parallelization of systems of partial differential equations on non-rectangular domains are considered, e.g., in [84].

For simplicity, it is assumed in this section that only one user process is mapped to each processor. Possible advantages attainable by mapping several (small) processes onto one processor are not considered.

2.11.1 Speed-Up and Efficiency

Important quantities for the evaluation of parallel applications are the *speed-up* and the *parallel efficiency*. The speed-up is usually defined as

$$S := \frac{\text{solution time using 1 processor}}{\text{solution time using } p \text{ processors}}. \tag{2.109}$$

Ideally, it should be close to p. The parallel efficiency

$$E := \frac{S}{p} \tag{2.110}$$

is a p-independent measure for the parallel quality of an application. Unfortunately, speed-up and efficiency cannot be determined via Equation (2.109) for large problems which do not fit on one processor due to storage limitations and/or to huge computing times. Another definition for the speed-up, which can be measured in parallel applications, is

$$S := \frac{\sum_{i=1}^{p} a_i}{\max_i (a_i + c_i)}, \tag{2.111}$$

where a_i is the time spent for arithmetic computations in processor i, and c_i denotes the sum of communication and idle time of processor i. The two definitions for S are equivalent under certain assumptions, including the independence of the time for floating point operations on the number of processes involved (no vectorization).

2.11.2 Criteria for Efficient Parallel Algorithm Development

Several basic criteria have to be considered in order to obtain high speed-ups and efficiencies, both for the development of new parallel algorithms and for the implementation of existing sequential codes on parallel machines: load-balancing, interprocess interferences, locality, granularity and scalability. Though they are of general importance, they are discussed here mainly with respect to distributed memory multiprocessors.

From Equation (2.111), it is clear that a satisfactory parallelization requires a careful *load-balancing* of the processors. Even if there were no communication nor idle times, the optimal speed-up of p (corresponding to an efficiency of 1) can be obtained only for $a_i = a_j$ for all i and j in $\{1,\ldots,p\}$, i.e. for an equally distributed load. Of course, this optimal case will hardly occur in any real application. But the crucial situation that one or a few processors have a significantly higher work load than the majority of processors, leads automatically to a very bad parallel efficiency and should thus be avoided.

Another requirement for high parallel efficiency concerns *interferences among processors such as synchronization or communication*. Dominating communication or dominating idle times may degrade the parallel behavior of an application severely. This situation (and a badly balanced load as well) has to be avoided by any means when developing efficient parallel algorithms.

In this connection, *locality* of an algorithm provides good preconditions for the parallelization. Working mainly with local data, i.e. data which are available in a processor's own memory, corresponds to a limited degree of communication with other processes. The grid partitioning approach described below makes use of this idea. It enlarges the data stored in the memory of each single processor to a small extent in order to reduce interprocess communication.

The *granularity* of an algorithm is a characterization for the independence of the single parallel processes with respect to each other. It has to be adjusted to the communication facilities and arithmetic performance of a particular parallel target machine under consideration. On message passing systems, the number of arithmetic operations to be performed between two communications steps may be a measure for the granularity.

The *scalability* of an algorithm describes its efficiency when increasing the number of available processors. The term scalability is commonly used in two different meanings:

- It may be based on a fixed total problem size reflecting Amdahl's law: The parallel speed-up is limited by the inverse of the sequential portion contained in the algorithm. Additionally, the granularity is refined when smaller and smaller portions of work are mapped to more and more processors.

- In a second approach the problem size per process is kept fixed. When employing more processors, the global problem size increases whereas the relative sequential portion of the algorithm is generally reduced. This possibility is suited for distributed memory multiprocessors since the available total memory increases with the number of processors. Here, a high scalability of an algorithm means that the efficiency is independent or only weakly dependent on the number of processors employed.

2.11.3 Grid Partitioning—An Efficient Parallelization Concept for Multigrid

The grid partitioning approach is essentially independent of the particular partial differential equation or system to be solved. It is applicable to general n-dimensional domains using block-structured grids [55, 84, 85]. In order to make the basic parallelization principle as easily intelligible as possible, we restrict ourselves here to the parallel treatment of partial differential equations such as Poisson's equation on a two-dimensional rectangular grid consisting of $N_x \times N_y$ points, which can be discretized by a five-point stencil. Generalizations to more complicated problems are straightforward. For the problems under consideration, each grid point corresponds to one discrete unknown (or a set of unknowns in the case of systems of partial differential equations). The parallelization can thus be described in terms of grid points, which corresponds to an intuitive understanding of locality. Grid partitioning distributes the grid points (and the corresponding discrete equations and variables) to the processors available for a given application. Obviously, this can be done in many different ways.

For an efficient parallelization, however, *load balancing and locality* have to be taken into account with high priority. A satisfactory load balancing presupposes that all processors are responsible for approximately the same number of grid points, i.e. approximately the same number of discrete equations and variables, since the overall arithmetic work is proportional to the number of grid points in typical multigrid applications. A high amount of locality is required in any case in order to keep the communication costs low. Since the difference equation for a variable at a given grid point involves variables which are located at neighbouring points, points at the interior boundaries of the subgrids, which have been mapped to the distinct processes, require information from other processes. Thus, low communication costs correspond to compact and simply-connected subgrids rather than fissured or multiply-connected ones. Rectangular grids are usually distributed to rectangular process grids; Figure 2.37, for instance, shows the mapping of a 12×9 grid to 4×3 processes.

In order to reduce the amount of communication, the use of an overlap area has proved to be successful, which makes available data from neighbouring processes. Figure 2.37 contains an example with an overlap width of one point.

In addition to its subgrid, each process has knowledge of current approximations at neighbouring points. Of course, these data have to be refreshed regularly at certain stages in the algorithm requiring interprocess communication. Provided that the data in the overlap area are up-to-date, the processes can, for instance, calculate the residuals at all points fully in parallel. Though the overlap belonging to a process in the interior of the rectangular process grid contains data from eight different other processes, communi-

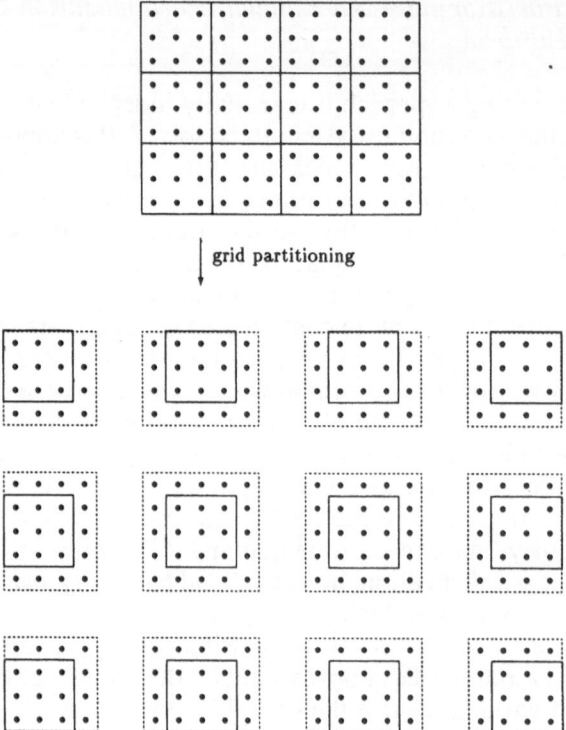

Fig. 2.37 Partitioning of a 12 × 9 grid to 3 × 3 subgrids including an overlap of width one

cation is required with four processes only. The data from diagonal neighbours can be obtained automatically if the data exchanges with respect to
the x- and the y-direction are performed one after the other.

Multigrid consists essentially of three components: smoothing, restriction
to a coarser grid, and interpolation of corrections back to the fine grid.
Typical multigrid components such as red-black point relaxation, full weighting, or linear interpolation of some coarse grid functions to the fine grid are
parallel by nature. Pure red-black point relaxation, for example, can be
carried out fully in parallel, provided that the overlap regions are updated
after every relaxation halfstep. Full weighting represents a local weighted
average of residuals, and linear interpolation can be performed independently
at all grid points, too.

However, a remaining difficulty in the parallelization of multigrid on highly
or massively parallel machines is the treatment of the very coarse grids. If
only a few grid points are available per process, the amount of communication may totally dominate the arithmetic work on that grid, possibly im-
plying a significant performance loss. Though the quantity of this effect is
dependent on the concrete machine parameters, it is of importance on all

current distributed memory machines. In many cases improvements can be achieved by agglomerating the problem on a coarse grid to fewer processes, resulting in a medium-sized problem per process on the agglomerated process grid; the decreased communication overhead on the current and still coarser grids outweighs the disadvantage of keeping the majority of the processes idle (see [54] for details).

2.11.4 Parallel Performance of Multigrid for Poisson's Equation

In this section some measurements concerning the performance of the parallel multigrid program MGDEMO [53, 54] on the Intel iPSC/2 and on the SUPRENUM machine are discussed. We concentrate on scalability aspects keeping the problem size per processor constant.

The numerical solution of Poisson's equation via multigrid represents a worst case problem for an efficient parallelization. Since it is a very simple partial differential equation, only a small number of operations is required for its numerical solution. The communication overhead in corresponding parallel multigrid programs is, however, hardly dependent of the concrete form of the partial differential equation. A very decisive criterion for the efficiency on parallel systems, i.e. the ratio of the times spent for arithmetic work and communication work, is thus particularly bad for the solution of Poisson's equation as compared with more complicated problems.

The multigrid algorithm applied in this section comprises

- two red-black point Gauss-Seidel relaxations for smoothing before and one after the coarse grid correction,
- half injection for the restriction, and
- linear interpolation of the corrections on the way back from coarse grids.

Data in the overlap areas are updated after every relaxation halfstep. Thus, it is guaranteed that the parallel algorithm is independent of the number of processors employed and is equivalent to its sequential counterpart.

The results in Table 2.11 are obtained on the Intel iPSC/2 in scalar mode.

Table 2.11. *MGDEMO benchmarks on the Intel iPSC/2; 256 × 256 grid on each processor; scalar execution*

Processor	MFLOPS	Time/Cycle (sec)	Efficiency (%)
1 = 1 × 1	0.3	6.6	100
2 = 2 × 1	0.5	6.8	98
4 = 2 × 2	1.0	6.9	97
8 = 4 × 2	2.0	7.0	96
16 = 4 × 4	3.9	7.0	95
32 = 8 × 4	7.8	7.1	95

Table 2.12. *MGDEMO benchmarks on the iPSC/2;*
192 × 192 grid on each processor; vector execution

Processor	MFLOPS	Time/Cycle (sec)
$1 = 1 \times 1$	0.8	1.2
$2 = 2 \times 1$	1.5	1.3
$4 = 2 \times 2$	2.9	1.4
$8 = 4 \times 2$	5.5	1.4
$16 = 4 \times 4$	10.3	1.5
$32 = 8 \times 4$	20.4	1.5

Agglomeration to one process keeping all other processors idle is carried out on the eighth grid. The coarse grid problem is solved by one relaxation step.

The vectorization facilities of this computer are not utilized. A 256×256 grid is mapped to each processor; thus, the largest application on 32 processors solves a 2048×1024 grid problem. The parallel efficiency proves to be excellent, yielding at least 95% in all cases. The small reduction of the parallel efficiency up to 16 processors is mainly caused by the fact that the full communication complexity of the overlap data exchange is not yet reached for smaller process grids: An exchange of the overlap data corresponds to communication with $3 (2, 1, 0)$ neighbouring processors if $8 (4, 2, 1)$ processors are employed. Communication with four neighbouring processors is required if 16 or more processors are utilized.

Table 2.12 is obtained upon switching on vectorization on the Intel iPSC/2. As compared with the scalar case, the grid size per processor has to be reduced to 192×192 since the available vector memory of the machine is only one Mbyte. Agglomeration to one process is performed on the seventh grid. The problem on the coarsest grid is solved by 10 relaxations.

Possible optimizations with respect to the partitioning of the grid, which may be important in connection with vectorization, were not taken into account. For example, four 192×192 subgrids, four 384×96 subgrids, or four 96×384 subgrids form a partitioning of a 384×384 grid problem. The effective vector length, which is very crucial for the performance of each processor, is, however, different in all cases. The grid level, at which agglomeration has to take place, depends on the structure of the grid partitioning, too.

As mentioned in Section 2.11.1, the speed-up definition (2.111) is different from definition (2.109) in cases utilizing vectorization. Thus, Tables 2.12 and 2.13 do not contain entries for the parallel efficiencies. The MFLOPS obtained and the times required per multigrid cycle indicate, how the system performance scales up with growing processor numbers and growing problem sizes. The optimal case for a fixed problem size per processor would show constant times per cycle and a doubling of the MFLOPS rate if the number of active processors is doubled.

Rating this optimal case as 100% efficient, the results in Table 2.12 are close
to 80% efficiency for 16 and 32 processors. Again, it is obvious that the increasing communication complexity up to 16 processors corresponds to a
certain loss of the obtained performance. It is not surprising that this loss is
larger than in the scalar case. Since execution in vector mode is significantly
faster, the time for arithmetic computations is lower whereas the time for
communication remains constant.

Before we discuss the results obtained on SUPRENUM, it should be mentioned that the SUPRENUM computer has a two-level connection network
(as compared to the hypercube structure of the Intel iPSC/2). 16 nodes are
combined into a cluster using the lower connection network, and 16 such
clusters form the whole machine. The interconnection of the clusters is
provided by an upper connection network.

If more than 16 processors are used, a first agglomeration step is carried out
on grid level 7. The result of this agglomeration is a process grid, which is
coarsened by a factor of 4 in each direction. For example, the 8×16 process
grid is reduced to 2×4 processes. Agglomeration to one processor is carried
out on the eighth grid in all tests contained in Table 2.13.

The MFLOPS rates on SUPRENUM are significantly higher than on the
Intel iPSC/2 because the vector unit is much more powerful. On the other
hand, communication between nodes is dominated severely by the high
communication start-up times for applications such as MGDEMO on
SUPRENUM. This is the reason for the high performance losses up to 16
processors where the full communication complexity is reached (communication with four neighbours is required). But then a similar effect is observed
for the increasing communication complexity among clusters. Employing 32
processors, the exchange of the overlap data requires communication with
one neighbouring cluster as compared to communication with 2 (3, 4)
clusters in case 64 (128, 256) processors are used. Nevertheless, in the range
from 16 to 128 processors the system performance increases nearly linearly
with the processors. More detailed measurements show that the relatively

Table 2.13. *MGDEMO benchmarks on SUPRENUM;*
512×512 grid on each processor; vector execution

Processor	MFLOPS	Time/Cycle (sec)
$1 = 1 \times 1$	4.7	0.7
$2 = 1 \times 2$	7.4	0.9
$4 = 2 \times 2$	12.4	1.1
$8 = 2 \times 4$	19.9	1.4
$16 = 4 \times 4$	33.8	1.6
$32 = 4 \times 8$	66.5	1.7
$64 = 8 \times 8$	125.0	1.8
$128 = 8 \times 16$	237.6	1.9
$256 = 16 \times 16$	391.1	2.3

bad performance on 256 processors is based on a bottleneck, caused by the upper connection network of the SUPRENUM machine, which limits the system performance severely.

This example demonstrates that scalability does not only depend on the algorithmic components of an application, but on the real parallel hardware available as well. Of course, for more complicated multigrid applications with a much larger amount of arithmetic work, the effect of the high communication start-up times on the performance is reduced significantly.

2.11.5 Other Parallelization Concepts

There are several other parallelization concepts which we will not discuss, here. Nevertheless, the most important of them will be mentioned in order to give the reader the opportunity of obtaining further information.

The *domain decomposition* approach (see, e.g., [69]), which has to be clearly distinguished from the approach described above, is also widely used. Recent results [117], however, indicate that its overall efficiency including the numerical efficiency seems to be inferior to grid partitioning, at least for some variants of this approach.

Sparse grids [46, 140] are a new discretization-based approach to obtain highly accurate numerical solutions of problems such as partial differential equations. It does not specify the solution algorithm (e.g., multigrid) for the various grid problems arising out of this method. In any case, the parallelization can be carried out very easily since the distinct discrete problems are fully decoupled.

For parabolic partial differential equations, *waveform relaxation methods* [136, 135] are another interesting algorithmic approach, particularly if vectorization *and* parallelization have to be taken into account.

2.11.6 Parallelization of Non-Local Smoothing Operators

The principle of grid partitioning leads in a natural way to the parallelization of multigrid methods. This is easily seen for multigrid components that involve only *local* grid operations (such as point relaxation, intergrid transfer operations and residual calculations). For an efficient error smoothing within the multigrid solution of anisotropic problems, however, *block relaxation methods* are used. These involve non local operations and hence are more difficult to parallelize.

Performing, for example, a step of line relaxation for the 2D-model problem (2.5) or (2.65) requires the solution of 'many' tridiagonal systems (each line corresponding to one system). As long as the parallelization of the multigrid method can be based on a stripwise partitioning of the computational grid in such a way that the lines are not intersected by artificial boundaries, the tridiagonal systems are independent from each other. In the general case, however, the choice of such a partitioning is not possible (what is easily seen for the case that *alternating* line relaxation steps have to be performed) and one has to solve tridiagonal systems that are distributed over several processors. For the treatment of such systems a solver with high multiprocessor speedup, based on cyclic reduction, has been implemented [70, 71]. Beyond parallel line relaxation methods, parallel block ILU smoothers (for 2D problems) and parallel plane relaxation methods (for 3D applications) based on parallel 2D multigrid methods have been implemented and applied successfully to problems with arbitrary anisotropies [38]. Important for the design of all these block smoothing methods is the following principle: 'Simultaneous messages for all of the blocks are collected and sent as a single message'. By this way, a high multiprocessor performance is reached also on machines with a relatively high startup time.

2.12 Standard Multigrid for Semiconductor Device Simulation

In this section it is pointed out, that even for more complicated systems of partial differential equations the standard multigrid components as described in Section 2.3 will work. System 4 of Example 2.23 is the classical drift-diffusion model for VLSI device simulation in the stationary case. A derivation and some remarks about other choices of variables may be found in [126].

It is well known [93, 126], that for computational reasons the system has to be scaled. Using the singular perturbation scaling [93], the scaled system has the following form:

$$\lambda^2 \Delta \psi - (n - p) = -N \tag{2.112}$$

$$\nabla(\mu_n(\nabla n - n\nabla\psi)) - R(\psi, n, p) = 0 \tag{2.113}$$

$$-\nabla(\mu_p(\nabla p + p\nabla\psi)) + R(\psi, n, p) = 0 \tag{2.114}$$

Looking at the determinant of the Jacobian matrix of this system, it can be seen that the highest partial derivatives have the factor λ^2. This makes the singular perturbed character of the system visible, if the factor is small compared to the factors of lower order derivatives. For MOS transistors a typical value for λ^2 at room temperature (300 K) is 10^{-7}. The independent variables ψ, n and p represent the electrostatic potential, electron and hole concentration, respectively. In addition μ_n, μ_p, N and R are the electron and hole mobilities, the net ionized impurity concentration and the net carrier recombination rate.

Some physical models have to be choosen for μ_n, μ_p and R. Here only the following should be mentioned [126]:

$$R(\psi, n, p) = R_{SRH}(\psi, n, p) = \frac{np - n_i^2}{\tau_p(n + n_i) + \tau_n(p + n_i)}$$

$$\mu_{n,p} = \mu_{min_{n,p}} + \frac{\mu_{max_{n,p}} - \mu_{min_{n,p}}}{1 + \left(\dfrac{N_i}{N_{ref_{n,p}}}\right)^{\alpha_{n,p}}}$$

$\tau_{n,p}, n_i$ and N_i are the carrier-lifetime, the intrinsic concentration and the total impurity concentration. $\mu_{min_{n,p}}, \mu_{max_{n,p}}, N_{ref_{n,p}}$ and $\alpha_{n,p}$ are temperature dependent constants.

A typical domain for a MOS transistor, building up a small p-channel between source and drain for some applied negative voltage at the gate, is shown in the following figure.

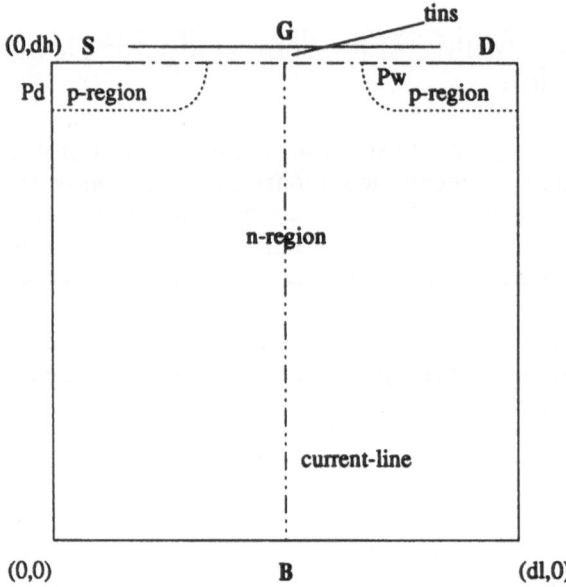

Fig. 2.38 Simulation area for the presented MOS example
S: source, D: drain, G: gate, B: bulk,
$\Omega_S = [0, dl] \times [0, dh]$: the semiconductor region
$\Omega_I = [S, dl - D] \times [dh + tins]$: the insulator region

Looking for boundary conditions describing the physical behavior is a nontrivial problem [126]. Here, as an example, Dirichlet conditions are assumed at source, drain and bulk. At the semiconductor-insulator interface a mixed boundary condition is assumed for (2.112) in case of a thin insulator.

For (2.113) and (2.114)

$$J_n \cdot \vec{n} = 0,$$
$$J_p \cdot \vec{n} = 0$$

is used. At the outside boundaries, symmetry conditions are valid:

$$J_p \cdot \vec{n} = 0,$$
$$J_n \cdot \vec{n} = 0$$
$$\nabla \psi \cdot \vec{n} = 0,$$

where $\vec{n}$ is the outward unit normal vector and J_n, J_p are given by:

$$J_n = \mu_n (\nabla n - n \nabla \psi),$$
$$J_p = -\mu_p (\nabla p + p \nabla \psi).$$

Discretizing the system (2.112)–(2.114) by the method of finite differences using central differencing, results in an unstable scheme if the singular perturbed character is significant. This is well known. In [114] it is shown for the one-dimensional case that using upwind differencing solves the stability problems, but now in an error estimate for n and p the constant depends on $\nabla \psi$. So the accuracy is not good if $\| \nabla \psi \| \gg 1$. This can be overcome by using the Scharfetter-Gummel discretization for (2.113) and (2.114), a special upwind differencing scheme.

On the given grid $G_h = \{(x_i, y_j) | x_i = i \cdot h_x, y_j = j \cdot h_y; h_x = dl/N, h_y = dh/M\}$ with $h = (h_x, h_y)$ all equations (2.112)–(2.114) have the same structure. Using the notation $U = (u_1, u_2, u_3)^T = (\psi, n, p)^T$, $U_h = (\psi_h, n_h, p_h)^T$ this is given by:

$$\frac{1}{h_x h_y} \begin{bmatrix} & s_{01} & \\ s_{-10} & -s_{00} & s_{10} \\ & s_{0-1} & \end{bmatrix}_h u_{k,h}(P) - g_{k,h}(U)(P) = f_{k,h}(P) \qquad (2.115)$$

(with $k = 1, 2, 3$), where P is an inner point of G_h and g_k, f_k are given functions, specified below for the example of this section. For (2.112) the coefficients $s_{i,j}$ are given by the usual 5-point stencil for $\lambda^2 \cdot \Delta$. For (2.113) and (2.114) the coefficients in (2.115) are summarized in Table 2.14. $g_k(U)$ and f_k for (2.112)–(2.114) are given by Table 2.15.

The discrete equations for boundary points are derived analogously [126]. The system (2.115) is solved by a nonlinear multigrid (FAS-method), as described by Algorithm 2.7. The smoothing process is done by "local collective" Gauss-Seidel relaxation with chequerboard ordering of the points, as explained in Section 2.9, here called lcGS-ch. At every gridpoint, the three nonlinear equations are solved approximately with a damped Newton method. Different strategies are applied:

1. Due to [24] a strategy is used which controls the arguments, whereas
2. following [1, 2] the residuals are controlled.

On the coarsest grid global linearization with a damped Newton method
was implemented. Full weighting (2.58) has been used for the restriction of
the residuals and injection for the restriction of the solution, while interpolation is done in bilinear manner. All these multigrid components are standard
as described in Section 2.3.
In the FAS-method, local linearization is done in every smoothing step. So
the Bernoulli-functions and their derivatives have to be calculated more

Table 2.14. *Bernoulli-Stencil*

	Equation for n	Equation for p
s_{00}	$\mu_n\|_{i+\frac{1}{2},j}B(\psi_{i,j}-\psi_{i+1,j})\frac{h_y}{h_x}$ $+\mu_n\|_{i-\frac{1}{2},j}B(\psi_{i,j}-\psi_{i-1,j})\frac{h_y}{h_x}$ $+\mu_n\|_{i,j+\frac{1}{2}}B(\psi_{i,j}-\psi_{i,j+1})\frac{h_x}{h_y}$ $+\mu_n\|_{i,j-\frac{1}{2}}B(\psi_{i,j}-\psi_{i,j-1})\frac{h_x}{h_y}$	$\mu_p\|_{i+\frac{1}{2},j}B(\psi_{i+1,j}-\psi_{i,j})\frac{h_y}{h_x}$ $+\mu_p\|_{i-\frac{1}{2},j}B(\psi_{i-1,j}-\psi_{i,j})\frac{h_y}{h_x}$ $+\mu_p\|_{i,j+\frac{1}{2}}B(\psi_{i,j+1}-\psi_{i,j})\frac{h_x}{h_y}$ $+\mu_p\|_{i,j-\frac{1}{2}}B(\psi_{i,j-1}-\psi_{i,j})\frac{h_x}{h_y}$
$s_{-1,0}$	$\mu_n\|_{i-\frac{1}{2},j}B(\psi_{i-1,j}-\psi_{i,j})\frac{h_y}{h_x}$	$\mu_p\|_{i-\frac{1}{2},j}B(\psi_{i,j}-\psi_{i-1,j})\frac{h_y}{h_x}$
$s_{1,0}$	$\mu_n\|_{i+\frac{1}{2},j}B(\psi_{i+1,j}-\psi_{i,j})\frac{h_y}{h_x}$	$\mu_p\|_{i+\frac{1}{2},j}B(\psi_{i,j}-\psi_{i+1,j})\frac{h_y}{h_x}$
$s_{1,0}$	$\mu_n\|_{i,j+\frac{1}{2}}B(\psi_{i,j+1}-\psi_{i,j})\frac{h_x}{h_y}$	$\mu_p\|_{i,j+\frac{1}{2}}B(\psi_{i,j}-\psi_{i,j+1})\frac{h_x}{h_y}$
$s_{-0,1}$	$\mu_n\|_{i,j-\frac{1}{2}}B(\psi_{i,j-1}-\psi_{i,j})\frac{h_x}{h_y}$	$\mu_p\|_{i,j-\frac{1}{2}}B(\psi_{i,j}-\psi_{i,j-1})\frac{h_x}{h_y}$
$B(z)=\dfrac{z}{e^z-1}$		

Table 2.15. *Definition of $g_k(U)$ and f_k*

	(2.112)	(2.113)	(2.114)
$g_k(U)$	$n-p$	$R(U)$	$R(U)$
f_k	$-N$	0	0

Table 2.16. *Parameters for the MOS-example*

	μm								
dl	dh	S	D	gate	Pd	Pw	tins	N_D	N_A
6	6	1	1	4	0.6	2.0	0.03	$3\cdot10^{16}$	$1.8\cdot10^{18}$

Table 2.17. *Empirical convergence rates,* $W(2,1)$, *lcGS-ch,* $V_S = V_D = V_B = 0.0V, V_G = -5.0V$

gridsize		$\frac{1}{3}\left(\sum_{\psi,n,p}\left(\frac{\|r_h^{n+l}\|}{\|r_h^n\|}\right)^{\frac{1}{l}}\right)$
finest	coarsest	ρ_{av}
$\frac{1}{48}$	$\frac{1}{3}$	0.44
$\frac{1}{48}$	$\frac{1}{6}$	0.46
$\frac{1}{96}$	$\frac{1}{3}$	0.44
$\frac{1}{96}$	$\frac{1}{6}$	0.44
$\frac{1}{128}$	$\frac{1}{4}$	0.45
$\frac{1}{128}$	$\frac{1}{8}$	0.44

Table 2.18. ρ_{av} *for different cycles, lcGS-ch,* $V_S = V_D = V_B = 0.0V, V_G = -5.0V$

cycle	$\frac{1}{3}\left(\sum_{\psi,n,p}\left(\frac{\|r_h^{n+l}\|}{\|r_h^n\|}\right)^{\frac{1}{l}}\right)$
$W(2,1)$	0.44
$F(2,1)$	0.49
$V(2,1)$	0.72

Table 2.19. *Influence of pre- and post-smoothing steps,* $W(v_1,v_2)$-*cycle, lcGS-ch,* $V_S = V_D = V_B = 0.0V, V_G = -5.0V$

v_1	v_2	$\frac{1}{3}\left(\sum_{\psi,n,p}\left(\frac{\|r_h^{n+l}\|}{\|r_h^n\|}\right)^{\frac{1}{l}}\right)$
1	1	*div.*
2	1	0.44
2	2	0.41
2	3	0.39
2	4	0.37

often than in the case of global linearization and the use of "linear multi-grid". This is very expensive and a crucial point in the algorithm. The machine accuracy and the possibility of vectorization has to be taken into account. But the function is smooth and monotone. So using

$$B(-z) = z + B(z), \quad \frac{d}{dz}B(-z) = -1 - \frac{d}{dz}B(z), \quad B(0) = 1,$$

$$\frac{d}{dz}B(0) = -\tfrac{1}{2}, \quad B(z) = o(\exp^{-z}) \quad \text{for} \quad z \to \infty,$$

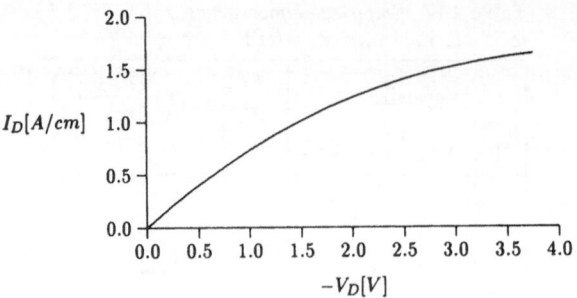

Fig. 2.39 MOS-characteristic, W(2, 1), lcGS-ch, $V_S = V_B = 0.0V$, $V_G = -5.0V$

Fig. 2.40 ψ and p for $V_D = -2.25V$ and $V_G = -5.0V$

the function can be approximated polygonally with a few nodes on a limited interval $[0, z_0]$. The values for B and $\frac{d}{dz} B$ at the nodes are stored in a table that is calculated once in advance. Typically 200 nodes are used.

Results are presented for an example which is described by the parameters of Table 2.16. The h-independency of the convergence rate can be seen from Table 2.17.

In practice, $\frac{1}{48}$ for the finest grid may be of no relevance, but neglecting special physical effects, a typical solution has been computed. With the described standard multigrid components, the experiments indicate that for this example the residuals on the coarsest grid have to be smaller than $5 \cdot 10^{-10}$ to get typical multigrid convergence.

The choice of the coarsest grid may not be obvious. Using a very coarse grid may result in incorrect representations of the fine grid problems. In the worst cases, the coarse-grid-solver did not converge. Table 2.18 shows the influence of the cycle type (compare Figure 2.19). The rate for the simple V-cycle is worse than for the other types. To find a good initial guess, W- and F-cycle schemes are more useful. The influence of pre- and post-smoothing is given by Table 2.19. lcGS-ch with $v_1 = 2$ and $v_2 = 1$ is the most efficient strategy.

Figure 2.39 shows the initial part of the characteristic for the calculated MOS transistor.

A simple continuation strategy has been used. Starting with $V_D = 0.0$ the solution serves as an initial guess for $V_D = -0.25$ etc. The current I_D, presented in Figure 2.39 is the flow of the majority carriers. For the p-channel transistor of the example, these are the holes. It is calculated on a 97×97 grid along the 'current-line' shown in the picture of the example. Notice, that all the current flowing through this line must flow through the drain. There is no flow of electrons. So in Figure 2.40 only the electrostatic potential ψ and the function p is shown. Here both the channel and the flow of holes are easily recognized.

3 Adaptive Multilevel Grid Selection Strategies for Process Simulation Evolution Problems

An adaptive simulation approach with grid selection strategies, which keeps the accuracy of a discrete approximation below a predetermined tolerance with a minimum of computational effort, is a highly desirable feature for an accurate analysis and an efficient simulation of processes. The adaptive grid selection strategies generally have to fulfill two basic goals. Firstly, they have to supply the discretization approach with a domain partitioning method capable of adapting the size of the discretization cells locally, and secondly, they should supply the adaptive simulation approach with a method for the evaluation and the control of the discrete approximation errors during simulation.

Besides these two general demands, there are also several special objectives that one should consider when designing efficient adaptive grid selection strategies for the simulation of evolution processes. Namely, the evolution character of the problem implies that the construction of adaptive grids and the control of the discrete approximation accuracy should be performed both in space and time dimensions. The spatial adaptive grid structure should be modified in order to follow the evolution of the discrete solution. Consequently, as the most important design requirements adaptive grid structures should

- allow for easy and efficient local modifications without introducing unnecessary refined domain areas;
- be as regular as possible in order to provide an acceptable order of consistency and to minimize the errors introduced with data transfers between different grid structures.

Additionally, for an efficient computation and control of the discrete approximation accuracy, adaptive grid selection strategies should be based on

- error estimators which can be reliably and accurately evaluated with computational cost negligible to the total simulation cost;
- compatible error estimators in space and time, controlled by the same predetermined tolerance parameters which are independent of the problem.

The problems of effectively meeting the above requirements with singlegrid
adaptive methods have proven to be extremely challenging. The usage of
nonuniform grid structures to achieve the adaptive discrete approximation
is a source of severe problems. It is well known that nonuniform discretization stencils may cause lower-order accuracy and that they typically require
huge information overheads for the representation and modification of grid
structures. The main problem with many singlegrid adaptive approaches is
the lack of efficient and accurate methods for the evaluation of the discretization errors. This is crucial for the reliable formulation of the local refinement criteria.

On the other hand, the multilevel adaptive concept is qualitatively different
from any singlegrid adaptive approach. It allows the construction of highly
flexible refinement structures based on the local uniform grids. In that way,
the multilevel adaptive concept eliminates most of the inherent weaknesses
of the singlegrid adaptive methods originating from the irregular grid
geometry. Moreover, the multilevel concept of adaptivity offers a large
freedom for the design of efficient strategies to control the accuracy according to particular goals of adaptive process simulation.

The presented adaptive multilevel grid selection strategies are designed
having in mind the idea to emphasize the basic principles which make them
advantageous to the adaptive singlegrid selection approaches in fulfilling
adaptive objectives for a wide class of general evolution problems. They
also give an indication for the effective programming of procedures for the
practical simulation of evolution processes (see Chapter 5).

3.1 Adaptive Multilevel Discrete Approximation

Starting from the generalized process simulation evolution problem, both
the basic steps to derive the adaptive discrete approximation in space and
time and the way in which these steps direct the adaptive simulation procedure are discussed. The adaptive multilevel discrete approximation of
evolution processes is considered in terms of grid structures originating
from the finite difference discretization and using the MLAT adaptive
concept.

3.1.1 General Process Simulation Evolution Problems

Let us consider an initial boundary value problem:

$$\frac{\partial u_j}{\partial t} - A_j(t, u_j) = f_j \text{ in } \Omega(t) \times \Omega_t$$

$$B_j(t, u_j) = g_j \text{ on } \partial\Omega(t) \times \Omega_t$$

$$u_j = u_j^0 \text{ in } \bar{\Omega}(0)$$

(3.1)

where the time-dependent nonlinear elliptic operator A_j contains no time derivatives. This property is used here to classify the initial boundary value problem (3.1) as an evolution problem. It is also used as the generalized mathematical model for evolution processes. The unknown u_j stands for the continuous space-time distribution of the j-th physical quantity of interest for the simulation. For example, in the case of a diffusion process, u_j may be the concentration profile of the j-th impurity. B_j is the nonlinear time-dependent operator of boundary conditions while f_j and g_j represent spatial distributions of the generation-recombination terms which do not depend on u_j. The time-dependence of the spatial differential operators A_j and B_j in evolution processes originates from the time varying processing conditions (e.g. variation of processing temperature). The space distribution u_j^0 defines an initial state for the evolution processing. Generally it is the result of both the previous evolution processes and intermediate auxiliary processes. For example, in the case of the diffusion process simulation, the initial state is typically formed of impurity profiles, resulting from ion implantation processes, superposed to impurity profiles resulting from the previous diffusion process.

The evolution problem (3.1) has to be solved in the space-time domain $\bar{\Omega}(t) \times \Omega_t$, where $\Omega(t) \subset \mathbb{R}^3$ is the time-dependent space domain with a non-planar and moving boundary $\partial\Omega(t)$. $\Omega_t = (0, T] \subset \mathbb{R}$ is a time domain, where T denotes the processing time. The boundary $\partial\Omega(t)$ can be generally split into

$$\partial\Omega(t) = \partial\Omega_p(t) \cup \partial\Omega_s \cup \partial\Omega_d$$

where $\partial\Omega_p(t)$, $\partial\Omega_s$ and $\partial\Omega_d$ are physical boundaries, symmetry lines and distant boundaries, respectively. The physical boundaries represent real material surfaces or interfaces. Their nonplanar nature and time-dependence are typically caused by auxiliary processes like deposition, selective etching or oxidation. The boundary conditions at the physical boundaries are typically of mixed type, relating the flux normal to the boundary to various physical mechanisms (segregation, volume expansion, etc.) which take place there. Apart from these "real" boundaries, the symmetry lines and distant boundaries are introduced artificially in order to avoid unnecessary computations in the case of symmetric problems and in order to cut out the areas which are not affected by the processing. To simplify the discretization on $\partial\Omega_s$ and on $\partial\Omega_d$ these boundaries are typically assumed to be planar. The symmetry lines are characterized by the Neumann (zero flux) boundary conditions, while a stationary Dirichlet boundary condition is usually posed at the distant boundary.

The design of a space-time discrete analogue for the evolution problem (3.1) is simplified by its evolution nature since discretization in time and space dimensions can be performed subsequently and independently. This fact has a significant influence on the formulation of grid refinement and adaptive simulation strategies.

3.1.2 Adaptive Multilevel Discrete Approximation in Space

Without loss of generality, the principles used to construct the spatial adaptive discrete approximation of (3.1) are considered in two dimensions. The basic constructive element for this is the infinite uniform lattice $\mathscr{G}_h$ with meshsize h, whose elements are considered here as nodes:

$$\vec{p} = (\alpha h, \beta h); \quad \alpha, \beta \in \mathbb{Z}.$$

A uniform grid is

$$G_h = \partial G_h \cup G_h^\circ$$

where ∂G_h represents the uniform discrete boundary and G_h° is an internal uniform grid defined by

$$G_h^\circ = \mathscr{G}_h \cap \Omega_{(h)}.$$

$\Omega_{(h)} \subseteq \Omega$ represents a continuous space area where the discrete approximation requires a grid resolution with at least the meshsize h. Generally it can be a set of disjoint subdomains of Ω. The discrete boundary $\partial G_h \subset \mathscr{G}_h$ is composed of lattice nodes laying outside $\Omega_{(h)}$ or at the continuous boundary $\partial \Omega_{(h)}$ which have at least one neighboring node inside $\Omega_{(h)}$. The term "neighboring" depends on the particular spatial discretization of the boundary conditions (see for example the set $\partial \Omega_{4h}$ in Section 4.2.2). However, it should be noted that whatever particular discretization approach is chosen for the boundary conditions, it should fit into the regular structure of G_h without introducing any interstitial grid nodes which would spoil the overall uniformity of the grid structures. The discrete boundary ∂G_h can be split into

$$\partial G_h = \partial G_h^i \cup \partial G_h^g$$

where ∂G_h^i is an internal discrete boundary, defined as a subset of ∂G_h lying inside Ω. The remaining boundary grid nodes belong to the global boundary grid ∂G_h^g. A uniform grid is local if it possesses nodes of an internal boundary. Otherwise it is global.

In the multigrid context, standard coarsening plays a significant role. It creates uniform grids with meshsizes

$$h_l = \frac{h_1}{2^{l-1}}; \quad l \in \mathbb{N}$$

where l denotes the discretization level and h_1 is the coarsest meshsize. Due to the above conventions, the grid $G_l \equiv G_{h_l}$ represents a uniform grid which covers the subdomain $\Omega_{(l)} \equiv \Omega_{(h_l)}$ on the l-th level. The complete multilevel adaptive grid structure is defined as a set of uniform grids

$$G = \{G_1, G_2, \ldots, G_M\}.$$

On the first m_g $(1 \leq m_g < M)$ levels the grids are global, while the other uniform grids on levels $l > m_g$ are local ones and defined on subdomains for which $\Omega_{(l)} \subset \Omega_{(l-1)}$ holds. M denotes the finest level. The local grids in the multilevel adaptive framework actually produce different levels of refinement. This is why the multilevel adaptive concept is qualitatively different from any singlegrid adaptive approach. Formulating the grid refinement (or coarsening) in terms of extending (or contracting) noncoextensive, and properly aligned uniform local grids on different levels, the globally required nonuniform resolution of the discrete approximation is obtained although exclusively uniform discretization stencils are used locally. This leads to important simplifications both for the discretization and for the handling of the adaptive grid structure.

The multilevel concept of adaptivity is rather general and independent of the particular discretization approach. The uniform grids are suitable for the direct implementation of the finite difference technique which is the basic discretization approach in the following chapters. In principle, other discretization approaches (finite element [52] or finite volume [96]) are possible, too. However, any particular discretization approach should be based on uniform local or global grids. For sake of convenience, when denoting discrete functions and operators, the index j which refers to the fact that (3.1) represents a system of equations, will be omitted in the sequel of this chapter. Replacing the differential terms A and B by corresponding discrete analogues A_h and B_h, converts the original evolution problem (3.1) into the semi-discrete problem:

$$\frac{du_h}{dt} - A_h(t, u_h) = f_h \text{ in } \overset{\circ}{G_h} \times \Omega_t$$

$$B_h(t, u_h) = g_h \text{ on } \partial G_h^g \times \Omega_t \tag{3.2}$$

$$u_h = I_{2h}^h u_{2h} \text{ on } \partial G_h^i$$

$$u_h = u_h^0 \text{ in } G_h \text{ at } t = 0,$$

which can be also considered as the set of ordinary differential equations for the unknown time-dependent components of the vector u_h. The semi-discrete problem (3.2) can be also expressed in the operator form

$$L_h(u_h) = f_h \text{ in } G_h \times \Omega_t \tag{3.3}$$

where it is assumed for convenience that L_h and f_h include the boundary conditions.

The complete multilevel discrete approximation in space is a set of both global and local problems (3.2), defined on the corresponding levels. Each local problem is coupled by an internal boundary condition $(I_{2h}^h u_{2h})$ to the discrete solution on the next coarser grid. This also implies the natural order in which the problems should be defined: sequentially, starting from the coarsest global problem towards the finest local problem using the coarser level discrete solution for the formulation of internal boundary conditions of the next finer local problem.

It is interesting to note that similar local adaptive concepts based on the formulation of internal boundary conditions for local problems have been also proposed in the singlegrid environment, as *grid chopping* adaptive strategies [119]. It was observed that in a global solution-refinement procedure an acceptable quality of both the grid resolution and the discrete solution is achieved in some regions before the refinement-solution cycle is completed. As a strategy to keep the computational work as low as possible, it is suggested to chop off those regions where the grid resolution and the solution are satisfactory. The locally refined problems defined on the remaining regions are then solved using the chopped off discrete solution for the formulation of their internal boundary conditions. It is obvious that the accuracy of the final discrete solution obtained by the grid chopping technique is determined by the error which is transported by the internal boundary conditions into the solution of the local discrete problems. This fact suppresses the general applicability of grid chopping techniques with internal boundary conditions of Dirichlet type. This problem can be handled by modifying internal boundary conditions so as to include the first derivative of the discrete solution and provide flux conservation [96]. However, this approach requires irregular discretization stencils at internal boundaries. In spite of the fact that, at first sight, the multilevel semi-discrete formulation (3.2) resembles the basic idea of grid chopping strategy, it has to be considered as an integral part of the adaptive multigrid solution method where both the accuracy of the internal Dirichlet boundary conditions and the conservation properties of the discretization scheme are controlled on the corresponding coarser levels.

An adaptive multilevel grid structure is used to approximate both the differential solution u and the operators A and B of the evolution problem (3.1) on different discretization levels. It is natural that the final discrete approximation of the differential solution has to be uniquely defined on the simulation domain Ω. Therefore, it is always considered on the composite grid structure

$$G_{\mathcal{H}} = \bigcup_{l=1}^{M} G_{h_l}. \tag{3.4}$$

From the structural point of view the composite grid $G_{\mathcal{H}}$ is very similar to some singlegrid adaptive structures (finite element [7, 62] or finite box grids [32]). However, in the classical multigrid context it is not used for the formulation of the discrete operators in space, but compare the FAC/AFAC [96] presentation in Section 2.7.4.

3.1.3 Adaptive Temporal Discrete Approximation

A natural extension of the multilevel adaptive concept into time dimension
is also possible. However, the multigrid methods, which exploit different
discretization levels in time at different spatial discretization levels, referred
to in the literature as *parabolic multigrid methods* [14, 48], are still under
active development and practically used exclusively for model problems
(see Section 2.8). Moreover, the parabolic multigrid methods also face some
conceptual problems concerning anisotropically distributed discretization
errors in space and time. In order to stay on the safe side and with the idea
to present here only adaptive techniques whose efficiency and robustness
has been practically confirmed, the conventional approach for the adaptive
time discretization is used (see Section 2.8). Adaptivity in time is achieved
by the nonuniform partition

$$\Omega_{\Delta t} = \{t^n | t^n = t^{n-1} + \Delta t^n; n \in \mathbb{N}, t^0 = 0\}$$

of the temporal domain Ω_t, where Δt^n is the time-step size which corresponds
to the n-th discrete temporal node. Consequently, the complete space-time
grid structure $G_{\Delta t}$ is considered as a sequence of multilevel adaptive grid
structures

$$G_{\Delta t} = \{G(t^0 = 0), G(t^1), G(t^2), \ldots, G(t^{N_t} = T)\}$$

defined at each temporal node in $\Omega_{\Delta t}$.
Restarting the formulation of a discrete analogue for the evolution problem
(3.1) by time discretization, the resulting semi-discrete (or semi-differential)
problem allows a qualitatively different insight. For the sake of simplicity,
the implicit Euler discretization is applied. The resulting semi-discrete
problem is:

$$\frac{u_{\Delta t}^n - u_{\Delta t}^{n-1}}{\Delta t^n} - A(t^n, u_{\Delta t}^n) = f_{\Delta t} \text{ in } \Omega \times \Omega_{\Delta t} \tag{3.5}$$

$$B(t^n, u_{\Delta t}^n) = g_{\Delta t} \text{ on } \partial\Omega \times \Omega_{\Delta t}$$

which can be considered as a sequence of local semi-differential incremental
elliptic problems with parameters Δt^n and $u_{\Delta t}^{n-1}$. In operator form it is
abbreviated by

$$L_{\Delta t}(u_{\Delta t}) = f_{\Delta t} \text{ in } \bar\Omega \times \Omega_{\Delta t}, \tag{3.6}$$

assuming that $L_{\Delta t}$ and $f_{\Delta t}$ include the boundary conditions. The local incremental elliptic problems are coupled through the previous temporal
node discrete solutions. Consequently, the local problems should be solved
sequentially at each temporal grid node starting from the initial state u^0.
The combination of both the multilevel space discretization of (3.5) and the
temporal discretization of (3.2) leads to the final form of the discrete approxi-

mation to the evolution problem:

$$\frac{u^n_{h,\Delta t} - u^{n-1}_{h,\Delta t}}{\Delta t^n} - A_h(t^n, u^n_{h,\Delta t}) = f_{h,\Delta t} \text{ in } G^\circ_h(t^n)$$

$$B_h(t^n, u^n_{h,\Delta t}) = g_{h,\Delta t} \text{ on } \partial G^q_h(t^n) \qquad (3.7)$$

$$u^n_{h,\Delta t} = I^h_{2h} u^n_{2h,\Delta t} \text{ on } \partial G^i_h(t^n)$$

$$u^0_{h,\Delta t} = u^0_h \text{ in } G_h(t^0).$$

The previous temporal node discrete solution defined on the multilevel adaptive grid structure $G(t^{n-1})$, which is generally different from $G(t^n)$, should be transferred to $G(t^n)$ prior to the formulation of the temporal discrete operators. The algorithmic details of this transfer are discussed in Section 5.3. The discrete operator formulation is

$$L_{h,\Delta t}(u_{h,\Delta t}) = f_{h,\Delta t} \text{ in } G_{h,\Delta t} \qquad (3.8)$$

and again it is assumed that (3.8) also includes the discrete boundary conditions.

The dual view of the discrete evolution problem (3.7), imposed by its semi-discrete formulations (3.2) and (3.5), implies the basic adaptive steps for the construction of the multilevel space grid structures at nonuniformly distributed temporal grid nodes. With a given time-step size at a certain temporal grid node, the first adaptive step is to construct a space discrete approximation to the discrete solution of the evolution problem (3.1), which is for that purpose considered as semi-differential incremental elliptic problem (3.5). This adaptive step is tightly coupled with an adaptive multigrid approach for solving (3.8). The result of this adaptive step is both an incremental discrete solution and a corresponding multilevel adaptive grid structure. The second step is an adaptive prolongation of the temporal grid structure. In this case, with the given spatial adaptive grid structure, the evolution problem (3.1) should be considered as a set of ordinary differential equations (3.2) for which an appropriate next time-step size should be anticipated. The complete adaptive simulation procedure recursively repeats these two basic adaptive steps starting from an initial state discrete approximation and an initial time-step size.

3.2 Discretization Errors and Their Properties

The accuracy of the discrete approximation is commonly expressed in terms of discretization errors. Starting from the basic definition of discretization errors for the discrete evolution problem (3.8), this section focuses on properties which are particularly important for the design of multilevel adaptive strategies.

3.2.1 Global and Local Errors

There are two basic types of discretization errors which can be considered
for the discrete evolution problem (3.8) on the spatial discretization level
with meshsize h and temporal grid structure $\Omega_{\Delta t}$. The *global error*, $e_{h,\Delta t}$, is
defined by

$$e_{h,\Delta t} = \hat{I}^{h,\Delta t} u - u_{h,\Delta t}. \tag{3.9}$$

$\hat{I}^{h,\Delta t}$ is a linear projection operator which transfers the continuous solution
into the discrete function space. Also, it describes the way how to compare
the discrete solution $u_{h,\Delta t}$ to the differential solution u. The common and the
most trivial choice for $\hat{I}^{h,\Delta t}$ is an injection operator defined by $\hat{I}^{h,\Delta t} u = (u)_{h,\Delta t}$,
which means evaluating u at the position of grid nodes. Then the global
error measures how well the discrete solution approximates the differential
solution evaluated at grid positions. For the fulfillment of other discrete
approximation goals, different projection operators can be formulated. For
example, the global error evaluated by an averaging projection operator
$\hat{I}^{h,\Delta t} \equiv F^{h,\Delta t}$ where

$$F^{h,\Delta t} u := \frac{\left(\int_{\Omega_h^{\vec{p}}} u \, d\Omega \right)_{h,\Delta t}}{|\Omega_h^{\vec{p}}|} \tag{3.10}$$

measures how well the discrete solution $u_{h,\Delta t}$ approximates an average value
of u over the elementary discretization cell $\Omega_h^{\vec{p}}$ surrounding the grid node $\vec{p}$.

The second discretization error type of interest is the *local error* (sometimes
called "truncation error") which for the discrete problem (3.8) is defined by

$$\tau_{h,\Delta t} = L_{h,\Delta t}(\hat{I}^{h,\Delta t} u) - f_{h,\Delta t}. \tag{3.11}$$

In other words, the local error is the defect of the discrete problem (3.8)
where the discrete solution $u_{h,\Delta t}$ is replaced by the projection $\hat{I}^{h,\Delta t} u$.

Remark 3.1. *For an explicitly known differential solution u the global error
can be evaluated only after "globally" solving the discrete problem. The local
error depends only on the "locally" given discrete operators and can be expressed in terms of high derivatives of u.*

It is obvious from (3.9) and (3.11) that the global and local errors are related
by

$$L_{h,\Delta t}(u_{h,\Delta t} + e_{h,\Delta t}) = f_{h,\Delta t} + \tau_{h,\Delta t}. \tag{3.12}$$

That means, the discrete solution of the discrete problem where the right
hand side is modified by adding the local error, coincides with the corresponding projection of the differential solution. On the other hand, if both

the local error and the discrete solution are given, the equation (3.12) can be solved for the unknown global error $e_{h,\Delta t}$, in the same way as the original discrete problem is solved for $u_{h,\Delta t}$. Using a physical analogy, with the right hand side f representing a generation-recombination term for the physical quantity u, the local error may be considered as an artificially introduced generation-recombination term, which governs the spatial distribution and evolution of the global error and can be used for its indirect control.

Example 3.1. *The following one-dimensional initial boundary value model problem*

$$\frac{\partial u}{\partial t} - D\frac{\partial^2 u}{\partial x^2} = 0, \quad (0 < x < l_x; 0 < t \le T)$$

$$\frac{\partial u}{\partial x} = 0, \quad (x = 0; 0 < t \le T) \tag{3.13}$$

$$u = 0, \quad (x = l_x; 0 < t \le T)$$

$$u = u^0, \quad (0 \le x \le l_x; t = 0)$$

originates from one-dimensional modeling of impurity diffusion processes. u represents the impurity concentration, D is the constant diffusion coefficient and l_x is the length of the space domain. In spite of the fact that the above linear diffusion problem is far away from being a proper representation of complex real life diffusion, it is very useful for the practical analysis of the discretization errors because the exact differential solution is known. This allows an easy understanding of the discretization errors and their properties.

The half-Gaussian implantation profile is assumed as an initial state $u^0(x) = u_{max} \cdot \exp(-x^2/(2\Delta R_p^2))$, where u_{max} is the maximum concentration of the Gaussian impurity profile at $x = 0$ and ΔR_p is the standard deviation. The differential solution of (3.13) is [120]

$$u = \frac{u_{max}}{\sqrt{1 + \dfrac{2Dt}{\Delta R_p^2}}} e^{-\frac{x^2}{2\Delta R_p^2 + 4Dt}} \tag{3.14}$$

The physical parameters used for the numerical experiments are $D = 10^{-14}\, cm^2/s$ and $l_x = 0.35\, \mu m$, while the Gaussian distribution has the standard deviation $\Delta R_p = 0.02 \cdot 10^{-4}\, cm$ and $u_{max} = 10^{20}\, cm^{-3}$.
In order to consider the discretization errors on different levels, the discretization is performed on a sequence of global grids

$$G_l = \{x_i | x_i = (i - 1) \cdot h_l; i = 1, \dots, N_l\},$$

where $N_l = 2N_{l-1} - 1$, $N_1 = 4$ and $h_l = l_x/(N_l - 1)$. The implicit Euler discretization in time and second order finite difference scheme in space with *"mirror imaging" for the Neumann boundary condition (see Sections 2.4, 4.2.2 and for instance [126]) give the discrete problem*

$$\frac{u_{h,\Delta t}^{i,n} - u_{h,\Delta t}^{i,n-1}}{\Delta t} - D \cdot \frac{u_{h,\Delta t}^{i-1,n} - 2u_{h,\Delta t}^{i,n} + u_{h,\Delta t}^{i+1,n}}{h^2} = 0 \ (i = 1, \ldots, N_l - 1; \ n = 1, \ldots, N_t)$$

$$u_{h,\Delta t}^{0,n} = u_{h,\Delta t}^{2,n} \ (n = 1, \ldots, N_t)$$

$$u_{h,\Delta t}^{N_l,n} = 0 \ (n = 1, \ldots, N_t)$$

$$u_{h,\Delta t}^{i,0} = (u^0)_h^i, \quad (i = 1, \ldots, N_l) \tag{3.15}$$

Due to (3.12), a similar discrete initial boundary value problem

$$\frac{e_{h,\Delta t}^{i,n} - e_{h,\Delta t}^{i,n-1}}{\Delta t} - D \cdot \frac{e_{h,\Delta t}^{i-1,n} - 2e_{h,\Delta t}^{i,n} + e_{h,\Delta t}^{i+1,n}}{h^2} = \tau_{h,\Delta t}^{i,n} (i = 1, \ldots, N_l - 1; \ n = 1, \ldots, N_t)$$

$$e_{h,\Delta t}^{0,n} = e_{h,\Delta t}^{2,n} (n = 1, \ldots, N_t)$$

$$e_{h,\Delta t}^{N_l,n} = 0 \quad (n = 1, \ldots, N_t)$$

$$e_{h,\Delta t}^{i,0} = \hat{e}_h^i \quad (i = 1, \ldots, N_l) \tag{3.16}$$

holds for the local and global errors. Here $\hat{e}_h$ is an initial global error distribution obtained with $(u^0)_h$ as a trivial discrete solution of the first time step for $\Delta t \to 0$. Actually, this particular and trivial case of the global error belongs to the more general formulation of the generalized local error (see Section 3.2.2). Assuming a uniform temporal grid, the accumulation of the global error in time is bounded by

$$\max_{1 \leq i \leq N_l} |e_{h,\Delta t}^{i,n}| \leq \max_{1 \leq i \leq N_l} |\hat{e}_h^i| + \Delta t \cdot n \cdot \max_{\substack{1 \leq i \leq N_l - 1 \\ 1 \leq n' \leq n}} |\tau_{h,\Delta t}^{i,n'}| \tag{3.17}$$

which can be proved by mathematical induction.

In order to demonstrate the validity of (3.17) in practice, the discrete diffusion problem is solved with parameters $l = 6$ and $\Delta t = 5s$ performing $N_t < 20$ time steps. The discrete solution and the exact differential solution supply the necessary information for an exact evaluation of both the global and the local error. For the analysis both the injection and the averaging projection operators are used. In order to make the local and global errors compatible for quantitative comparison, the local error is considered in its "undivided" form

$$\tilde{\tau}_{h,\Delta t} = \tau_{h,\Delta t} \cdot \Delta t.$$

For convenience, the discretization errors are scaled with the constant concentration $u_s = 10^{18} \, cm^{-3}$. The evolution of the discretization errors is considered in terms of the "scaled L_∞-norm"

$$\| \cdot \| \equiv \frac{\max_{1 \leq i \leq N_l} |\cdot|}{u_s}.$$

This convention is used also for other examples of this chapter referring to the considered model problem.

Figures 3.1 and 3.2 show the evolution of both the global error and the local error in terms of time-step numbers N_t. The discretization errors in Figure 3.1 are evaluated by injection and in Figure 3.2 by averaging projection operators. In both cases the evolution of the discretization errors is compared to the right hand side of (3.17) denoted in Figures 3.1 and 3.2 by dotted lines.

It can be seen that in both cases the inequality (3.17) holds and approximately follows the accumulation of the global error at the very beginning of the time integration. However, it also significantly overestimates the global error accumulation for large integration times and actually represents its worst case estimate. More closer control of the global error by its corresponding local error requires additional assumptions on the properties of the discrete solution [23].

Note from Figure 3.1 that, using the injection operator both for the formulation of the discrete initial state and for the evaluation of the global error, the trivial global error $\hat{e}_h$ vanishes. Consequently, at $N_t = 1$ $\|e_{h,\Delta t}\| \cong \|\bar{\tau}_{h,\Delta t}\|$ holds. On the other hand, it is evident from Figure 3.2 that the trivial global error evaluated by the averaging projection operator has finite value and represents an initial state for the global error evolution. This observation is very important for the proper selection of an error estimator which is based on the global

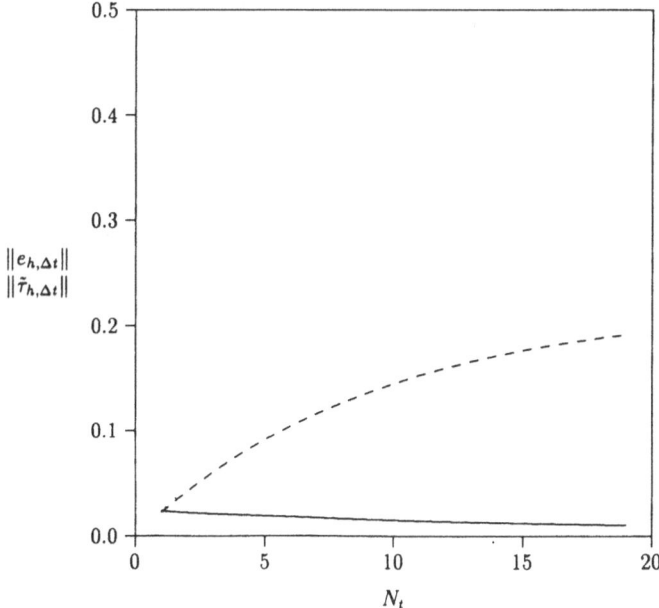

Fig. 3.1. The evolution of the local (solid line) and the global (dashed line) errors obtained by the injection

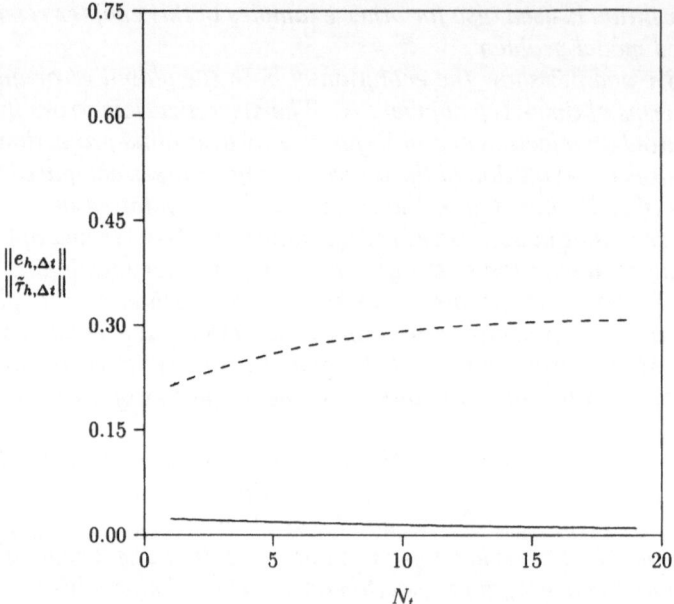

Fig. 3.2 The evolution of the local (solid line) and the global (dashed line) errors obtained by the averaging projection

error and which is used to define a spatial local refinement criterion. Using for that purpose the global error obtained by injection will result in extremely coarse grids at the initial stage of the transient simulation. But extremely coarse spatial grid structures lead to an irreversible loss in the conservation of the local doses inside elementary discretization cells which may results in an nondesired rate of the global error accumulation in the later simulation phases.

Both the global and the local errors (3.9) and (3.11) are introduced so far for the completely discretized evolution problem (3.8). However, they can be defined for the semi-discrete problems (3.3) and (3.6) analogously. The two step discretization of the evolution problem (3.1) together with the relation (3.12) allows the local and global errors to be split into

$$e_{h,\Delta t} = \hat{I}^{\Delta t} e_h + \hat{I}^h e_{\Delta t} \tag{3.18}$$

$$\tau_{h,\Delta t} = I^{\Delta t} \tau_h + I^h \tau_{\Delta t} \tag{3.19}$$

where e_h, τ_h and $e_{\Delta t}$, $\tau_{\Delta t}$ are global and local errors with respect to space and time for the semi-discrete problems (3.3) and (3.6), respectively. $\hat{I}^h$, I^h and $\hat{I}^{\Delta t}$, $I^{\Delta t}$ are corresponding semi-discrete projection operators satisfying

$$\hat{I}^{h,\Delta t} = \hat{I}^{\Delta t} \cdot \hat{I}^h$$

and

$$I^{h,\Delta t} = I^{\Delta t} \cdot I^h.$$

The obvious benefit of the discretization error splitting is the fact that the local and global errors (3.11) and (3.9), corresponding to the full time-space discrete approximation, can be controlled by the corresponding local and global errors of the semi-discrete problems (3.3) and (3.6) separately and independently in time and space.

3.2.2 A Generalized Local Error

Starting from the basic formulation of the local error (3.11), it is possible to define an important generalization. The right-hand side of the discrete problem (3.8) can be generally expressed in the form

$$f_{h,\Delta t} = I^{h,\Delta t} f$$

where $I^{h,\Delta t}$ defines the discretization of the right-hand side of the continuous problem $L(u) = f$. The operator $I^{h,\Delta t}$ may be different from $\hat{I}^{h,\Delta t}$ since they are generally defined on different function spaces. Then the local error can be rewritten as

$$\tau_{h,\Delta t} = L_{h,\Delta t}(\hat{I}^{h,\Delta t} u) - I^{h,\Delta t} L(u).$$

Thus, the local error is expressed in terms of the discrete analogue $L_{h,\Delta t}$ and the right hand side, hidden in the projection of the differential operator L applied to u. However, this formulation applies even in the case of more general operators W and $W_{h,\Delta t}$ which do not originate from evolution problems but which are defined on the same function space.

The generalized local error $\tau_{h,\Delta t}(W_{h,\Delta t})$ is of particular importance to control the discretization errors for discrete approximations of continuous distributions. In this case the choice of the operator W and its discrete analogue $W_{h,\Delta t}$ depends on the particular approximation goal of the adaptive simulation in which the discrete approximation of the continuous distribution participate.

Example 3.2. *An example, which demonstrates the usage of the generalized local error is its formulation for the discrete approximation of a compact distribution w with the goal to control the local dose conservation inside the elementary discretization cells. The local dose of w inside an elementary discretization cell $\Omega_h^{\vec{p}}$ is expressed by the generalized integral operator*

$$W^{\Omega_h^{\vec{p}}}(w) = \int_{\Omega_h^{\vec{p}}} w \, d\Omega. \tag{3.20}$$

The discrete analogue of (3.20) in finite difference formulation is

$$W_h^{\Omega_h^{\vec{p}}}((w)_h) = |\Omega_h^{\vec{p}}| \cdot (w)_h \tag{3.21}$$

and the generalized local error in space is

$$\tau_h(W_h^{\Omega_h^{\vec{p}}}) = |\Omega_h^{\vec{p}}| \cdot (w)_h - \left(\int_{\Omega_h^{\vec{p}}} w d\Omega \right)_h,$$

using the injection for both I^h and $\hat{I}^h$. Due to (3.10), the spatial average projection operator can be expressed as $F^h \equiv (W^{\Omega_h^{\vec{p}}}(\cdot))_h / |\Omega_h^{\vec{p}}|$. If w is chosen to be an initial state of an evolution problem $w = u^0$ then

$$|\tau_h(W_h^{\Omega_h^{\vec{p}}})| = |\hat{e}_h| \cdot |\Omega_h^{\vec{p}}|$$

holds, where $\hat{e}_h$ is the trivial global error evaluated by the average projection operator (see Example 3.1).

3.2.3 Local-Global Error

Besides the local and global errors with clearly distinguished local and global properties, there is another type of discretization error to be defined for the semi-discrete problem (3.5) using implicit time discretization. Considering the single time-step integration starting from an exact previous time-step differential solution, the temporal global error (using the injection operator) is then given by [23]

$$l_{\Delta t} = (u)_{\Delta t} - \hat{u}_{\Delta t} \tag{3.22}$$

where $\hat{u}_{\Delta t}$ is the semi-discrete solution of (3.5), which in the case of the implicit Euler time-discretization is obtained from the equation

$$\hat{u}_{\Delta t}^n = (u)_{\Delta t}^{n-1} + \Delta t^n \cdot (A(t^n, \hat{u}_{\Delta t}^n) + f_{\Delta t}).$$

In spite of the fact that the error (3.22) is similar to the definition of the global error, it is qualitatively different since it does not take into account the accumulation of the global error along the integration path. This fact gives definition (3.22) local properties, too. Having in mind the dual character of this particular type of discretization error, in the sequel it is referred to as a *local-global error*.

In order to relate the local-global and the global error the temporal grid is assumed to be uniform. From (3.9) and (3.22) we have

$$e_{\Delta t}^n = \hat{u}_{\Delta t}^n - u_{\Delta t}^n + l_{\Delta t}^n.$$

If the relationship

$$\| \hat{u}_{\Delta t}^n - u_{\Delta t}^n \| \le c \cdot \| \hat{u}_{\Delta t}^{n-1} - u_{\Delta t}^{n-1} \|, \quad \hat{u}_{\Delta t}^{n-1} = (u)_{\Delta t}^{n-1}$$

holds for a given norm $\| \cdot \|$ in $\mathbb{R}^2$ with $0 < c \in \mathbb{R}$ then after n integration steps the global error satisfies the inequality

$$\| e_{\Delta t}^n \| \le c \cdot \| e_{\Delta t}^{n-1} \| + \| l_{\Delta t}^n \| \le c^n \cdot \| e_{\Delta t}^0 \| + \rho \cdot \sum_{i=0}^{n-1} c^i$$

where $\rho = \max_{1 \le n' \le n} \| l_{\Delta t}^{n'} \|$. This inequality demonstrates the importance of "unconditional contractivity", that means $c < 1$, in order to suppress the propagation of the initial error and to provide for the stable accumulation of the global error. For $e_{\Delta t}^0 = 0$ the inequality reduces to

$$\| e_{\Delta t}^n \| < \rho \cdot \frac{1}{1-c}, \tag{3.23}$$

showing that with unconditional contractivity the temporal global error can be controlled by the local-global error.

Example 3.3. *The evolution of the temporal global error is practically demonstrated using the model problem from Example 3.1. The discrete solution is computed with parameters $l = 10$ and $\Delta t = 5s$ performing $N_t < 60$ time-steps. The extremely fine space grid is chosen in order to make the spatial errors negligible compared to temporal ones. Consequently, the choice of $\hat{I}^h$ has no influence on the discretization error.*
The evolution of both the temporal global and the local-global errors are shown in Figure 3.3. Note that the local-global error decreases monotonely while, due to the contractivity, the global error shows a stable accumulation. For $N_t \le N_t'$ where $\| l_{\Delta t}^{N_t} \| \approx (1-c) \cdot \| e_{\Delta t}^{N_t - 1} \|$ the global error norm increases due to the dominant accumulation of the local-global errors. For $N_t > N_t'$, it decreases due to the dominant influence of the contractivity parameter c.

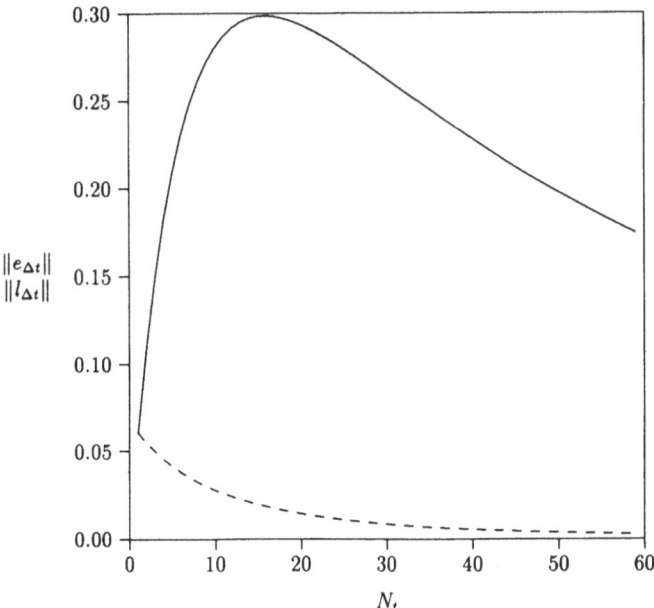

Fig. 3.3 The evolution of the temporal global (solid line) and the local-global (dashed line) errors

On the other hand, the local-global error must not be confused with an ordinary local discretization error for the semi-discrete problem (3.5) given in its undivided form:

$$\tilde{\tau}_{\Delta t} = (L_{\Delta t}((u)_{\Delta t}) - f_{\Delta t}) \cdot \Delta t. \tag{3.24}$$

Apart from the local-global error this quantity does not depend on the discrete solution $u_{\Delta t}$. From (3.22), (3.24) and (3.5) these two discretization errors are related by

$$l_{\Delta t} - \tilde{\tau}_{\Delta t} = \Delta t \cdot (A((u)_{\Delta t}) - A(\tilde{u}_{\Delta t})). \tag{3.25}$$

Remark 3.2. *The quantitative difference between the local error in undivided form and the local-global error exists in the case of the implicit time discretization (3.5) and obviously vanishes for $\Delta t \to 0$. In the case of an explicit time discretization, the local-global error and the local error in the undivided form are equivalent.*

Example 3.4. *Using again the model problem from Example 3.1 both the undivided local error and the local-global error are evaluated on the space level $l = 10$ for different sizes of the first time-step. The dependence of both errors on the initial time-step size is shown in Figure 3.4.*

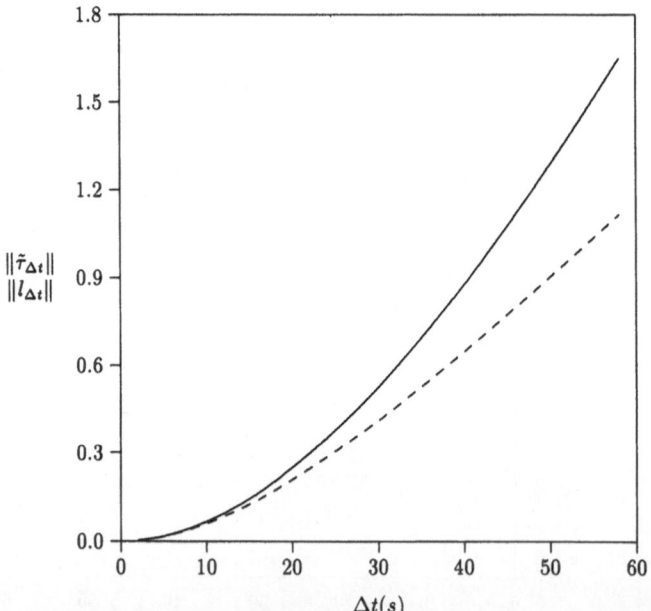

Fig. 3.4 The local-global (dashed line) and the local (solid line) errors with respect to the initial time-step size

It is observed that the discrepancy between local and local-global errors becomes of minor importance as the time-step size approaches the magnitude which is required to solve the problem at hand with a reasonable accuracy. This fact implies that both local-global and local errors may be used as an error estimator for an adaptive control of temporal discretization.

3.3 Evaluation of the Discretization Errors by Two-Level Extrapolation Techniques

The essential phase in an adaptive control of discretization errors is their evaluation for the given discrete evolution problem. However, the formulation of a reliable technique to evaluate the discretization errors, as the constituent part of an adaptive simulation procedure, is a critical task, due to the *a posteriori* nature of this problem. At the same time, it is one of the most difficult challenges for an adaptive treatment of differential equations. Since the exact differential solution is unknown, the evaluation of both local and global (or local-global) errors is always an approximate evaluation. It can only be based on some discrete approximation of the differential solution and operators. Moreover, local refinement criteria based on discretization errors, cannot be applied at the very beginning of an adaptive simulation. Typically, some intuitively posed initial grid structure is used to start a solution-refinement cycling, which converges both to the final accuracy of the discrete solution and to the corresponding adaptive grid structure. For the evaluation of discretization errors at a certain stage of this procedure only an intermediate adaptive grid structure and some incomplete discrete approximation of the differential solution is available. Thus a numerical technique to evaluate the discretization errors should be effective and reliable even in the case of coarse grids and corresponding intermediate discrete solutions.

The presence of the discrete operators and solutions on different discretization scales for space grids as well as time discretization levels provides an excellent framework for a cheap estimate of the discretization errors. It is based on two-level extrapolation techniques which use asymptotic expansions of the discretization errors.

3.3.1 Asymptotic Dependence of Discretization Errors on the Discretization Scale

The evaluation and the control of discretization errors on an adaptive grid structure becomes possible if they are related to the discretization scale. In order to derive such relations a general operator W and its discrete analogue $W_{\Delta x}$ along a generalized coordinate x, with a uniform discretization scale defined by Δx, is considered. It is assumed that the asymptotic expansion

$$W_{\Delta x}(\hat{I}^{\Delta x} u) = I^{\Delta x} W(u) + \sum_{j=1}^{k} \Delta x^j \cdot I^{\Delta x} b_j(u) + R_{\Delta x} \qquad (3.26)$$

is given. The remainder term $R_{\Delta x}$ is estimated by

$$\| R_{\Delta x} \| \le c_1 \cdot \Delta x^{k+1} \qquad (3.27)$$

where the constant c_1 and the functions b_j do not depend on Δx.
Applying (3.26) to the semi-discrete operators of the evolution problem
(3.1), the generalized step-size Δx is replaced by h or Δt in the corresponding
semi-discrete operators (3.2) and (3.5), respectively.

Example 3.5. *Let us consider the time derivative $\dfrac{\partial u}{\partial t}$ within the evolution*

problem (3.1) and its discrete analogue obtained by implicit Euler discretiza-
tion. Using Taylor's expansion of u in the vicinity of the temporal node t^n we
have

$$\frac{u(t^n) - u(t^{n-1})}{\Delta t^n} = \dot{u}(t^n) + \sum_{j=1}^{k} (\Delta t^n)^j \cdot \frac{(-1)^j u^{(j+1)}(t^n)}{(j+1)!} + R_{\Delta t}(t^n) \qquad (3.28)$$

where

$$|R_{\Delta t}(t^n)| \le \frac{\Delta t^{k+1}}{(k+2)!} \max_{t^0 \le t \le t^n} |u^{(k+2)}(t)|.$$

It is obvious that the expansion (3.28) resembles to (3.26) and (3.27) with

$$b_j(u) = \frac{(-1)^j u^{(j+1)}(t^n)}{(j+1)!},$$

$$c_1 = \frac{\max_{t^0 \le t \le t^n} |u^{(k+2)}(t)|}{(k+2)!}$$

and the injection operators for $\hat{I}^{\Delta t}$ and $I^{\Delta t}$.

For linear semi-discrete problems which satisfy the asymptotic expansion
for discrete operators and which possess additional properties such as
stability conditions for the $u_{\Delta x}$ and both the existence and the uniqueness of
the differential solution u the asymptotic expansion

$$\hat{I}^{\Delta x} u = u_{\Delta x} + \sum_{j=1}^{k} \Delta x^j \cdot \hat{I}^{\Delta x} a_j(u) + \hat{R}_{\Delta x} \qquad (3.29)$$

with the residual term

$$\| \hat{R}_{\Delta x} \| \le c_2 \cdot \Delta x^{k+1}$$

is proven in [92]. The constant c_2 and the functions a_j do not depend on Δx. In spite of the fact that (3.29) is valid particularly for linear problems, it is justified to assume in the sequel that it approximately holds also for the quasi-linear and (not strongly) nonlinear operators as they typically appear in process simulation evolution problems.

The most important benefit from the asymptotic expansions (3.26) and (3.29) is the fact that both the local and the global errors (3.11) and (3.9) of the semi-discrete operators (3.2) and (3.5) can be approximated by

$$e_h \simeq \hat{I}^h a_p \cdot h^p \tag{3.30}$$

$$e_{\Delta t} \simeq \hat{I}^{\Delta t} a_q \cdot \Delta t^q \tag{3.31}$$

$$\tau_h(L_h) \simeq I^h b_p \cdot h^p \tag{3.32}$$

$$\tau_{\Delta t}(L_{\Delta t}) \simeq I^{\Delta t} b_q \cdot \Delta t^q \tag{3.33}$$

which means that the discretization errors are dominated by their principal truncation terms, that is, by the first non-vanishing terms under the sum of the corresponding asymptotic expansions. In (3.30) and (3.31) p and q are the orders of convergence while in (3.32) and (3.33) they represent the approximation orders of space and time discretization, respectively. It is assumed that the corresponding convergence and approximation orders are equal.

It is important to note that if the semi-discrete operator (3.6) is given in its undivided operator form (that means multiplied by Δt), the principal truncation term for the local error in time is

$$\tilde{\tau}_{\Delta t} \simeq I^{\Delta t} b_q \cdot \Delta t^{q+1}. \tag{3.34}$$

The expansion (3.26) cannot be directly applied to evaluate the principal truncation term of the local-global error because it depends on the discrete solution $\hat{u}_{\Delta t}$. However, it follows from (3.25) that [23]

$$l_{\Delta t} \simeq I^{\Delta t} b_q' \cdot \Delta t^{q+1} \tag{3.35}$$

with $b_q' \simeq b_q$ for $\Delta t \to 0$.

3.3.2 Richardson Extrapolation Technique

Two subsequent discretization levels along the generalized coordinate x with step-sizes Δx and $\Delta X = 2\Delta x$, respectively, are considered. Assuming that the discrete solutions $u_{\Delta x}$ and $u_{\Delta X}$ satisfy the governing semi-discrete problem for the corresponding dimension, the global errors on these two discretization levels can be expressed as

$$\hat{I}^{\Delta x} u - u_{\Delta x} = \hat{I}^{\Delta x} a_r \cdot \Delta x^r \tag{3.36}$$

$$\hat{I}^{\Delta X} u - u_{\Delta X} = \hat{I}^{\Delta X} a_r \cdot \Delta X^r \tag{3.37}$$

where r stands for the order of convergence. Applying the relative restriction operator $\hat{I}^{\Delta X}_{\Delta x}$ which satisfies

$$\hat{I}^{\Delta X} = \hat{I}^{\Delta X}_{\Delta x} \cdot \hat{I}^{\Delta x},$$

to both sides of (3.36) and subtracting the result and (3.37), the unknown projection $\hat{I}^{\Delta X}u$ is eliminated and the global errors on these two levels are given by

$$e_{\Delta X} = \frac{2^r}{2^r - 1} e^{\Delta x}_{\Delta X} \tag{3.38}$$

$$e_{\Delta x} = \frac{1}{2^r - 1} I^{\Delta x}_{\Delta X} e^{\Delta x}_{\Delta X} \tag{3.39}$$

where

$$e^{\Delta x}_{\Delta X} = \hat{I}^{\Delta X}_{\Delta x} u_{\Delta x} - u_{\Delta X}$$

is the relative global error of the coarse grid solution with respect to the fine grid solution and $I^{\Delta x}_{\Delta X}$ is a discrete prolongation operator. The special relation between ΔX and Δx has been used to obtain (3.38) and (3.39).

This technique to extrapolate the global error from its relative value obtained by the discrete solutions on two subsequent discretization levels is well known as Richardson extrapolation. It is also proposed as a technique to increase the convergence order of the discrete solution. This is simply done by

$$\bar{u}_{\Delta x} = u_{\Delta x} + e_{\Delta x}. \tag{3.40}$$

The global error of the resulting discrete solution $\bar{u}_{\Delta x}$ is at least of order $O(\Delta x^{r+1})$. Unfortunately, it is not possible to exploit this increased accuracy of the discrete solution in the local grid refinement processes because there is no way to quantitatively estimate the global error of $\bar{u}_{\Delta x}$. Therefore, the extrapolated global errors serve only for the construction of error estimators (see Section 3.4).

Example 3.6. *The Richardson extrapolation is now applied to the model problem (3.13–3.15). The spatial global error is extrapolated after $N_t = 6000$ time-steps with size $\Delta t = 0.1s$. The time-step size is chosen sufficiently small in order to make the temporal part of the global error negligible to the spatial one. The number of time-steps corresponds to the simulation time at which the global errors obtained by injection and averaging projections have comparable values. Figure 3.5 shows the spatial distribution of the exact global errors obtained by both the injection and the averaging projection operators on level $l = 5$ as well as their extrapolated spatial distributions evaluated by (3.39) using discrete injection and full weighting restriction operators, respectively. The corresponding spatial distributions on level $l = 4$ are shown in Figure 3.6.*

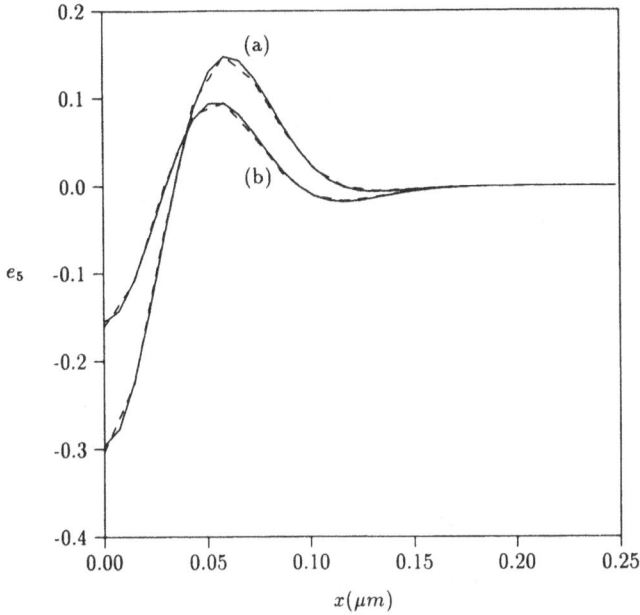

Fig. 3.5 Spatial distribution of the exact (solid lines) and the extrapolated (dashed lines) global errors for $l = 5$ ((a) full weighting, (b) injection)

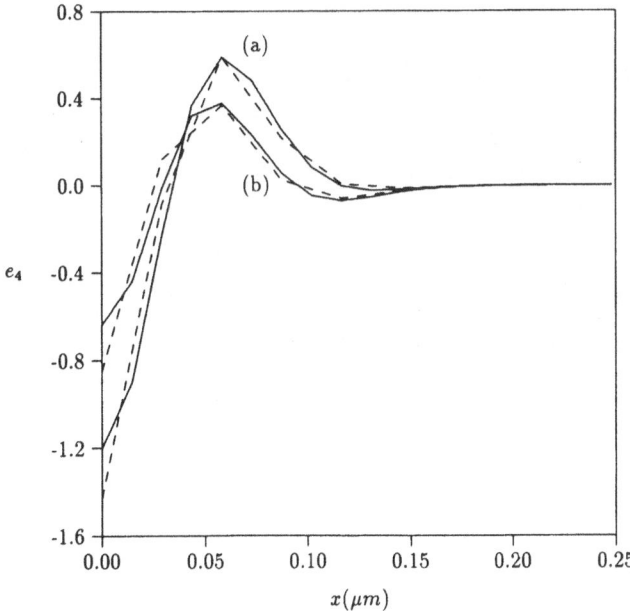

Fig. 3.6 Spatial distribution of the exact (solid lines) and the extrapolated (dashed lines) global errors for $l = 4$ ((a) full weighting, (b) injection)

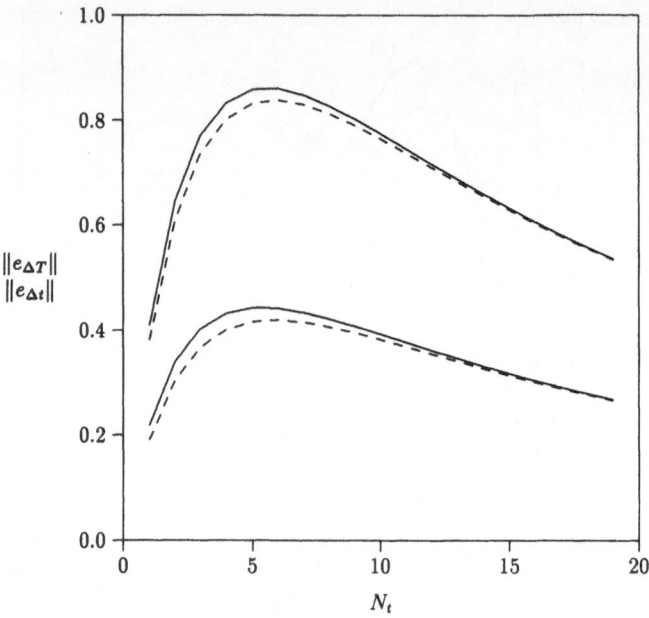

Fig. 3.7 The evolution of the exact (solid lines) and the extrapolated (dashed lines) global errors

The temporal global errors are extrapolated using two time discretization levels with the time-step size $\Delta t = 15s$ on the finer time level while the spatial level $l = 10$ is chosen to make the spatial part of the global error unimportant. The evolution of the exact and the extrapolated global errors on both levels evaluated at grid nodes $1 \leq N_t < 20$ corresponding to the coarse temporal level is shown in Figure 3.7.

Richardson extrapolation technique is so far considered as a powerful strategy for the evaluation of global errors. However, it can be used also for the estimation of the local-global and the generalized local error.

The Richardson extrapolation of the local-global error is considered now using two uniform temporal grids with time-steps Δt and $\Delta T = 2\Delta t$. In order to handle the local properties of the local-global error in the Richardson extrapolation procedure, some assumptions are made. First, it is assumed that

$$e_{\Delta t}^n = e_{\Delta t}^{n-1} + l_{\Delta t}^n \tag{3.41}$$

which is equivalent to the assumption that the contractivity parameter c is close to one. Additionally, it is assumed that the local-global error on the Δt-level does not change drastically within two subsequent Δt integration steps inside a single ΔT-step, i.e. $l_{\Delta t}^n \approx l_{\Delta t}^{n+1}$.

In order to extrapolate the local-global error it is necessary to perform the time stepping on the ΔT-level in such a way that each step uses as "previous discrete solution" those obtained on the Δt-level. Consequently, for all coarse time-steps, preceding the one for which local-global error should be extrapolated, holds

$$e_{\Delta T} = (e_{\Delta t})_{\Delta T}. \tag{3.42}$$

Then, the global errors on the ΔT- and Δt-levels expressed with the aid of the corresponding principal truncation terms for the local-global error and the previous time-step global errors are

$$(u)^m_{\Delta T} - u^m_{\Delta T} = (b'_q)_{\Delta T} \cdot \Delta T^{q+1} + (e^{n-2}_{\Delta t})^{m-1}_{\Delta T}$$

$$(u)^n_{\Delta t} - u^n_{\Delta t} = 2 \cdot (b'_q)_{\Delta t} \cdot \Delta t^{q+1} + e^{n-2}_{\Delta t}$$

where $n = 2m, m \in \mathbb{N}$. The same procedure as above, but using the injection for the relative restriction operator and the special choice $\Delta T = 2\Delta t$ leads to

$$l^m_{\Delta T} = \frac{2^q}{2^q - 1}((u^n_{\Delta t})^m_{\Delta T} - u^m_{\Delta T}) \tag{3.34}$$

$$l^n_{\Delta t} = \frac{1}{2(2^p - 1)}((u^n_{\Delta t})^m_{\Delta T} - u^m_{\Delta T}). \tag{3.44}$$

Remark 3.3. *The two-level time stepping which fulfills condition (3.42) eliminates the accumulation of the global error on the coarse temporal level. This makes the Richardson extrapolation of the local-global error also applicable to coarse temporal grids with nonuniform time intervals and fine time grids obtained by halving each of the coarse time-steps (see Subsection 3.5.2).*

Example 3.7. *The Richardson extrapolation is now applied to the local-global error of the model problem (3.13–3.15).*
The experiments again use $l = 10$ and two temporal levels with the time-step size $\Delta t = 15s$ on the finer one. The evolution of the exact and extrapolated local-global error norms at grid nodes $1 \le N_t < 20$ corresponding to the coarse temporal level is shown in Figure 3.8.

Finally, the Richardson extrapolation is used to evaluate the generalized local error of the discrete operator $W_{\Delta x}(I^{\Delta x}u)$. Due to the asymptotic expansion (3.26) the local errors of the discrete operator $W_{\Delta x}$ can be expressed in terms of its principal truncation error terms on Δx- and ΔX-levels as:

$$W_{\Delta x}(\hat{I}^{\Delta x}u) - I^{\Delta x}W(u) = I^{\Delta x}b_r \cdot \Delta x^r \tag{3.45}$$

$$W_{\Delta X}(\hat{I}^{\Delta X}u) - I^{\Delta X}W(u) = I^{\Delta X}b_r \cdot \Delta X^r, \tag{3.46}$$

where r now stands for the order of discrete approximation. Applying again

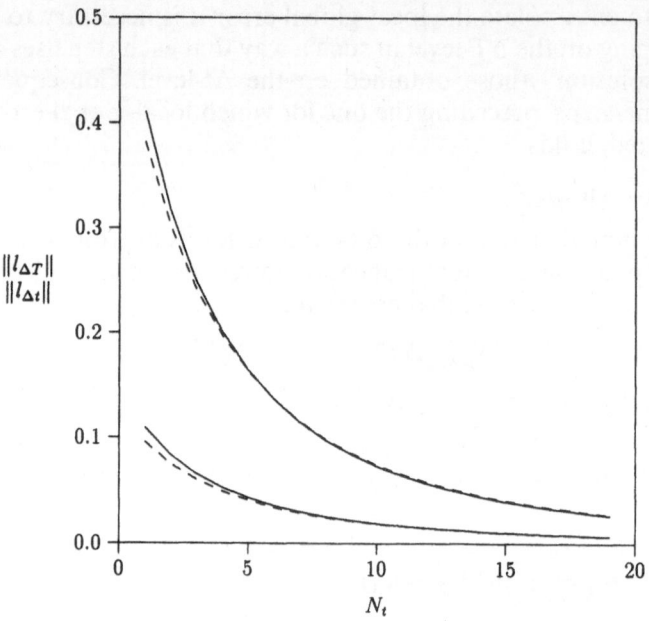

Fig. 3.8 The evolution of the exact (solid lines) and the extrapolated (dashed lines) local-global errors

a relative restriction operator $I_{\Delta x}^{\Delta X}$ which satisfies

$$I^{\Delta X} = I_{\Delta x}^{\Delta X} \cdot I^{\Delta x}, \tag{3.47}$$

and using a similar technique as above, the local error is expressed in terms of discrete operators on two subsequent levels and corresponding projections $\hat{I}^{\Delta x} u$ and $\hat{\hat{I}}^{\Delta X} u$ by

$$\tau_{\Delta x}(W_{\Delta x}) = \frac{2^r}{(2^r - 1)} \cdot \tau_{\Delta X}^{\Delta x}(W_{\Delta x}) \tag{3.48}$$

$$\tau_{\Delta x}(W_{\Delta x}) = \frac{1}{2^r - 1} \cdot I_{\Delta X}^{\Delta x} \tau_{\Delta X}^{\Delta x}(W_{\Delta x}) \tag{3.49}$$

where

$$\tau_{\Delta X}^{\Delta x}(W_{\Delta x}) = W_{\Delta x}(\hat{I}^{\Delta X} u) - I_{\Delta x}^{\Delta X} W_{\Delta x}(\hat{I}^{\Delta x} u)$$

is the relative local error of the ΔX-level discrete operator with respect to the Δx-level. However, a practical usage of the extrapolation formulae (3.48) and (3.49) requires the knowledge of the projections $\hat{I}^{\Delta x} u$ and $\hat{\hat{I}}^{\Delta X} u$. Since the averaging projection operator generally cannot be evaluated in the closed form (even with explicitly known u) the injection is the most convenient choice for this projection operator. The most important benefit of this technique is the possibility to evaluate the generalized local error for some discrete functional $W_{\Delta x}((u)_{\Delta x})$ when it is not possible to evaluate $W(u)$ although the continuous function u is explicitly given.

Example 3.8. *Using 3.49 and the discrete formulation (3.21) of the integral operator (3.20) leads to the extrapolated generalized local error*

$$\tau_h\left(\frac{W_h^{\Omega_h^{\vec{p}}}}{|\Omega_h^{\vec{p}}|}\right) = \frac{1}{2^p - 1} I_H^h((u)_H - F_h^H(u)_h) \tag{3.50}$$

where $p = 2$ is the approximation order of the discrete operator (3.21). The restriction is performed with the full weighting operator F_h^H which satisfies (3.47), while in case of the injection operator the generalized local error vanishes.

Remark 3.4. *Note that the extrapolation of the generalized local error is equivalent to the extrapolation of the global error applied to some "discrete solution" $v_{\Delta x} := W_{\Delta x}(\hat{I}^{\Delta x}u)$. On the other hand, the extrapolation of the global error can be considered as the extrapolation of the "generalized local error" of the operator $L_{\Delta x}^{-1}(I^{\Delta x}f)$. The right hand side f takes the role of u in the above formulae.*

3.3.3 τ-Extrapolation Technique

It is obvious that the extrapolation formulae (3.48) and (3.49) for the generalized local error cannot be directly applied to the local errors (3.11) of the semi-discrete evolution problems (3.2) and (3.5) since the differential solution and its projections $\hat{I}^{\Delta x}u$ and $I^{\Delta x}u$ are not available. On the other side $L(u)$ is explicitly given and is equal to f. However, the relative local error for the semi-discrete problem $L_{\Delta x}$ can be rewritten as

$$\tau_{\Delta X}^{\Delta x}(L_{\Delta x}) = L_{\Delta x}(\hat{I}_{\Delta x}^{\Delta X}u_{\Delta x} + \hat{I}_{\Delta x}^{\Delta X}e_{\Delta x}) - I_{\Delta x}^{\Delta X}L_{\Delta x}(u_{\Delta x} + e_{\Delta x})$$
$$\approx L_{\Delta x}(\hat{I}_{\Delta x}^{\Delta X}u_{\Delta x}) - I_{\Delta x}^{\Delta X}L_{\Delta x}(u_{\Delta x}). \tag{3.51}$$

Then the extrapolated local errors on two subsequent discretization levels are given by

$$\tau_{\Delta x} = \frac{2^r}{2^r - 1} \cdot \tau_{\Delta X}^{\Delta x}(L_{\Delta x}) \tag{3.52}$$

$$\tau_{\Delta x} = \frac{1}{2^r - 1} \cdot \tau_{\Delta X}^{\Delta x}(L_{\Delta x}). \tag{3.53}$$

This particular type of Richardson extrapolation is called τ-extrapolation technique (see Sections 2.5 and 2.7.2). It obviously resembles the Richardson extrapolation, but it can be profitably done even in the case when the approximation order changes its value over the grid since it only exploits local properties of the discrete operators.

Example 3.9. *The local error extrapolation formula (3.52) and (3.53) are verified with the model problem (3.13–3.15). The spatial local errors are evaluated with a sufficiently small time-step size $\Delta t = 0.1s$ and after $N_t = 6000$ time-steps. Figure 3.9 shows both the exact and the extrapolated local error evaluated on levels $l = 4$ and $l = 3$.*
The temporal local error is extrapolated on space level $l = 10$ using two temporal levels with the time-step size $\Delta t = 15s$ on the finer one. The evolution of both the exact and the extrapolated local errors at grid nodes $1 \le N_t < 20$ of the coarse time level is shown in Figure 3.10.

Similarly to the possibility to use the global error for achieving the higher order of convergence, the extrapolated local error can be used to increase the approximation order of discrete equations. This is simply done by adding the extrapolated local error to the right hand side of the corresponding discrete equation. Namely, the discrete evolution problem (3.8) with a new right hand side

$$\bar{f}_{h,\Delta t} = f_{h,\Delta t} + \tau_{h,\Delta t}.$$

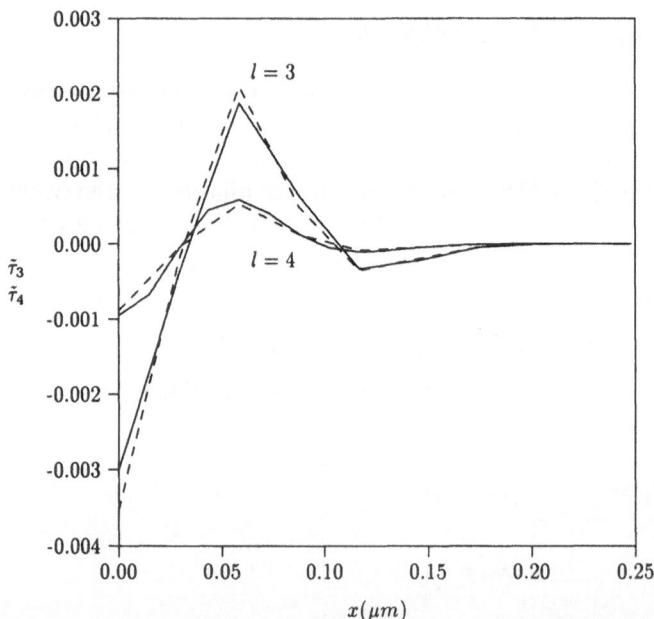

Fig. 3.9 Spatial distribution of the exact (solid lines) and the extrapolated (dashed lines) local errors

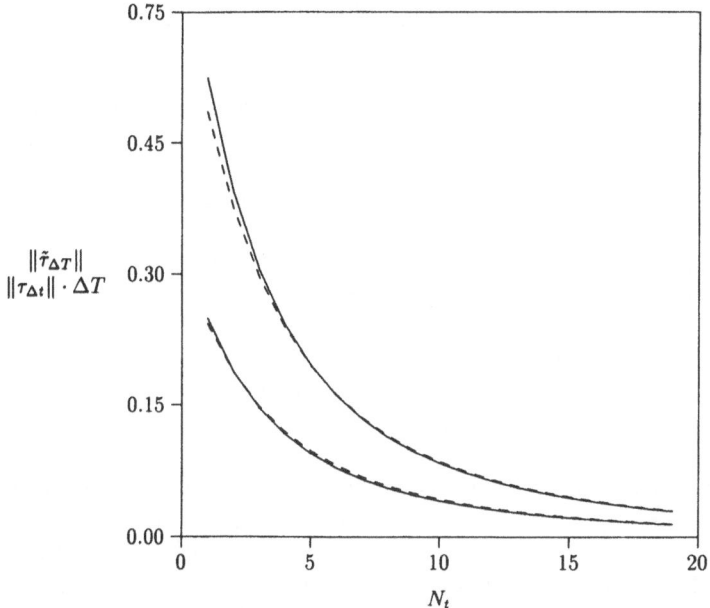

Fig. 3.10 The evolution of the exact (solid lines) and the extrapolated (dashed lines) local errors

has the local error which is at least $O(\Delta t^{q+1})$ and $O(h^{p+1})$. But, again, the local refinement method cannot benefit from this improved approximation order.

3.4 Local Refinement Criteria Based on Discretization Errors

The adaptive treatment of the discrete problem (3.8) includes the following grid adaptation steps in space and time:

1. The construction of the adaptive multilevel spatial grid structure at the beginning of the transient simulation according to an initial distribution u^0.
2. The refinement of the space grid structure during transient simulation with a refinement strategy which controls a space discrete approximation of the incremental elliptic problems (3.5).
3. The construction of an adaptive temporal grid structure according to the time stepping scheme which controls the temporal discrete approximation of the semi-differential problem (3.2).

To implement any of these grid adaptation steps, reliable refinement criteria are needed in order to locally identify domain areas where the grid modification is required.

The formulation of the local refinement criteria for a given discrete problem should be based on realistic objectives of adaptive simulation. This requires both the specification of such quantities which represent the most appropriate measure for the simulation accuracy and a way to monitor and control them.

The general local refinement criterion for the discrete evolution problem (3.8) has the following formulation:

$$\varepsilon_{h,\Delta t}(\vec{p}, t^n) > Tol_{h,\Delta t}(\vec{p}, t^n). \tag{3.54}$$

This means practically: if (3.54) holds introduce a new local grid structure above node $\vec{p}$ on $h/2$ discretization level and find a better guess for the time-step size at node t^n. $\varepsilon_{h,\Delta t}$ is the discrete error estimator while $Tol_{h,\Delta t}$ represents the discrete tolerance vector which is an upper limit for $\varepsilon_{h,\Delta t}$.

The ultimate objective of all adaptive techniques for differential problems is to control the relative and absolute errors of the discrete solution in comparison to some projection of the differential solution. The error estimator and the tolerance which are appropriate to fulfill this goal are

$$\varepsilon_{h,\Delta t}(\vec{p}, t^n) = |e_{h,\Delta t}(\vec{p}, t^n)|$$

$$Tol_{h,\Delta t}(\vec{p}, t^n) = Rel \cdot |\hat{I}^{h,\Delta t} u(\vec{p}, t^n)| + Abs$$

where $Rel > 0$ denotes a control parameter which defines the required relative error of the discrete solution and the parameter $Abs > 0$ defines the allowed absolute error of the discrete solution. The absolute error parameter can be considered as optional. Its practical role is to suppress the grid refinement in those areas where an extremely fine resolution is irrelevant. For example, in diffusion process simulation Abs is typically chosen to be equal to the minimum impurity concentration of interest for the simulation. However, since the projection of the differential solution is generally unknown, it is reasonably to replace its value with the extrapolated discrete solution $\bar{u}_{h,\Delta t}$ and to use

$$Tol_{h,\Delta t}(\vec{p}, t^n) = Rel \cdot |\bar{u}_{h,\Delta t}(\vec{p}, t^n)| + Abs \tag{3.55}$$

as a tolerance vector. This tolerance vector makes only sense for non-oscillating functions. This problem can be resolved using the scalar tolerance

$$Tol = Rel \cdot \max_{\vec{p}, t^n} |\bar{u}_{h,\Delta t}(\vec{p}, t^n)| \tag{3.56}$$

instead of (3.55).

In the adaptive discrete simulation of the evolution problems, the local refinement steps are separated in space and time dimensions. Therefore, having in mind the splitting property stated in (3.18) and (3.19) as well as the applicability of the extrapolation techniques to both spatial and temporal discretization error, the refinement criterion (3.54) is split into

$$\varepsilon_h(\vec{p}) > Tol_h(\vec{p}) \tag{3.57}$$

$$\varepsilon_{\Delta t}(t^n) > Tol_{\Delta t}(t^n) \tag{3.58}$$

For the corresponding tolerance vectors $Tol_h(\vec{p}) = Tol_{\Delta t}(t^n) = Tol_{h,\Delta t}(\vec{p}, t^n)/2$, is chosen, while $\varepsilon_h(\vec{p})$ and $\varepsilon_{\Delta t}(t^n)$ are given by

$$\varepsilon_h(\vec{p}) = |e_h(\vec{p})| \tag{3.59}$$

$$\varepsilon_{\Delta t}(t^n) = |e_{\Delta t}(t^n)|. \tag{3.60}$$

The most important feature of discrete error estimators for the grid adaptation in time and space is their dependence on the local meshsize. Consequently, the global error can be used as an error estimator only if it is evaluated on a uniform grid or on a uniform local subgrid. This condition is naturally satisfied in the adaptive multilevel spatial grid structure because the local grid refinement is based on a sequence of uniform grids. This fact, along with the possibility to exploit the Richardson extrapolation technique, represents an important advantage of the multilevel adaptive methods over conventional singlegrid adaptive approaches where global error estimators are very rarely used. Most of the singlegrid adaptive approaches [113] are based on the derivative methods [119] which use

$$\varepsilon_h = h^k |u^{(k)}| \tag{3.61}$$

to estimate the global error and which try to equidistribute an approximation to some derivative of u computed with the aid of u_h. However, the approximations of higher derivatives obtained from the discrete solution on irregular grid structures can be quite inaccurate. Moreover, this approach is far from being a general principle. An artificial multileveling to evaluate the global error can be achieved in singlegrid adaptive approaches [62], too. The discrete problem is solved twice, once with the regular discretization and then using higher-order discretization, where the difference of these two discrete solutions is used as an approximation of the global error. The obvious disadvantage of this method is the fact that an additional discrete problem with higher-order discretization requires greater computational time than the calculation of the original discrete solution and usually rules out the possibility of obtaining faster solutions using mesh adaptations. However, the advantage of multilevel adaptive methods to effectively exploit asymptotic expansions of the global error for its extrapolation and formulation of the error estimators sometimes have to be treated carefully. There are two main problems to be distinguished. First, the assumption that the principal truncation term of the global error is given in the form $\hat{I}^h a_p \cdot h^p$ may be invalidated by the insufficient smoothness of the solution. However, this problem has a local nature occurring typically in the vicinity of internal discontinuities or near nonplanar boundaries. The second problem is that the Richardson extrapolation of the global error can be rather inaccurate on extremely coarse discretization levels. It is particularly critical if the extrapolated global error severely underestimates the exact error. That is, in the multilevel refinement approach, where finer discretization levels are subsequently added, the erroneously extrapolated global error on the very coarse discretization levels results in local grids of irreversibly reduced size, which can seriously spoil the structure of the local grids on the finer discretization levels. Exploiting the local nature of both problems an obvious remedy is to use as a global error estimator

$$\varepsilon_h(\vec{p}) = |e_h(\vec{p})| \cdot w(h, \vec{p}) \tag{3.62}$$

where w is a problem dependent weighting function which is designed so as to locally modify the value of the global error estimated in the vicinity of critical domain areas and for large meshsizes, while in the greatest part of the simulation domain and for reasonably fine discretization levels the global error estimator is unmodified (see Section 2.7.3 or [12]).

Apart from the multilevel spatial grid structures, the adaptive temporal grid structure with nonuniformly distributed time-step sizes, inherently suppresses the use of the global temporal error estimators to formulate the local refinement criteria. That is, since the temporal global error reflects the accumulation of the discrete solution error all along the simulation path, it cannot be related to the local time-step size. Moreover, the Richardson extrapolation of the temporal global error is a reasonable approach only for uniform temporal grids. However, the control of relative and absolute errors of the discrete solution can be performed in the framework of the single time-step using the error estimator

$$\varepsilon_{\Delta t}(t^n) = |l_{\Delta t}(t^n)| \tag{3.63}$$

based on the local-global error, which can also be associated with nonuniform temporal grid structures.

Another generally accepted way to control the relative error of the discrete solution in comparison to the differential solution is to do it indirectly by using error estimators based on the local errors. This technique, well known as *local error equidistribution* is formally justified by stability-consistency theory [93] which states that

$$\| e_{h,\Delta t} \| \leq \xi \cdot \| \tau_{h,\Delta t} \| \tag{3.64}$$

where the constant ξ does not depend on the grid spacings. The local error equidistribution is unavoidable if the global error estimates require extensive external tailoring by weighting functions or if it is not possible at all to evaluate the global errors.

In order to properly formulate a local error estimator it should be noted that local errors corresponding to discrete evolution problem operators, given in divided form, are not compatible with the discrete tolerance vectors (3.55) and (3.56). The local error has the same dimension as the discrete operator and not the discrete solution. Having in mind the evolution character of the problem this is easily resolved by error estimators formulated in terms of the spatial local error multiplied by the time-step size and the undivided form of the temporal local error:

$$\varepsilon_h(\vec{p}) = |\tau_h(L_h(\vec{p}))| \cdot \Delta t \tag{3.65}$$

$$\varepsilon_{\Delta t}(t^n) = |\tilde{\tau}_{\Delta t}(L_{\Delta t}(t^n))|. \tag{3.66}$$

The usage of the error estimators based on the local error is fairly safe in almost any case including the nonuniform temporal grid structure. However, in spite of the fact that the local error estimators effectively distinguish areas with "large" and "small" global errors, they fail to yield accurate global error estimates and typically underestimate these values. The possible remedy is to introduce local error estimators as

$$\varepsilon_h(\vec{p}) = K_1 \cdot |\tau_h(L_h(\vec{p}))| \cdot \Delta t$$

$$\varepsilon_{\Delta t}(t^n) = K_2 \cdot |\tilde{\tau}_{\Delta t}(L_{\Delta t}(t^n))|$$

where the introduced constants $K_1 > 1$ and $K_2 > 1$ compensate for the low level of the local error estimators. Unfortunately, these constants are problem-dependent and in general unpredictable.

In addition to error estimators based on local errors for the evolution problem (3.1) they may also apply the generalized local error for generalized discrete operators which dimensionally corresponds to the discrete solution. This is of special importance for the adaptive grid refinement according to an initial state distribution in compact form. In this case an error estimator

$$\varepsilon_h = \left| \tau_h \left(\frac{W_h^{\Omega_h^{\vec{p}}}}{|\Omega_h^{\vec{p}}|} \right) \right| \tag{3.67}$$

obtained by the Richardson extrapolation formula represents the most convenient choice since it is compatible with the global error estimators.

Example 3.10. *The applicability of the global and local error based error estimators for the selection of grid parts which require higher resolution is demonstrated again for the simple diffusion model problem (3.13–3.15). In order to compare the space refinement criteria based on global and local error estimators and the discrete tolerance (3.55), the following error estimators are defined:*

$$r_l^e = \frac{|e_l|}{Rel \cdot u_l + Abs} \tag{3.68}$$

$$r_l^\tau = \frac{|\tau_l| \cdot \Delta t}{Rel \cdot u_l + Abs} \tag{3.69}$$

with the spatial global and local errors evaluated by Richardson and
τ-extrapolation technique, respectively. The space grid nodes on the l-th level
which require higher grid resolution are simply selected by $r_l^e(x_i) > 1$ or
$r_l^\tau(x_i) > 1$.
The fixed simulation parameters are $l = 6$, $\Delta t = 10s$, $Rel = 10^{-2}$ and
$Abs = 10^{13}\,cm^{-3}$. The transient simulation is performed for $1 \leq N_t \leq 60$ and
the refinement criteria are compared at time-steps $N_t = 1$ and $N_t = 60$. For the
extrapolation of the global and local errors both injection and full weighting
restriction operators are used. The distributions of the global and local error
estimators are shown in Figures 3.11–3.14.
Comparing the error estimators at the beginning and at the end of the
transient simulation shows that the position of the maximum components of
the global and local error estimators are shifted towards the semiconductor
substrate actually tracing the steep front of the diffusion profile. However, it is
much more important to note the difference between the detected critical areas
which require better grid resolution. It can be seen that for $N_t = 1$, the critical
areas which require better grid resolution strongly depend on the selection of
the restriction operators. The local and the global error estimators give almost
the same result in the case of injection while for the full weighting restriction the
local error estimators significantly underestimate the global one. On the other

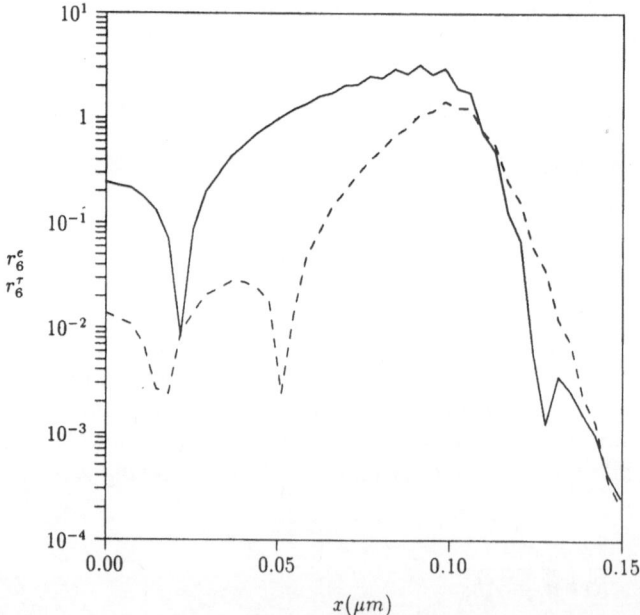

Fig. 3.11 Spatial distribution of the global (solid line) and the local (dashed line) error
estimators ($N_t = 1$, full weighting)

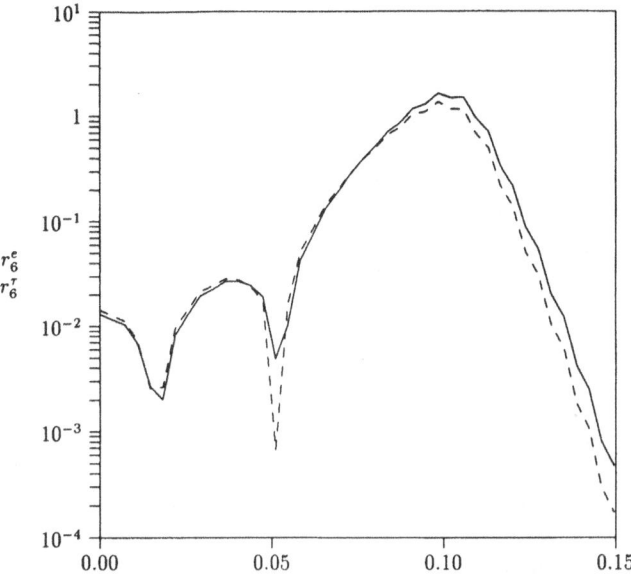

Fig. 3.12 Spatial distribution of the global (solid line) and the local (dashed line) error estimators ($N_t = 1$, injection)

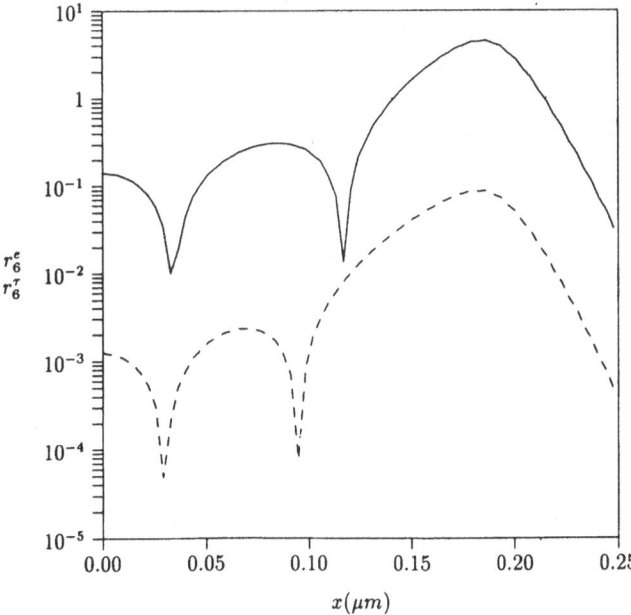

Fig. 3.13 Spatial distribution of the global (solid line) and the local (dashed line) error estimators ($N_t = 60$, full weighting)

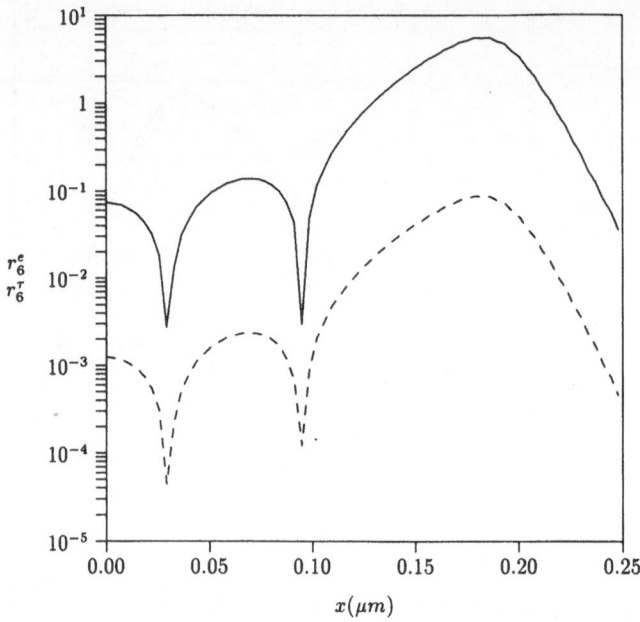

Fig. 3.14 Spatial distribution of the global (solid line) and the local (dashed line) error estimators ($N_t = 60$, injection)

side, for $N_t = 60$, the local error estimator distinguishes areas requiring different grid resolution in the similar way as the global error estimator does but significantly underestimates the global error both in case of the injection and the full weighting restriction operators. Note, also that for $N_t = 60$ the choice of the restriction operator has almost no influence on the selection of critical areas.

3.5 Two-Level Time Stepping Schemes Based on Extrapolation Techniques

The multilevel space grid structure offers a natural environment for the evaluation of the spatial discretization errors by the extrapolation techniques. On the other hand, since the adaptivity in time dimension is introduced by the singlegrid structure $\Omega_{\Delta t}$, the extrapolation technique for temporal discretization errors requires an auxiliary time level. This additional temporal level is introduced only to extrapolate temporal discretization errors and thus becomes an integral part of the two-level adaptive time-stepping scheme. Having in mind that with respect to a nonuniform temporal grid structure only the local-global or the local error can be used to control the time-step size, the two-level time-stepping scheme should be designed so to enable the application of the corresponding extrapolation techniques in the framework of the nonuniform temporal grid structure $\Omega_{\Delta t}$.

3.5.1 Technology of Adaptive Time-Step Size Control

In this section, a strategy to construct the nonuniform temporal grid $\Omega_{\Delta t}$ during transient simulation is derived. Suppose that an error estimator $\varepsilon_{\Delta t}^n$ (which may concern the local-global or the local error), is evaluated for the semi-discrete problem (3.6) at the n-th integration step ($n \geq 1$). For a given tolerance $Tol_{\Delta t}^n$, a new spatial distribution of the time-step size

$$\Delta t_{\Omega}^{new} = \left(\left(\frac{Tol_{\Delta t}^n}{\varepsilon_{\Delta t}^n} \right)^{\frac{1}{q+1}} \right) \cdot \Delta t^n \tag{3.70}$$

is obtained using the principal truncation term of the local-global error or the temporal local error in the undivided form and exploiting the condition that the desired error should be equal to the tolerance. A new integral time-step size which is appropriate to the whole spatial domain $\bar{\Omega}$ is

$$\Delta t^{new} = \frac{1}{\left\| \dfrac{1}{\Delta t_{\Omega}^{new}} \right\|}. \tag{3.71}$$

If it is required that the temporal error estimator is below the respective tolerance in the whole domain $\bar{\Omega}$, the norm in the denominator of (3.71) should be the L_∞-norm. However, it can be replaced by the L_2 norm as a tradeoff between the allowed tolerance and the controlled discretization error [3].

The refinement criterion (3.58) is rewritten in an integral form

$$\left\| \frac{\varepsilon_{\Delta t}^n}{Tol_{\Delta t}^n} \right\| \leq 1.$$

If it is satisfied then the n-th time-step is accepted and $\Delta t^{n+1} = \Delta t^{new}$ is used as the next time-step size. Otherwise, the current time-step is rejected and repeated with $\Delta t^n = \Delta t^{new}$. The evaluation of the refinement criterion uses the same norm $\| \cdot \|$ as above for the integral time-step size.

A robust automatic time-step size control code also requires some additional care of the time-step size increase or decrease rate in order to avoid both repeated time-step rejections and an abrupt change in the time-step size at the end of a transient simulation. The first problem is usually handled by replacing

$$\Delta t^{new} \leftarrow \Delta t^n \cdot \min\left(Facmax, \max\left(Facmin, Safety \cdot \frac{\Delta t^{new}}{\Delta t^n} \right) \right).$$

The typical values of the parameters *Facmax* and *Facmin* are 2 and 0.5, respectively. The *Safety* parameter is introduced in order to increase the probability for the next time-step to be accepted with typical value 0.9 [3]. In order to solve the second problem it is convenient to make a further modification of the time-step size due to [3]

$$\Delta t^{new} \leftarrow \frac{T - t^n}{\left\lceil \dfrac{T - t^n}{\Delta t^{new}} - \epsilon \right\rceil} \tag{3.72}$$

where ϵ is a small multiple of the machine epsilon and $\lceil \cdot \rceil$ denotes the integer ceiling function.

The selection of the first time-step size is much more difficult to handle automatically. Fortunately, a wrongly predicted initial time-step sizes will be quickly repaired by the automatic time-step size control procedure itself together with the help of a rough idea for the initial time-step size which is usually present.

3.5.2 Two-Level Step-Doubling Strategy

One of the most popular time stepping schemes in CAD transient simulation tools is the two-level step-doubling strategy or Milne's device [74]. It is based on the Richardson extrapolation technique for the local-global error corresponding to the current time-step. The extrapolated value is then used to predict the size of the next time step.

The basic structure of the two-level step-doubling strategy is shown in Figure 3.15. An additional fine discretization level is introduced by halving each time-step of the basic nonuniform grid structure $\Omega_{\Delta t}$. The time integration is performed on both levels, once with a single step on the coarse level, then, independently, as two half-steps on the fine level, resulting in discrete solutions $u_{\Delta t}$ and $u_{\Delta t/2}$. The time integrations on coarse and fine

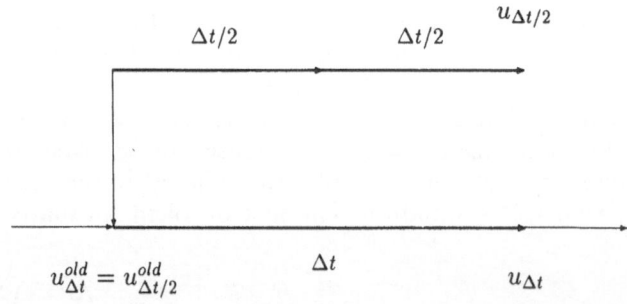

Fig. 3.15 Time stepping scheme for two-level step-doubling strategy

levels are independent only inside the single coarse integration step, since both integration paths share the same starting discrete solution $u_{\Delta t/2}^{old}$, which can be the initial state of the transient simulation or the discrete solution of the previous time-step obtained on the fine level.

The local-global errors of the coarse and fine time-steps obtained by Richardson extrapolation are from (3.43 and 3.44):

$$l_{\Delta t} = \frac{2^q}{2^q - 1}(u_{\Delta t/2} - u_{\Delta t}) \tag{3.73}$$

$$l_{\Delta t/2} = \frac{1}{2(2^q - 1)}(u_{\Delta t/2} - u_{\Delta t}). \tag{3.74}$$

Due to the artificially mixed local and global properties of the controlled discretization error, there are different practical implementations of the two-level step-doubling time stepping strategy which may lead to confusion. For example, if the global character of the discretization error on the fine level is neglected, that is, expressing the global error after two halfsteps on the fine level only by the local-global error of the second halfstep [139], the resulting extrapolation expression is analogous to (3.73) but with $q + 1$ instead of q. The extrapolated local-global error (3.73) on the coarse level has the same form as the corresponding extrapolated global error obtained by Richardson extrapolation using two globally uniform levels with time-step sizes Δt and $\Delta t/2$, where the time integration on coarse and fine discretization levels are performed from the very beginning of the transient simulation independently of each other. If, due to the similarity of the extrapolated global and local-global error expressions, the global convergence relation (3.31) is used instead of the local-global principal truncation term for the anticipation of a new time-step size [72], then an expression is obtained which is analogous to (3.70) but with q instead of $q + 1$.

In spite of the fact that (3.73) represents the correctly extrapolated local-global error on the coarse discretization level with time-step size Δt, it should not be used in (3.70) as an error estimator $\varepsilon_{\Delta t} = |l_{\Delta t}|$ because only the fine discretization level provides the required accuracy: two halfstep integrations generate a more accurate discrete solution than one single coarse step.

Remark 3.5. *From the discrete solution accuracy point of view the fine temporal level represents the actual adaptive integration path while the coarse level is considered only as an auxiliary level which is necessary for the extrapolation of the halfstep local-global errors.*

Consequently, the tolerance is formulated by the fine-step solution as

$$Tol_{\Delta t} = Rel \cdot (u_{\Delta t/2})_{\Delta t} + Abs$$

while for an error estimator there are two equivalent possibilities at our disposal. The first one is to use

$$\varepsilon_{\Delta t} = |l_{\Delta t}|$$

as the error estimator with the modified tolerance vectors

$$Tol_{\Delta t} \cdot 2^q \quad \text{or} \quad Tol_{\Delta t} \cdot 2^{q+1}$$

depending on the choice to control either the global error after two half-steps or the individual local-global errors of single halfsteps. The second possibility avoids the artificial modification of the tolerance vectors but uses either the error estimator

$$\varepsilon_{\Delta t} = 2 \cdot |l_{\Delta t/2}|$$

to control the global error after two halfsteps or the error estimator

$$\varepsilon_{\Delta t} = |l_{\Delta t/2}|$$

if the local-global error of a single halfstep is controlled.

The two-level step-doubling strategy suffers from two inherent weaknesses which affect the efficiency of transient simulation. The first originates from the fact that this strategy actually controls the error of the discrete solution on the fine level which is quasi-nonuniform since timeintegration proceeds with coupled pairs of halfsteps equal in size. Consequently, the accuracy of the extrapolation procedure depends on the assumption of approximately equal-sized local-global errors for the two halfsteps and on the contractivity parameter c close to one. Moreover, the anticipation of the next time-step (or a better guess for the current time-step) by two-level step-doubling strategy actually means the anticipation of a new couple of halfsteps instead of a single one, which limits the flexibility of the time-step size modification. The second but much more important problem is the fact, that the extrapolation procedure for the local-global error requires extra computational work for the coarse time integration step, which may be a high price for the introduction of temporal adaptivity when large process simulation problems are considered.

3.5.3 A Two-Level Strategy Based on τ-Extrapolation

Now the two-level time stepping scheme for the two-level step-doubling strategy, as shown in Figure 3.15 is revisited. Having in mind that for time-step sizes which correspond to the realistic range of the relative and absolute error control parameters, the difference between local and local-global errors becomes negligible,

$$\varepsilon_{\Delta t} = |\tau_{\Delta t/2}| \cdot \Delta t/2$$

may be used as an estimator. This results in almost the same temporal adaptive grid structure at the end of transient simulation, as in the case of the same time stepping scheme based on the local-global error.

In order to evaluate the local error estimator the Richardson extrapolation of the local-global error can be simply replaced by the basic τ-extrapolation of the local error on the same time stepping scheme. This also leads to a significant reduction of time stepping computational cost because the local error of the second halfstep obtained by τ-extrapolation is

$$\tau_{\Delta t/2} = \frac{1}{2^q - 1}(L_{\Delta t}(u_{\Delta t/2}) - f_{\Delta t}). \tag{3.75}$$

Remark 3.6. *Comparing the error estimates based on the local error with those based on the local-global error shows that only the solution of the fine temporal level is involved in the formulation of the error estimator based on the local error.*

This actually means that with τ-extrapolation based step-doubling time stepping strategy, the discrete solution procedure on the coarse discretization level is completely avoided. The auxiliary coarse time-step is here involved only in order to estimate the defect of the coarse grid operator for the fine level discrete solution. Since the computational cost for the evaluation of the coarse discrete equation defect is negligible to the complete solution procedure of the coarse time-step discrete problem, the application of τ-extrapolation eliminates the computational overhead introduced by the two-level step-doubling strategy based on Richardson extrapolation of the local-global error.

Remark 3.7. *The τ-extrapolation based approach for the local error estimation generally does not pose any limitation to the ratio of the coarse and half time-step sizes on the two temporal levels.*

Following Remark 3.7, the time stepping scheme from Figure 3.15 can be also profitably modified in order to optimally fit into the nonuniform time-stepping procedure with no additional computational cost. The idea to do this is to replace the finer auxiliary time level which is required for the extrapolation of the local error by another auxiliary time level which is coarser than $\Omega_{\Delta t}$. The improved time stepping scheme is shown in Figure 3.16.

The real fine level integration path now allows flexible step-by-step modifications of the time-step sizes. The fictitious coarse step (Δt_c) is formulated as common for the previous and the current time-step on the fine level and it is used only to evaluate the defect of the corresponding coarse level operator injecting the fine level discrete solution. In that way, the limited flexibility of the time-step size evolution due to the step-doubling is avoided without any additional computational cost. The doubling of the time-step size is required only for the first two time-steps at the very beginning of the adaptive

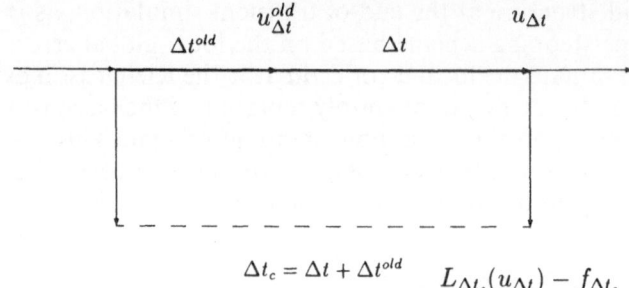

$$\Delta t_c = \Delta t + \Delta t^{old} \qquad L_{\Delta t_c}(u_{\Delta t}) - f_{\Delta t_c}$$

Fig. 3.16 The improved time stepping scheme

transient simulation. The local temporal error of the current time-step Δt on the fine (and, for the solution procedure, the only existing) integration path is

$$\tau_{\Delta t} = \frac{1}{\left(\dfrac{\Delta t_c}{\Delta t}\right)^q - 1}(L_{\Delta t_c}(u_{\Delta t}) - f_{\Delta t_c}) \tag{3.76}$$

and

$$\varepsilon_{\Delta t} = |\tau_{\Delta t}| \cdot \Delta t \tag{3.77}$$

is the error estimator which should be used in (3.70). The application of a similar strategy along with Richardson extrapolation of the local-global error is also possible but it leads to an additional increase in the computational cost of the time-stepping procedure since the coarse level incremental elliptic problem should be solved along with each fine level one.

It is difficult to accurately predict the computational gain that could be achieved by using the improved time stepping strategy based on τ-extrapolation technique. However, even neglecting the benefits of the more flexible time step size modifications and assuming equal computational costs for each of the incremental problems (3.8), approximately 1/3 of the transient simulation computational cost may be saved with the improved time-stepping strategy.

Example 3.11. *In order to make a practical comparison between the two-level step-doubling time stepping strategy with the local-global error control and the improved time stepping strategy with local error control, the diffusion model problem (3.13–3.15) is considered again.*

Using the automatic time-step size control technique as described in the previous section, the transient simulation of (3.13) is performed using both of the explained two-level time stepping strategies. The simulation parameters are $T = 600\,s$, $Rel = 10^{-2}$ and $Abs = 10^{13}$. The time-step size evolutions for the two-level step-doubling and τ-extrapolation based strategies are shown in Figure 3.17 and Figure 3.18, respectively. The more flexible and the more

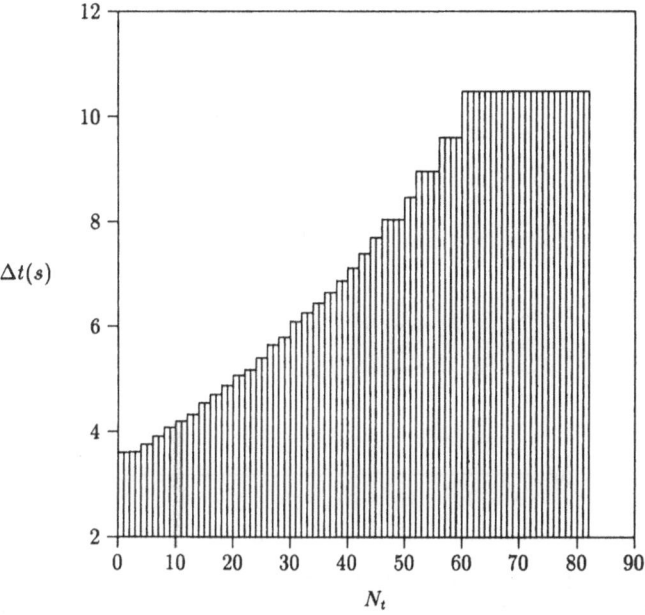

Fig. 3.17 The evolution of the time-step size with the two-level step-doubling time stepping strategy

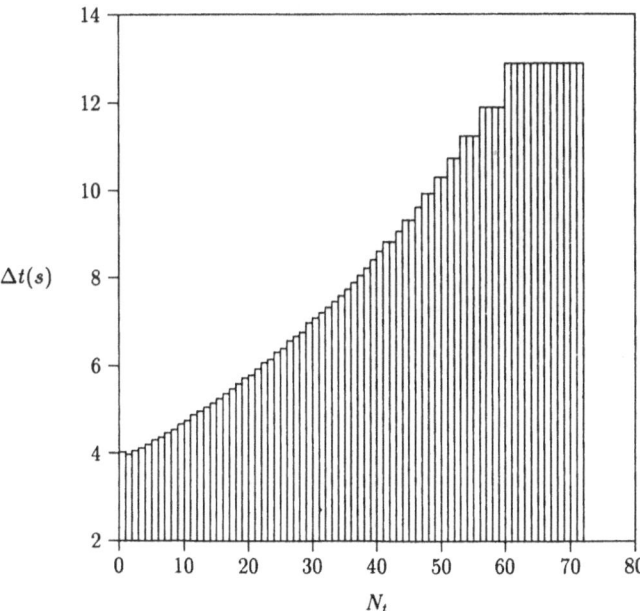

Fig. 3.18 The evolution of the time-step size with the improved two-level time stepping strategy

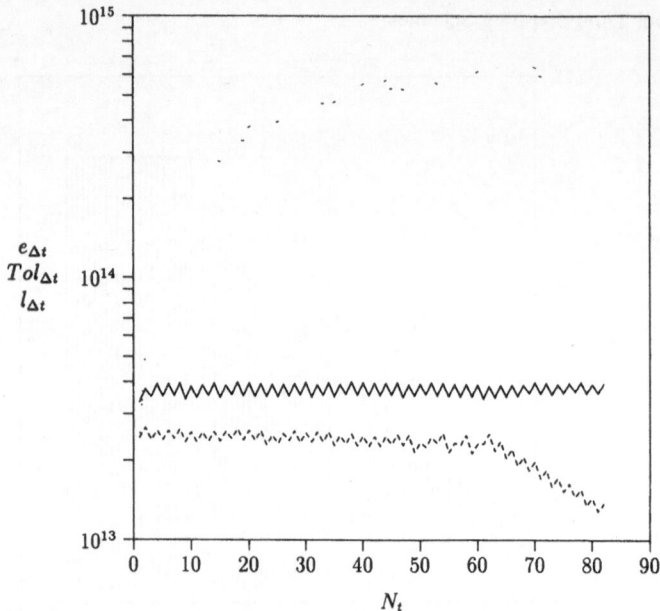

Fig. 3.19 The evolution of the discretization errors with the two-level step-doubling time stepping strategy (the solid line denotes the tolerance level, the dashed line denotes the exact local-global error and the dotted line denotes the exact global error)

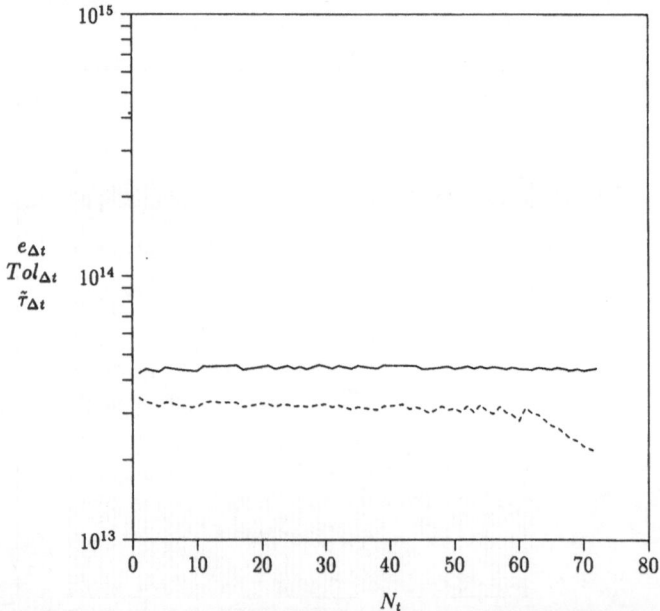

Fig. 3.20 The evolution of the discretization errors with the improved two-level time stepping strategy (the solid line denotes the tolerance level, the dashed line denotes the exact local error and the dotted line denotes the exact global error)

moderate changes in the time-step sizes by the improved strategy results in approximately 10% less integration steps. The improved time stepping strategy is also less sensitive to the influence of the finishing device (3.72), which restricts the rise of the time-step size near the end of the transient simulation.

Finally, Figure 3.19 and Figure 3.20 show the evolution of the exact local-global and local errors during transient simulation along with the evolution of the tolerance levels and the exact global errors. Both the discretization errors and the tolerances are evaluated at the spatial grid nodes where Δt_Ω^{new} takes the smallest value. It is evident from Figures 3.19 and 3.20 that both the exact local and the exact local-global errors are below the corresponding tolerance levels. The exact global error in both cases shows the stable accumulation which saturates towards a small multiple of the average tolerance level for the local or the local-global error. Again, the smooth evolution of the discretization errors is evident in case of the improved time stepping strategy based on τ-extrapolation.

In addition to the smooth time-step size evolution and the proper behavior of the discretization errors during transient simulation, the most important benefit of the improved time stepping strategy is the elimination of the computational work which is necessary to solve the additional incremental problems on the coarse temporal level in the step-doubling time stepping strategy.

4 Tayloring Multigrid Components for a Diffusion Model Problem

The main difficulty to construct efficient algorithms for application problems consists in the careful selection of those components a multigrid algorithm is composed of. The components have to be tuned carefully, because the proper choice decides on the efficiency of the multigrid solver. For realistic applications this problem is complemented by particular technical questions, which sometimes prohibit the use of standard components. A serious and often heard advice, not to say theorem, of the multigrid philosophy is: *treat local problems locally*. Having this in mind, the presented approach for the model problem of this chapter will differ by several details from other singlegrid <u>and</u> multigrid approaches [89, 90, 88, 99, 124]. The most significant of them are summarized in the following list:

1. A standard second order discretization technique is used for the differential equation in a time-dependent domain without coordinate transformation (computational domain = original domain).
2. A special second order formula is derived to discretize boundary conditions with normal derivatives on curved boundaries (Taylor series expansion on a cartesian grid).
3. Implicit time discretization schemes are combined with a time stepping procedure which is based on extrapolation techniques.
4. The incremental problem in time is solved by use of a multigrid method within every time-step.
5. Because of the special shape of the profiles, high order monotone and shape preserving multigrid interpolation techniques are applied.
6. The adaptive local refinements within the multigrid approach are controlled by an estimation of the local discretization error.

Most of the algorithmic components are chosen with the intention to make them useful for similar applications—and not only within a multigrid program. The monotone and shape preserving interpolation technique is an excellent example for this idea. The application of interpolation methods which maintain special properties of the interpolant, improves, in this case,

Table 4.1. *Number of points on grids G_1 upto $G_7, t = 0$*

G_1	G_2	G_3	G_4	G_5	G_6	G_7
5×7	9×13	17×25	33×49	65×97	129×193	257×385
(35)	(117)	(425)	(1.617)	(6.305)	(24.897)	(98.945)

the convergence behavior of the algorithm and, additionally, offers a solution to the often mentioned demand for nonoscillating interpolations which are interesting for the interchange of data between different simulation tools and different data structures.

All experiments of this chapter are executed on a domain being initially a rectangle of $3.0\,\mu m \times 2.0\,\mu m$. The multigrid algorithm uses a sequence of grids $G_{h_1}, \ldots, G_{h_7}, h_1 = \frac{1}{2}\,\mu m$. Their respective number of points, considering the grids as global ones, is for $t = 0$ given in Table 4.1. The interpretation of both the physical and numerical parameters is given in the "Notation". Times are given in terms of minutes.

4.1 The Physical Problem and the Mathematical Representation

Because this chapter is still "numerically" oriented and under the condition that characteristic features of "real" problems are present, the not highly sophisticated modeling of implantation, oxidation and diffusion is justified. The actually used approaches are explained only so far as it is necessary to understand the initial boundary value problem which has to be solved.

4.1.1 Implantation

From the mathematical point of view, the doping profile after ion implantation is not more than the initial condition of the evolution problem. Therefore, the basic model which is used here, may be replaced by any other more sophisticated approach (Pearson distributions or even Monte Carlo simulations, [126], see also Chapter 5). For simplicity, perpendicular mask edges as well as constant layers of deposited material are assumed. This is, in principle, no disadvantage for the numerical experiments with the model problem.

Taking into account the special symmetries of the model problem at hand, the formal representation of the field implant looks like

$$N(x, y, 0) = N_d(f_{0a}(x, y) + f_{a\infty}(x, y)) \tag{4.1}$$

where the functions $f_{0a}(x, y)$ and $f_{a\infty}(x, y)$ depend on the geometry and are responsible for the representation of the profile within selected parts of the geometry $(0 \le x < a, a \le x < \infty)$. More complex geometries are resolved similarly.

4.1.2 Diffusion

"Diffusion" will, in this chapter, denote the distribution of an existing impurity within the substrate without regarding the origin of the profile. The rate of the dopant movement within the semiconductor is given by the diffusion coefficient. Having in mind the construction and the investigation of a numerical algorithm, it is not intended to go into a discussion of effects which have to be regarded for a realistic modeling of the diffusion mechanism. Therefore a concentration-dependent (that means nonlinear) diffusion coefficient due to [88, 89] is used. With the "normalized" concentration $\tilde{N} := N/2n_i$ and a dopand-dependent fitting parameter γ, the diffusion coefficient is

$$D(\tilde{N}) = \frac{1 + \gamma(\tilde{N} + \sqrt{\tilde{N}^2 + 1})}{1 + \gamma}\left(1 + \frac{\tilde{N}}{\sqrt{\tilde{N}^2 + 1}}\right) D^0. \qquad (4.2)$$

Remark 4.1. *The above choice of the diffusion coefficient is not the most general one. Therefore it should be noted that any other diffusion coefficient (see for instance Section 5.3.4) may be used instead of (4.2). But the analytic representation allows an easy determination of the derivative with respect to N for nonlinear relaxation schemes.*

4.1.3 Oxidation

The two-dimensional version of the analytic oxidation model [89, 126] which is used for the present model problem is a generalization of the well-known one-dimensional oxide growth model due to Deal and Grove [22]. The initial geometry is selected by a proper choice of geometry parameters l_x, l_y, a and the mask profile including the initial oxide thickness U_0. It is sketched in Figure 4.1 at the beginning of the oxidation $t = 0$ and with a grown up oxide for $t > 0$. The oxide thickness $U(x, t)$ for a time $t > 0$ at the lateral position x is given by

$$U(x, t) = \frac{U(t) - U_0}{2} \operatorname{erfc}\left(\frac{x - a + \delta(U(t) - U_0)}{\sqrt{2R_0 U(t)}}\right) + U_0 \qquad (4.3)$$

with the grown up oxide $U(t)$ for $t > 0$ due to the one-dimensional Deal-Grove oxidation model where $\operatorname{erfc}(\xi) = 1 - \operatorname{erf}(\xi) = \frac{1}{\sqrt{\pi}} \int_{\xi}^{\infty} e^{-x^2} dx$ denotes

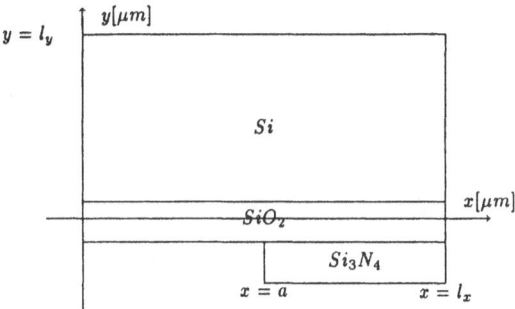

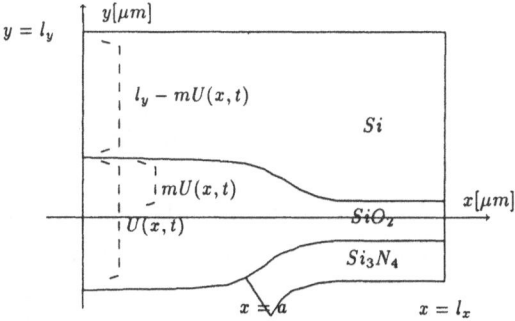

Fig. 4.1 Initial geometry and the situation for $t > 0$

the complementary error function. R_0 is the ratio between the lateral and the vertical oxide growth. δ is a fitting parameter to model the shape of the oxide.

For a given set of parameters the boundary of the physical domain is determined for each $t > 0$. This information is necessary to calculate the outward normal unit vector $\vec{n}(t)$ as well as the oxide growth rate perpendicular to the moving boundary, $\dot{U}_{\vec{n}}(x, t)$ which is used to set up the initial boundary value problem.

4.1.4 The Mathematical Formulation of the Model Problem

Using the previously described models for implantation, diffusion and oxidation (k and m are the segregation coefficient and the amount of consumed silicon to produce a volume unit of silicon dioxide, $m = 0.44$), the initial boundary value problem which has to be solved for the doping concentration $N(x, y, t)$ is represented by

$$\nabla(D(N)\nabla N) = \frac{\partial N}{\partial N} \text{ within } \Omega(t) \tag{4.4}$$

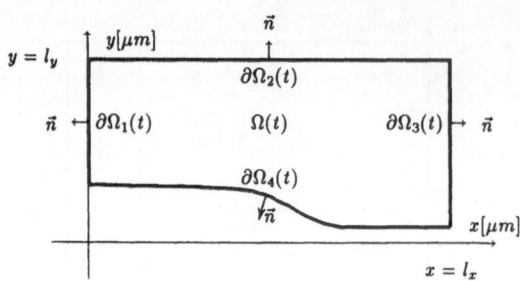

Fig. 4.2 Description of the initial boundary value problem

$$\frac{\partial N}{\partial n} = 0 \text{ at } \partial\Omega_1(t), \ \partial\Omega_2(t) \text{ and } \partial\Omega_3(t) \qquad (4.5)$$

$$D(N)\frac{\partial N}{\partial n} - (k-m)\dot{U}_n N = 0 \text{ at } \partial\Omega_4(t) \quad \text{for} \quad t > 0. \qquad (4.6)$$

$$N(x, y, 0) \quad \text{due to (4.1)} \quad \text{for} \quad (x, y)\in\overline{\Omega(0)}, \qquad (4.7)$$

$$\text{where} \quad \Omega(t) = \{(x, y)|0 < x < l_x, mU(x, t) < y < l_y\}$$

$$\text{and} \quad \partial\Omega(t) = \bigcup_{i=1}^{4} \partial\Omega_i(t)$$

$$\text{with} \quad \partial\Omega_1(t) = \{(0, y)|mU(0, t) < y < l_y\}$$

$$\partial\Omega_2(t) = \{(x, l_y)|0 \le x \le l_x\}$$

$$\partial\Omega_3(t) = \{(l_x, y)|mU(l_x, t) < y < l_y\}$$

$$\partial\Omega_4(t) = \{(x, y)|0 \le x \le l_x, y = mU(x, t)\}$$

4.2 The Discretization

Most process simulation tools map the physical domain onto a fixed rectangle. The usually chosen transformation techniques base on conformal
mappings [86, 87] or on a stretching into the direction of a selected coordinate
[88, 89, 90, 99, 124]. For a wide class of domains such a transformation is
useful, without any doubt. But the change of the coordinate system often
introduces mixed derivatives, first order derivatives and time-dependent
coefficients. The conformal mapping approach which avoids mixed derivates
is limited to two-dimensional applications.
In contradiction to the previously mentioned "*treat local problems locally*",
a local feature influences, by such transformations, the solution process on
the whole (global) computational domain. If local peculiarities are trans-

ported into the whole computational domain it is worth thinking about alternative approaches which circumvent this disadvantage. As there is no physically motivated reason to use the often recommended transformations, the decision to compute the solution on the original geometry (that means "computational domain = original domain") is a consequence of strictly following the above principle. As a result, the differential equation remains in its original form and standard finite difference or finite volume methods may be applied to define the discrete problem on regular grids.

4.2.1 Numerical Consequences of a Domain Transformation

An often used stretching in y-direction

$$t \mapsto \tau(t) = t$$

$$x \mapsto \xi(x) = x$$

$$y \mapsto \eta(x, t) = \frac{y - mU(x, t)}{l_y - mU(x, t)} l_y,$$

changes the partial differential equation (4.4) of the model problem into the general form

$$\frac{\partial N}{\partial \tau} = \frac{\partial}{\partial \xi} \left(D(N) \frac{\partial N}{\partial \xi} \right) + \Theta \frac{\partial}{\partial \eta} \left(D(N) \frac{\partial N}{\partial \eta} \right)$$

$$- \Xi \left(\frac{\partial}{\partial \xi} \left(D(N) \frac{\partial N}{\partial \eta} \right) + \frac{\partial}{\partial \eta} \left(D(N) \frac{\partial N}{\partial \xi} \right) \right) + \Phi \frac{\partial N}{\partial \eta} \qquad (4.8)$$

where the coefficients Θ, Ξ and Φ depend on the oxidation model $U(\xi, \tau)$ and derivatives of U with respect to τ and ξ.

The boundary conditions (4.5)–(4.6) are transformed similarly. The coefficients corresponding to the discrete version of (4.4) on the original domain are abbreviated here by $\mathcal{W}, \mathcal{S}, \mathcal{N}, \mathcal{E}$ and $\mathcal{L} = \mathcal{W} + \mathcal{S} + \mathcal{N} + \mathcal{E}$. Discretizing (4.8) in the same way introduces terms which correspond to the additional coefficients of (4.8). Denoting them in the sequence of their occurance again by Θ_h, Ξ_h and Φ_h, the five-point stencil for (4.4) is transformed into a nine-point formula:

$$\begin{bmatrix} & \mathcal{N} & \\ \mathcal{W} & -\mathcal{L} & \mathcal{E} \\ & \mathcal{S} & \end{bmatrix}_{(x,y,t)} \mapsto \begin{bmatrix} \Xi_h & \Theta_h \mathcal{N} + \Phi_h & \Xi_h \\ \mathcal{W} & -(\mathcal{W} + \mathcal{E} + \Theta_h(\mathcal{S} + \mathcal{N})) & \mathcal{E} \\ \Xi_h & \Theta_h \mathcal{S} + \Phi_h & \Xi_h \end{bmatrix}_{(\xi,\eta,\tau)}$$

For the transformed boundary conditions the discretization by introducing ghost points outside of the domain results in more complex stencils, too. The additional numerical effort to solve the transformed problem is mainly determined by the evaluation of the more complex stencils and is approximately doubled.

Conclusion 4.1. *As long as the boundary segments are smooth, it is not necessary at all to apply the transformation technique. The investigated problem with the evolving bird's beak geometry certainly belongs to such a class of problems. Alternatively one may think on the use of composite grids with orthogonal grids near the boundary and regular cartesian grids throughout the major part of the region. Such a strategy should be used adaptively and self-controlled, leading to the problem to find robust and reliable criteria which are not in sight, yet.*

4.2.2 The Discretization of the Diffusion Equation Including the Homogeneous Neumann Boundary Conditions

As the domain $\Omega(t)$ and the boundary segments $\partial\Omega_1(t), \partial\Omega_3(t)$ and $\partial\Omega_4(t)$ depend on time, the following considerations are valid for every $t \geq 0$. Only for simplification the specification "(t)" is omitted.

The discretization of the partial differential equation is performed on an equispaced grid where the boundary segments $\partial\Omega_1, \partial\Omega_2$ and $\partial\Omega_3$ coincide with grid lines. To approximate the normal derivative, so called "ghost points" are introduced (marked by $\circ$ in Figure 4.3). For the second order approximation of the boundary condition on $\partial\Omega_4$ another set of additional points, $\partial\Omega_{4h}$, is necessary:

$$\partial\Omega_{4h} := \{z \,|\, z = (ih, jh) \in \mathscr{G}_h - (\bar{\Omega} - \partial\Omega_4) \wedge \mathrm{dist}(z, \partial\Omega_4) < \sqrt{2}h,$$

$$\text{where } j := \max_{kh \leq mU(ih,t)} k \text{ or } ((i+1)h, (j+1)h) \in \Omega \cap \mathscr{G}_h\}$$

$$(4.9)$$

For those points in $\partial\Omega_{4h}$ where no connection to direct neighbouring grid points with distance h intersects the boundary $\partial\Omega_4$ (marked in Figure 4.3 by

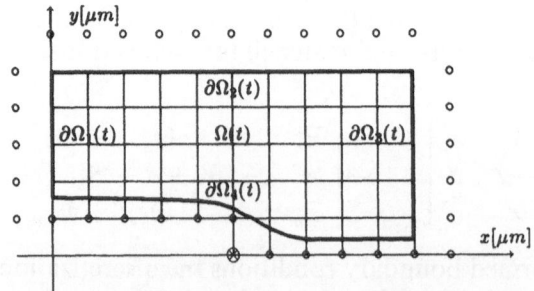

Fig. 4.3 Computational domain with grid, including ghost points

$\otimes$), necessary values are extrapolated from values at neighbouring inner points. The left hand side of the differential equation is approximated by standard discretization techniques (finite volume or finite difference approach) and results in a second order five-point formula [8,94]. In more detail: the space operator of the differential equation is discretized on the boundary pieces $\partial\Omega_1 \cap \mathcal{G}_h$, $\partial\Omega_2 \cap \mathcal{G}_h$ and $\partial\Omega_3 \cap \mathcal{G}_h$, too. The boundary conditions are used to eliminate the equations for the introduced ghost points (mirror imaging). The stencil which represents the space discrete problem including the eliminated equations of the homogeneous Neumann boundary conditions is

$$\frac{1}{h^2}\begin{bmatrix} 0 & (1-\delta_{jN_y})D_{i,j+1/2} & 0 \\ (1+\delta_{iN_x}-\delta_{i0})D_{i-1/2,j} & \Sigma & (1-\delta_{iN_x}+\delta_{i0})D_{i+1/2,j} \\ 0 & (1+\delta_{jN_y})D_{i,j-1/2} & 0 \end{bmatrix}$$

(4.10)

with

$$\Sigma = -((1+\delta_{iN_x}-\delta_{i0})D_{i-1/2,j} + (1+\delta_{jN_y})D_{i,j-1/2}$$
$$+ (1-\delta_{jN_y})D_{i,j+1/2} + (1-\delta_{iN_x}+\delta_{i0})D_{i+1/2,j}),$$

where δ_{kl} denotes Kronecker's symbol and

$$D_{i\pm1/2,j}:=\tfrac{1}{2}(D(N_{i\pm1,j})+D(N_{i,j})) \quad D_{i,j\pm1/2}:=\tfrac{1}{2}(D(N_{i,j\pm1})+D(N_{i,j})).$$

(4.11)

Remark 4.2. *Denoting the index J_i to be the largest J of column i with $Jh \leq mU(ih,t)$, it becomes obvious that the time-dependency of the domain induces that of $J_i(t)$. In case of the model problem in combination with the chosen coordinate system the moving boundary (decreasing domain) guarantees the condition $J_i(t) \leq J_i(t+\Delta t)$ for $\Delta t > 0$ and all grid columns $0 \leq i \leq N_x$. In (4.10) the nonlinearity of the problem is hidden in $D_{i-1/2,j}$ which really means $D_{i-1/2,j} = D_{i-1/2,j}(N_{i-1,j}, N_{i,j})$ due to (4.11). Similar dependencies are valid for the other coefficients.*

4.2.3 The Discrete Boundary Condition on the Time-Dependent Boundary

The second order difference approximation of the boundary condition

$$D(N)\frac{\partial N}{\partial \vec{n}} - (k-m)\dot{U}_{\vec{n}}N = 0$$

(4.12)

on the curved and moving boundary $\partial\Omega_4(t)$ is obtained by Taylor's series expansion. Again, the indication "(t)" is omitted for notational simplicity, because it is easy to recognize where it should appear.

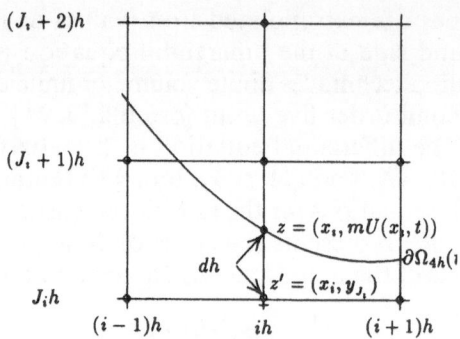

Fig. 4.4 Discretization of the normal derivative on a curved boundary

For a $z := (x_i, y + dh) = (ih, J_i h + dh) \in \partial\Omega_4$ the discretization of the Neumann part of the boundary condition (4.12) starts from the representation

$$D(N)\frac{\partial N}{\partial \vec{n}}\bigg|_z = D(N)\vec{n}\cdot\nabla N|_z = D(N)\left(n_x \frac{\partial N}{\partial x} + n_y \frac{\partial N}{\partial y}\right)\bigg|_z.$$

n_x, n_y and $d := \dfrac{mU(ih, t) - J_i h}{h}$ denote the x- and y-components of the normal unit vector $\vec{n}$, and the relative distance between a point $(ih, J_i h)$ and $\partial\Omega_4(t)$, respectively. Following [6], the functions $\dfrac{\partial N}{\partial x}$ and $\dfrac{\partial N}{\partial y}$ are expanded around the center $z' := (x_i, y_{J_i}) = (ih, J_i h)$ (denoted by $\bigcirc$ in Figure 4.4). To achieve the desired second order approximation, $\dfrac{\partial N}{\partial x}\bigg|_{z'}$ and $\dfrac{\partial N}{\partial y}\bigg|_{z'}$ are approximated with second order, while their derivatives only need a first order formula. This finally leads to a seven-point stencil to approximate the Neumann part of the boundary condition (4.12):

$$\frac{D(N)|_z}{h}\left(\frac{n_x}{2}\begin{bmatrix} 0 & 0 & 0 \\ -d & 0 & d \\ -(1-d) & \underline{0} & (1-d) \end{bmatrix} + n_y \begin{bmatrix} 0 & d-\frac{1}{2} & 0 \\ 0 & 2(1-d) & 0 \\ 0 & \underline{d-\frac{3}{2}} & 0 \end{bmatrix}\right)$$

(4.13)

The "center" $z' = (x_i, y_{J_i}) = (ih, J_i h)$ of the stencil is underlined and corresponds to the underlined circle in Figure 4.4.

Remark 4.3. *Commuting the x- and y-direction yields a rotated stencil which may be used for curved boundaries with a more extreme curvature. The previously defined as $\partial\Omega_{4h}$ has, in such situations, to be modified in a straightforward manner.*

Passing to the limits $d = 0$ and $d = 1$ shows that the x- and y-components tend to frequently used second order approximations of first derivatives using differences of size h and 2h, respectively. The general situation $0 < d < 1$ is an interpolation of the limiting special cases.

The discretization process of (4.12) is completed by the approximation

$$
\begin{bmatrix}
0 & \frac{1}{2}d(d-1) & 0 \\
0 & -d(d-2) & 0 \\
0 & \frac{1}{2}(d-1)(d \times 2) & 0
\end{bmatrix}
\tag{4.14}
$$

for $N|_z = N|_{(x_i, y_J, + dh)}$. With $\dot{f}_{\bar{n}} := (k - m)\dot{U}_{\bar{n}}|_z$ the discrete version of the boundary condition at the curved boundary segment in stencil notation reads

$$
\bar{D}
\begin{bmatrix}
0 & \frac{n_y}{h}(d-\frac{1}{2}) & 0 \\
-\frac{n_x}{2h}d & \frac{n_y}{h}2(1-d) & \frac{n_x}{2h}d \\
-\frac{n_x}{2h}(1-d) & \frac{n_y}{h}(d-\frac{3}{2}) & \frac{n_x}{2h}(1-d)
\end{bmatrix}
- \dot{f}_{\bar{n}}
\begin{bmatrix}
0 & \frac{1}{2}d(d-1) & 0 \\
0 & -d(d-2) & 0 \\
0 & \frac{1}{2}(d-1)(d-)2 & 0
\end{bmatrix},
$$

$$\tag{4.15}$$

where $\bar{D}$ is any reasonable approximation to $D(N)|_z$. Due to the boundary conditions at the symmetry lines ($i = 0$ and $i = N_x$) the stencil (4.15) is modified similar to (4.10).

Remark 4.4. *The relative distance to the boundary $d = d(x_i, t)$ is computed and stored for every time-step. The same computation provides $J_i = J_i(t)$, $n_x = n_x(t)$, and $n_y = n_y(t)$ which can be determined analytically with the information of Section 4.1.3. These quantities describe, from the programming technique point of view, the time-dependent domain $\Omega(t)$.*

Due to the "conservation law of difficulty" the global advantage has to be paid for locally. The relative distance d is an important parameter of the discrete boundary condition because it expresses the movement of the boundary. The varying d influences the numerical behavior in a non-ignorable way. The discrete boundary condition tends towards

$$
\bar{D}
\begin{bmatrix}
0 & -\frac{n_y}{2h} & 0 \\
0 & 2\frac{n_y}{h} & 0 \\
-\frac{n_x}{2h} & -3\frac{n_y}{2h} & \frac{n_x}{2h}
\end{bmatrix}
- \dot{f}_{\bar{n}}
\begin{bmatrix}
0 & 0 & 0 \\
0 & 0 & 0 \\
0 & 1 & 0
\end{bmatrix}
\tag{4.16}
$$

with $d \to 0$ and towards

$$
\bar{D}
\begin{bmatrix}
0 & \dfrac{n_y}{2h} & 0 \\[6pt]
-\dfrac{n_x}{2h} & 0 & \dfrac{n_x}{2h} \\[6pt]
0 & -\dfrac{n_y}{2h} & 0
\end{bmatrix}
- \dot{f}_{\bar{n}}
\begin{bmatrix}
0 & 0 & 0 \\
0 & \underline{1} & 0 \\
0 & \underline{0} & 0
\end{bmatrix}
\tag{4.17}
$$

for $d \to 1$. The first case shows, that no difficulties should be expected, while $d \to 1$ may produce extremely unbalanced stencils for $\| \dot{f}_{\bar{n}} \|$ large compared to $\| \bar{D} n_y/(2h) \|$. Such a situation may, for the bird's beak geometry, not only occur for a single point but even for a set of points. This requires a modification to reestablish the diagonal dominance of the corresponding matrix. A reasonable approach leaves the Neumann part unchanged, while

$$
\dot{f}_{\bar{n}}
\begin{bmatrix}
0 & \tfrac{1}{2}d(d-1) & 0 \\
0 & -d(d-2) & 0 \\
0 & \tfrac{1}{2}(d-1)(d-2) & 0
\end{bmatrix}
\quad \text{is replaced by} \quad
\dot{f}_{\bar{n}}
\begin{bmatrix}
0 & \tfrac{1}{2}d & 0 \\
0 & 0 & 0 \\
0 & 1-\tfrac{1}{2}d & 0
\end{bmatrix}
$$

$$\tag{4.18}$$

for all d with $1 > d \geq \varepsilon_d > 0$ and a given ε_d. Although this reduces the $O(h^2)$ order of consistency to an only first order approximation, damaging disadvantages are not observed in practice.

Remark 4.5. *The above discussion of the varying $d(t)$ is confirmed for $d \to 1$ by a significant deterioration of both smoothing and convergence properties of the used relaxation schemes what even causes the multigrid algorithm to diverge. This is demonstrated by numerical experiments to find a standard value for ε_d, the switching parameter. $\varepsilon_d = 0.5$ is an adequate choice, because $\varepsilon_d = 0.25$ results in approximately the same behavior, while $\varepsilon_d = 0.9$ often produces divergence.*
The above formula which involves not more than three points of a grid column is advantageous even in the case of column relaxation because the structure of the problem is unchanged: it remains tridiagonal with only one additional entry for the equation which corresponds to the point with the index J_i. Nevertheless, for block relaxation schemes a decomposition into two triangular systems is possible.

4.2.4 Time Discretization

Because diffusion periods in the range of minutes to several hours have to be considered, the implicit time discretization is a practical requirement.

The time-step restriction of the explicit Euler scheme for the actually regarded differential equation is for all (x_i, y_j) of the computational domain given by

$$\Delta t < \frac{h^2}{D_{i-\frac{1}{2},j} + D_{i,j-\frac{1}{2}} + D_{i,j+\frac{1}{2}} + D_{i+\frac{1}{2},j}}. \tag{4.19}$$

Depending on the doping concentration this enforces step sizes in the range of seconds for a long initial phase. In spite of the simple problem for every time-step an enormous computational effort is invested whenever the time period to be simulated lies in the range of hours.

Implicit schemes are usually combined with time stepping procedures to control the marching in time by a proper selection of the time-step size. This is possible because most of the implicit schemes for parabolic partial differential equations don't have to satisfy a stability condition which often restricts the time-step size dramatically. Although the Crank-Nicolson scheme for nonlinear differential operators has to satisfy such a condition, it allows considerably larger time-step sizes than explicit schemes would do. Fortunately, the convergence of multigrid methods does not depend in an extreme way on the increasing time-step size and they become competitive for large time-steps and even on coarse space grids. This has been shown for a heat transfer problem in [128]. The discretization due to Section 2.8 leads to a representation of the discrete problem due to (2.85) with L_h given by (4.10).

4.2.5 Time Stepping

The time stepping method used here estimates the temporal discretization error with the aim to predict the next time-step size in order to keep the error below a certain level. The necessary theoretical background is explained in Section 3.5.2. Controlling the local-global error determines the new time-step to be

$$\Delta t^{new} = \begin{cases} \Delta t \sqrt[3]{\dfrac{Tol}{\frac{4}{3}\varepsilon(t^n + \Delta t)}} & \text{for} \quad CN \\[4mm] \Delta t \sqrt[2]{\dfrac{Tol}{2\varepsilon(t^n + \Delta t)}} & \text{for} \quad BE \end{cases}$$

Unfortunately, there is no indication for an appropriate choice of Δt^1, the initial time-step size, within this procedure. Small time-steps are required at the beginning of the simulation to avoid the oscillating behavior of the CN-scheme, respectively to be sufficiently exact, when a first order scheme (BE) is used. By experience, the step sizes of the explicit Euler scheme have these properties. Although, therefore starting with very small values, the above procedure soon leads to considerably larger time-steps as it is expected for implicit approaches.

4.2.6 An Estimation of the Stability Condition for CN

The stability analysis for the CN scheme is due to (2.90)–(2.92) founded on the knowledge of eigenvalues of the operators $L_h(N)$ at different time levels $L_h(N_h^{n+1})$ and $L_h(N_h^n)$. This nonlinear dependency on N and especially on the time levels t^n and t^{n+1} complicates the exact computation of the stability condition. To get an estimation, the approximation $\Lambda_h(N_h^n) := \bar{D}^n \Delta_h$ to the spatial operator $L_h(N_h^n)$ for every t^n with $\bar{D}^n := \max_{x \in \Omega_h} D(N_h^n)$ is used. The formal representation of the Crank-Nicolson-scheme with respect to L_h is then replaced by that for Λ_h,

$$\left(I_h - \frac{\Delta t}{2}\Lambda_h\right)N_h^{n+1} = \left(I_h + \frac{\Delta t}{2}\Lambda_h\right)N_h^n,$$

and

$$\frac{\left|1 + \frac{\Delta t}{2}\lambda_{l,m}^h(N_h^n)\right|}{\left|1 - \frac{\Delta t}{2}\lambda_{l,m}^h(N_h^{n+1})\right|} < 1$$

becomes the local stability condition. For $k = n, n+1$ the $\lambda_{l,m}^h(N_h^k)$ are the eigenvalues of $\Lambda_h(N_h^k)$. Using the definition of Λ_h and inserting the eigenvalues $\lambda_{l,m}^h$ of the Laplacian leads to

$$\frac{2\Delta t}{h^2}(\bar{D}^n - \bar{D}^{n+1}) < 1. \tag{4.20}$$

After straight forward calculations the difference $\bar{D}^n - \bar{D}^{n+1}$ is estimated by

$$\bar{D}^n - \bar{D}^{n+1} \approx -\overline{D'(N_h^n)\nabla_h(D(N_h^n)\nabla_h N_h^n)}\Delta t =: \zeta \Delta t,$$

which in combination with (4.20) finally provides the "approximated stability condition"

$$-2\Delta t^2 \overline{D'(N_h^n)\underbrace{\nabla_h(D(N_h^n)\nabla_h N_h^n)}} < h^2 \quad \text{or} \quad \Delta t < \frac{h}{\sqrt{2\zeta}} \tag{4.21}$$

Due to the diffusion process ζ is positive. The bar again denotes taking the maximum value for all $x \in \Omega_h$ and the underbraced part of (4.21) is $L_h(N_h^n)$. Therefore this condition is numerically cheaply tested, depends only on the actual time level and is consequently well suited to be an additional quantity for the time-step size control.

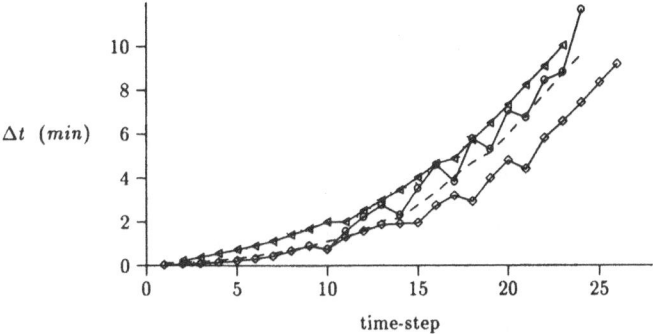

$\Delta t \ (min)$

time-step

Fig. 4.5 Time stepping (cntrl)/CN-stability condition (stab), ○ ≙ cntrl, ◇ ≙ cntrl/stab, ◁ ≙ stab

This is demonstrated by numerical experiments, whose results are summarized in Figure 4.5. Marching in time under the exclusive control of condition (4.21) is a natural choice. The respective time-step size development is marked by ◁ in Figure 4.5. For comparison, the proposed values of the time stepping procedure are indicated by the dotted line. The exclusive use of the time stepping procedure is another basic choice. Starting with the same initial time-step size as in the previous experiment, the time-step sizes (marked by ○) do in most situations satisfy the CN stability condition (dashed line). Only in the second half of the simulation the condition (4.21) is sometimes violated by the proposed Δt. But the time-step sizes are controlled back to approximately the size of the stability condition. The time stepping procedure does not automatically run into time-step sizes which are too large compared with the condition (4.21). A suitable combination of both methods, using the respective minimum value, results in a "save" time-step size development, indicated by ◇.

Conclusion 4.2. *A combination of the above discussed mechanisms with different time discretization schemes for special phases of the simulation offers a large variety of possibilities to control the time-step size. This means: during the first time-steps with small Δt because of the stiffness of the problem an absolutely stable scheme like the implicit Euler (BE) is recommended to smooth out the initially dominating high-frequent components very fast. As the stiffness requires small initial time-steps anyhow, it does not matter to use a scheme which is only first order in time. Later on, the low-frequent components dominate and do allow growing time-steps. Then it is advisable to use a second order implicit scheme (here CN).*

4.3 Relaxation Methods

Because the FAS-scheme needs no global linearization only nonlinear variants of Gauss-Seidel methods are presented for the considered diffusion model problem. The local linearization is based on Newton's method [109]. The point oriented schemes use a lexicographically or chequerboarded ordering of grid points, while the block Gauss-Seidel methods scan the columns lexicographically, respectively even-odd and odd-even. They serve as iterative coarse grid solver, too.

4.3.1 Gauss-Seidel-Newton Relaxations

The Newton method to solve the M-dimensional problem $F(x) = 0$

$$F: \mathbb{R}^M \rightarrow \mathbb{R}^M \quad \text{with} \quad x \mapsto F(x) = \begin{pmatrix} f_1(x_1, \ldots, x_M) \\ \vdots \\ f_M(x_1, \ldots, x_M) \end{pmatrix}$$

leads to the iteration

$$F'(x^{(k)})\delta x = - F(x^{(k)}) \quad \text{with} \quad x^{(k+1)} = x^{(k)} + \delta x. \tag{4.22}$$

F' is the Jacobian and the superscript (k) indicates the approximation to the solution after k steps of the Newton iteration. To solve the problem (4.22) the Jacobian $F'(x^{(k)})$ has to be computed. The point- and column-Gauss-Seidel-Newton methods (PGSN, CGSN) don't require the determination of the complete Jacobian $F'(x^{(k)})$. For CGSN it is only necessary to compute $F'(x)$ with respect to those unknowns which belong to grid points of the actually scanned grid column, while for PGSN the Jacobian reduces to a 1×1-matrix.

At the time level $t^{n+1} = t^n + \Delta t$ the discrete problem for an inner point of the domain is given by (2.85), where L_h simplifies from (4.10) to

$$\frac{1}{h^2} \begin{bmatrix} 0 & D_{i,j+\frac{1}{2}} & 0 \\ D_{i-\frac{1}{2},j} & \tilde{\Sigma} & D_{i+\frac{1}{2},j} \\ 0 & D_{i,j-\frac{1}{2}} & 0 \end{bmatrix} \tag{4.23}$$

and Σ to $\tilde{\Sigma} := - (D_{i-\frac{1}{2},j} + D_{i,j-\frac{1}{2}} + D_{i,j+\frac{1}{2}} + D_{i+\frac{1}{2},j})$. A reformulation leads to the equivalent problem to solve:

$$\mathscr{L}_h N_h^{n+1} := \begin{bmatrix} 0 & D_{i,j+\frac{1}{2}} & 0 \\ D_{i-\frac{1}{2},j} & \tilde{\Sigma} - \dfrac{h^2}{\beta \Delta t} & D_{i+\frac{1}{2},j} \\ 0 & D_{i,j-\frac{1}{2}} & 0 \end{bmatrix} N_h^{n+1} - F_{h,\Delta t}^n = 0. \tag{4.24}$$

Omitting the time level indicators and considering the equation which is associated to the unknown $N_{i,j}$, the Newton method has, within the $l+1$-th

Gauss-Seidel iteration, to compute the solution η of

$$\mathcal{L}_{ij}(N_{i-1,j}^{l+1}, N_{i,j-1}^{l+1}, \eta, N_{i,j+1}^{l}, N_{i+1,j}^{l}) = 0 \tag{4.25}$$

and to accept this value as the $l+1$-th iterate $N_{i,j}^{l+1}$. Step k of the iteration scheme to compute $N_{i,j}^{l+1} = \eta$ with the aid of the Newton scheme is

$$\eta^{(k)} = \eta^{(k-1)} - \frac{\mathcal{L}_{ij}(N_{i-1,j}^{l+1}, N_{i,j-1}^{l+1}, \eta^{(k-1)}, N_{i,j+1}^{l}, N_{i+1,j}^{l})}{\dfrac{\partial \mathcal{L}_{ij}}{\partial \eta}\Big|(N_{i-1,j}^{l+1}, N_{i,j-1}^{l+1}, \eta^{(k-1)}, N_{i,j+1}^{l}, N_{i+1,j}^{l})} \tag{4.26}$$

for $k, l = 1, \ldots$, using an initial guess $\eta^{(0)}$. Experiences with multigrid algorithms and nonlinear relaxation schemes, the quadratic convergence order of the Newton method and the usually good initial guess $\eta^{(0)} = N_{i,j}^{l}$, especially for time-dependent problems show, that it is sufficient to perform only one Newton step to get $N_{i,j}^{l+1} = \eta^{(1)}$, thus replacing equation (4.26) by

$$N_{i,j}^{l+1} = N_{i,j}^{l} - \frac{\mathcal{L}_{ij}(N_{i-1,j}^{l+1}, N_{i,j-1}^{l+1}, N_{i,j}^{l}, N_{i,j+1}^{l}, N_{i+1,j}^{l})}{\dfrac{\partial \mathcal{L}_{ij}}{\partial \eta}\Big|(N_{i-1,j}^{l+1}, N_{i,j-1}^{l+1}, N_{i,j}^{l}, N_{i,j+1}^{l}, N_{i+1,j}^{l})} \quad \text{for} \quad l \geq 1$$

$$\tag{4.27}$$

Collecting terms and using the abbreviations

$$a := D_{i-\frac{1}{2},j}(N_{i-1,j}^{l+1}, N_{i,j}^{l}), b := D_{i,j-\frac{1}{2}}(N_{i,j-1}^{l+1}, N_{i,j}^{l}),$$

$$c := D_{i,j+\frac{1}{2}}(N_{i,j+1}^{l}, N_{i,j}^{l}), d := D_{i+\frac{1}{2},j}(N_{i+1,j}^{l}, N_{i,j}^{l}), \tag{4.28}$$

$$z := \tilde{\Sigma}(N_{i-1,j}^{l+1}, N_{i,j-1}^{l+1}, N_{i,j}^{l}, N_{i,j+1}^{l}, N_{i+1,j}^{l}) - \frac{h^2}{\theta \Delta t} \quad \text{and}$$

$$\partial \psi := \frac{\partial \psi}{\partial \eta}\Big|_{(N_{i,j}^{l})}$$

the equation which has to be solved for the unknown $N_{i,j}^{l+1}$ reads

$$(a - N_{i,j}^{l}\partial a)N_{i-1,j}^{l+1} + (b - N_{i,j}^{l}\partial b)N_{i,j-1}^{l+1}$$
$$+ (c - N_{i,j}^{l}\partial c)N_{i,j+1}^{l} + (d - N_{i,j}^{l}\partial d)N_{i+1,j}^{l}$$
$$+ \{z + \partial a N_{i-1,j}^{l+1} + \partial b N_{i,j-1}^{l+1} + \partial z N_{i,j}^{l} + \partial c N_{i,j+1}^{l} + \partial d N_{i+1,j}^{l}\}N_{i,j}^{l+1}$$
$$= \partial z N_{i,j}^{l}N_{i,j}^{l} + F_{i,j}. \tag{4.29}$$

The corresponding stencil

$$\begin{bmatrix} & & c - N_{i,j}^{l}\partial c \\ & & \partial c \\ a - N_{i,j}^{l}\partial a & z + \begin{bmatrix} & \partial c & \\ \partial a & \partial z & \partial d \\ & \partial b & \end{bmatrix}^{l} & d - N_{i,j}^{l}\partial d \\ & \partial b & \\ & b - N_{i,j}^{l}\partial b & \end{bmatrix}^{l+1} \tag{4.30}$$

shows the general structure which is very convenient for the smoothing analysis in practice. The superscripts l and $l+1$ indicate the Gauss-Seidel step which is responsible for the selection of the center element $N_{i,j}$ of the stencil, respectively substencil.

With only minor modifications a similar representation for the column Gauss-Seidel-Newton method can be derived. Instead of considering (4.25) to be an equation for the unknown corresponding to (x_i, y_j), $\mathscr{L}_{ij}$ is now considered with respect to all the unknowns ξ, η and ζ which belong to the column i:

$$\mathscr{L}_{ij}(N_{i-1,j}^{l+1}, \xi, \eta, \zeta, N_{i+1,j}^{l}) = 0 \tag{4.31}$$

Exactly the same steps as previously performed lead to a stencil whose structure is similar to that of PGSN and which has to be interpreted analogously.

4.3.2 Relaxation Schemes

With point relaxations it is advantageous to satisfy the boundary condition on $\partial\Omega_4$ first, to relax all the other points afterwards and to finish the scheme with a sweep over boundary points _and_ additionally over a small strip of inner points. The additional boundary relaxation usually includes all the points with index J_i to J_i+2. This are exactly those points which are involved in the discrete boundary condition (4.15). Such a procedure is called PGS-lex or PGS-ch in dependency from the chosen relaxation method for the inner points. The column schemes (CGS-lex, CGS-eo and CGS-oe) treat all the points simultaneously which belong to the same grid column. An additional sweep over the boundary and some inner points near the boundary is added, too.

4.3.3 Empirical Convergence Rates

In Section 2.4 the numerical work has been considered assuming standard coarsening, that means using the relation $\mathscr{N}_k = 2^d \mathscr{N}_{k-1}$ for the number of points on level k. In practice, and especially in combination with refined grids, this assumption is no longer valid. A better approach uses the real number of grid points per level: the level-dependent factor

$$f_k = \frac{\text{number of points on level } k}{\text{number of points on the finest global grid}}$$

replaces $\left(\dfrac{1}{2^d}\right)^{l-k}$ in (2.68) and improves the determined W_l.

When refined grids are encountered within the algorithm, the finest grid level is, in general not a global grid at all. Therefore, the number of points on a global grid with the same meshsize is difficult to obtain. But an estimation (extrapolation) of this value may serve to get an approximate measure for the relative work on each grid level. This "extrapolated global grid RU" permits a more direct comparison of global grid experiments with those using locally refined meshes.

Example 4.1. *A computation for the initial geometry $(t = 0)$ shows the effect of this modification: four global grids with standard coarsening, V-cycling with totally v relaxations per grid and v iterations to solve on the coarsest grid result in $W_4 = 1\frac{21}{64}vRU \approx 3.98\,RU$ with $v = 3$ (excluding the effort for grid transfer). The modified calculation leads to the more exact value of $W_4 \approx 4.07$ RU, while an empirically determined value was $W_4 \approx 4.1\,RU$. The neglectable difference originates in additional relaxation sweeps to solve the coarsest grid problem which are counted, too.*
If there are only three global grids and one refined grid instead of four global ones, the number of points belonging to the fourth hypothetical global grid is extrapolated from the number of points which belong to the three existing global grids: 35, 117 and 425 points. If 1530 points are estimated to belong to the hypothetical finest global grid and if the refinement consists of only 385 points, the relative numerical work on the refined grid is estimated by $\frac{385}{1530}$. In a similar way the quotients for the other grid levels are computed. Finally, for a $V(2,1)$-cycle $W_4 \approx 1.88\,RU$ is stated for such a grid configuration.

In theory, the multigrid convergence does not depend on the meshsize h. Because this statement is an "asymptotic" one, this property must not be observed with meshsizes used in practice. Nevertheless, with a given set of admissable meshsizes and with different components of the algorithm the defect reduction and related quantities like empirical convergence rates are of interest. The time-dependent problem requires an investigation of several snap shots with special consideration of different time-step sizes.
The empirical convergence rates result from calculations which reduce an initial residual by a factor of at least 10^{-14}. All values are given in terms of the discrete L_2-norm. Before discussing the empirical convergence rates of the multigrid algorithm, the relaxation schemes are used as iterative single-grid solver on a grid with the same meshsize. This experiment demonstrates the h-dependent and poor convergence of these iterative singlegrid solvers. The iterations on single grids produce, after a certain number (Σit) of

Table 4.2. *Asymptotic singlegrid convergence rates on grids G_5 and G_6*

scheme	G_5		G_6	
	ρ_{av}	Σit	ρ_{av}	Σit
PGS-lex	0.91	≈ 400	0.98	≈ 300
PGS-ch	0.92	≈ 100	0.98	≈ 300
CGS-lex	0.74	≈ 100	0.94	≈ 100

iterations, asymptotic singlegrid rates ρ_{av}. The calculations have been performed for $T = 80$ using $\Delta t = 15$ to allow the direct comparison with the empirical convergence rates per RU in Tables 4.3 and 4.4. A first discussion of the stencil in (4.24) with respect to Δt suggests the better smoothing property for a given relaxation scheme in combination with CN than in combination with BE. The observed algorithms really do behave in this way. The influence of different time-step sizes is shown in Table 4.3 where empirical convergence rates for $\Delta t = 0.15$ and $\Delta t = 0.25$ on a fine grid at the initial phase of diffusion are given. The extreme dependence of the convergence rates on the time-step size is caused by the small Δt which has not yet reached the magnitude to be in the region of the asymptotic behavior with respect to Δt (Section 4.4).

Table 4.3. ρ_{av} and ρ_{rv} for $t = 0.05$ on G_6, $V(2,1)$-cycle, BE/CN

cycle : $V(2,1)$ time discr. : BE/CN		G_6		
		$\Delta t = 0.15$		$\Delta t = 0.25$
Scheme		ρ_{av}	ρ_{ru}	ρ_{av}	ρ_{ru}
PGS-lex	CN	0.0572	0.4918	0.0668	0.5111
PGS-ch	CN	0.0272	0.4092	0.0309	0.4224
CGS-lex	CN	0.0348	0.4349	0.0424	0.4567
PGS-lex	BE	0.0584	0.4944	0.0695	0.5163
CGS-lex	BE	0.0435	0.4597	0.0458	0.4656

Table 4.4. Empirical convergence rates, CN

cycle : $V(2,1)$	G_4		G_5		G_6	
	$T = 80$ $\Delta t = 15$		$T = 80$ $\Delta t = 15$		$T = 80$ $\Delta t = 15$	
Scheme	ρ_{av}	ρ_{ru}	ρ_{av}	ρ_{ru}	ρ_{av}	ρ_{ru}
PGS-lex	0.0445	0.4431	0.0816	0.5126	0.1032	0.5489
PGS-ch	0.0375	0.4302	0.0514	0.4684	0.0543	0.4779
CGS-lex	0.0109	0.3370	0.0303	0.4619	0.0811	0.5236
cycle : $V(2,1)$	$T = 300$ $\Delta t = 100$		$T = 300$ $\Delta t = 100$		$T = 300$ $\Delta t = 100$	
PGS-lex	0.0814	0.5191	0.1054	0.5560	0.1178	0.5701
PGS-ch	0.0447	0.4521	0.0688	0.4957	0.0935	0.5321
CGS-lex	0.0546	0.4691	0.0998	0.5487	0.0805	0.5265
grid : G_5	$V(1,1)$		$W(1,1)$		$W(2,1)$	
	$T = 80$ $\Delta t = = 15$		$T = 80$ $\Delta t = 15$		$T = 80$ $\Delta t = 15$	
PGS-lex	0.1325	0.4600	0.1324	0.5684	0.0794	0.6203
PGS-ch	0.0952	0.4125	0.0953	0.5334	0.0505	0.5775
CGS-lex	0.0867	0.4290	0.0864	0.5466	0.0316	0.5780
grid : G_6	$V(1,1)$		$W(1,1)$		$W(2,1)$	
	$T = 300$ $\Delta t = 100$		$T = 300$ $\Delta t = 100$		$T = 300$ $\Delta t = 100$	
PGS-lex	0.1764	0.5025	0.1893	0.6141	0.1146	0.6573
PGS-ch	0.1408	0.4577	0.1481	0.5933	0.1011	0.6587
CGS-lex	0.1571	0.4802	0.1719	0.6254	0.0933	0.6535

Multigrid cycling on G_6 with PGS-ch smoothing shows the best convergence both with respect to ρ_{av} and to ρ_{ru}. The time-step sizes of 0.15 and 0.25 have been chosen obeying practical requirements of accuracy and stability. They are, on grid G_6, more than one hundred times as large as the stability condition of the explicit Euler method would allow. Even the stability condition for the CN-scheme is, for G_6, more then ten times smaller than the values used for these convergence tests.

In addition to Table 4.3 the Table 4.4 contains a collection of empirical convergence rates for different cycle types on grids $G_4 - G_6$ for large T and Δt. The CGS-lex relaxation shows a better convergence than the PGS-lex. Similar to elliptic Poisson-like problems for the diffusion model problem PGS-ch turns out to produce well-converging multigrid algorithms on fine grids and for large times. These situations are characterized by concentrations without large variation over the length of $2h$ and so avoiding the problem of extremely unbalanced stencils. The Δt used additionally lie within the range of the asymptotic behavior (see Section 4.4).

4.4 Results of the Smoothing Analysis with Respect to Time-Step Size

The nonlinearity combined with concentrations changing their magnitude extremely over small distances induces numerical difficulties. For the discrete equations this means strongly varying coefficients of the difference equations: there are stencil entries of different order of magnitude which disturb the smoothing properties of relaxation methods. Such situations have to be detected and to be analyzed in combination with an increasing time-step size. An easily applied tool to do this is the smoothing analysis. Some results for PGSN and CGSN are presented and discussed in this section.

The concrete formula for $D_{i\pm\frac{1}{2},j}$ and $D_{i,j\pm\frac{1}{2}}$ (Section 4.2.2) are used to reformulate (4.30). This ends with a compact five-point stencil whose entries are given by

$$\mathscr{A} := \tfrac{1}{2}\{D(N_{i-1,j}^{l+1}) + D(N_{i,j}^{l}) - N_{i,j}^{l}D'(N_{i,j}^{l})\}$$

$$\mathscr{B} := \tfrac{1}{2}\{D(N_{i,j-1}^{l+1}) + D(N_{i,j}^{l}) - N_{i,j}^{l}D'(N_{i,j}^{l})\}$$

$$\mathscr{L} := -\tfrac{1}{2}\{D(N_{i-1,j}^{l+1}) + D(N_{i,j-1}^{l+1}) + 4D(N_{i,j}^{l}) + D(N_{i,j+1}^{l}) + D(N_{i+1,j}^{l})$$

$$- D'(N_{i,j}^{l})(N_{i-1,j}^{l+1} + N_{i,j-1}^{l+1} - 4N_{i,j}^{l} + N_{i,j+1}^{l} + N_{i+1,j}^{l})\} - \frac{h^2}{\theta\Delta t}$$

$$\mathscr{C} := \tfrac{1}{2}\{D(N_{i,j+1}^{l}) + D(N_{i,j}^{l}) - N_{i,j}^{l}D'(N_{i,j}^{l})\}$$

$$\mathscr{D} := \tfrac{1}{2}\{D(N_{i+1,j}^{l}) + D(N_{i,j}^{l}) - N_{i,j}^{l}D'(N_{i,j}^{l})\}$$

for PGSN (similar for CGSN).

A first impression of what can be expected is based on the assumption to have a function N which is approximately constant over an h-surrounding of (x_i, y_j). With $\tilde{D} := D(N_{i,j}) \approx D(N_{i-1,j}) \approx D(N_{i,j-1}) \approx \ldots$, the stencil reduces to:

$$
\tilde{D} \begin{bmatrix} & 1 & \\ 1 & -\left(4 + \dfrac{h^2}{\tilde{D}\theta\Delta t}\right) & 1 \\ & 1 & \end{bmatrix}
$$

As constant factors cancel out, the smoothing rates will, in this situation, probably not be worse than those of point- and column-relaxation schemes for the standard five-point formula of the Poisson equation. Especially for decreasing h or increasing Δt the smoothing rates should asymptotically approximate $\mu^*_{PGS-lex,\Delta_h} = 0.5$ for PGSN and $\mu^*_{CGS-lex,\Delta_h} = \frac{1}{5}\sqrt{5} \approx 0.4472$ for CGSN. In Figures 4.7–4.8 these values are marked by the dashed, respectively dotted lines parallel to the horizontal axis.

For nonlinear problems the smoothing rate depends on the variable. To achieve a complete overview, the smoothing rates have to be computed on the whole range of characteristic values. Therefore, a subdomain $\tilde{G}_h \subseteq G_h$ is chosen, covering regions with an initially strong variation of the profile. A smoothing factor is computed for all points of $\tilde{G}_h$. The maximum of all these values is then referred to as the smoothing rate. As this value additionally depends on time, the calculations are carried out for different times with varying time-steps and on a set of grids with meshsizes which are relevant for practical applications. For $T = 0.05$ a sequence of experiments with time-step sizes $\Delta t \in \{0.05, \ldots, 1.05\}$ on $\tilde{G}_4 - \tilde{G}_6$ determines the smoothing rates. One background to vary the Δt is the search for an appropriate initial time-step size, which has to be properly chosen both for the CN scheme and for the BE scheme (to avoid the "oscillations" of CN and taking into account the only first order approximation in time of BE). Additionally, it is of interest to notice the influence of the time-step size on the smoothing property.

The upper two parts of Figure 4.6 present from the left to the right characteristic distributions of the smoothing rates, CGSN with BE on the subdomain $\tilde{G}_6$ and CGSN with BE on $\tilde{G}_4$. The results demonstrate the dependency on Δt: the smoothing worsens with increasing Δt. The first example (left) makes the structure of the underlying approximation to the solution N_h transparent. For an approximately constant doping profile the predicted effect of the asymptotic behavior towards the reference value $\mu^*_{CGS-lex,\Delta_h}$ is observed in those parts of $\tilde{G}_6$ which are away from the dominant profile. Obviously, for large Δt, the smoothing rates within the "silent" part of the simulation area worsen faster than in regions where a relevant concentration exists. The second part (right), CGSN with BE on $\tilde{G}_4$, shows extremely

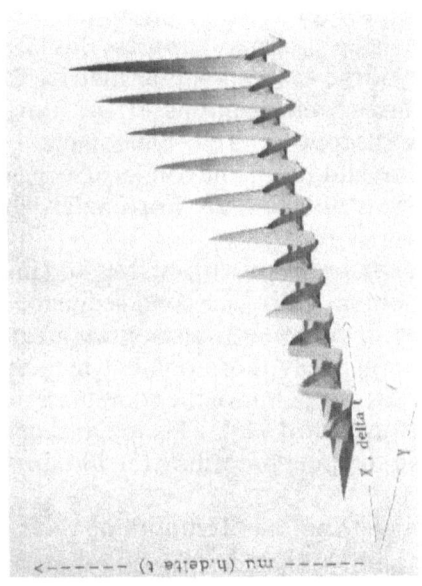

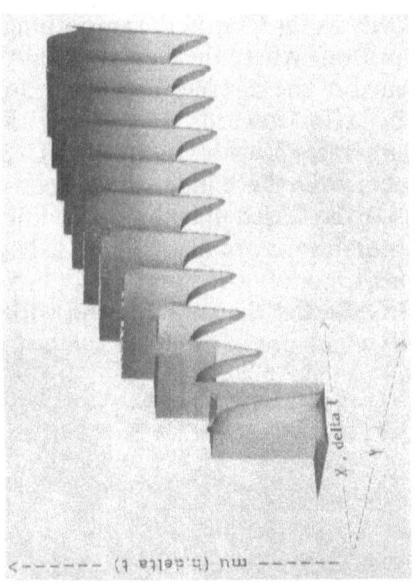 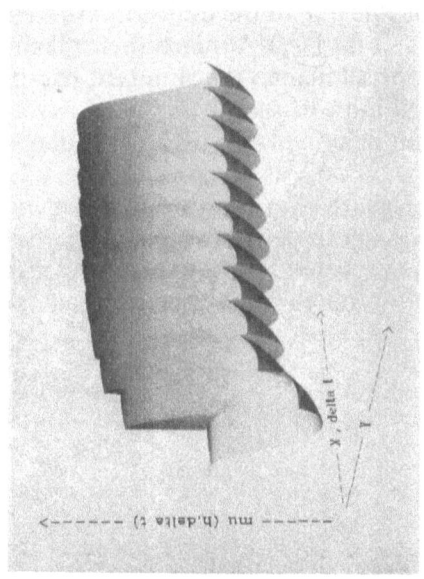

Fig. 4.6 Smoothing factors depending on Δt, CGSN, $T = 0.05$ and $T = 80$

bad values at some positions. There the strong variation of N between (x_i, y_{j-1}) and (x_i, y_{j+1}) carries over to the diffusion coefficient and produces stencil elements of an extremely different magnitude. This ruins the smoothing properties at this position. Small time-step sizes may suppress this effect by improving the diagonal dominance of the corresponding matrix. But with increasing time-step sizes the unfavourable relations of the entries dominate and the smoothing property disappears. This phenomenon is typically observed immediately after implantation and on coarse grids whenever the solution "jumps" on intervals of length $2h$. Fortunately this appearance is restricted to an isolated set of points.

As already mentioned, the time-dependent problem carries over to time-dependent smoothing rates. This is demonstrated by a second sequence of experiments at $T = 80$. The subdomains to compute the smoothing rates correspond to the previously used, excluding only those points which are, due to the movement of the boundary, no longer points of the computational domain. The time-step sizes now vary from 1.0 to 21.0. This magnitude is reached without any difficulty by the time stepping procedure for 80-minute simulations.

The lower left part of Figure 4.6 shows a distribution of smoothing rates at $T = 80$ for CGSN on $\tilde{G}_6$ combined with the CN-scheme. Again, the underlying discrete solution is transparent. The influence of the strongly varying coefficients is no longer observed. This is the benefit of the already flattened profile due to the diffusion. As well for PGSN as for CGSN the smoothing rates for large Δt reach their maxima at positions where the stencils locally look similar to the standard five-point stencil of the discrete Laplacian. In addition to the pronounced asymptotic behavior towards $\mu^*_{CGS-lex,\Delta_h}$, an impression of the still h-dependent smoothing rates $\tilde{\mu}(h)$ for CGSN with CN on $\tilde{G}_4 - \tilde{G}_6$ is given in Figure 4.7 (remember that the h-independent convergence is an asymptotic statement for $h \to 0$; the h used may not yet be fine enough to detect this property). Similar observations are made for PGSN. As expected by the experience of many linear model problems, the CGSN smoothing rates are better than those of PGSN and the combination with CN is superior to that with BE for all cases which are of practical interest.

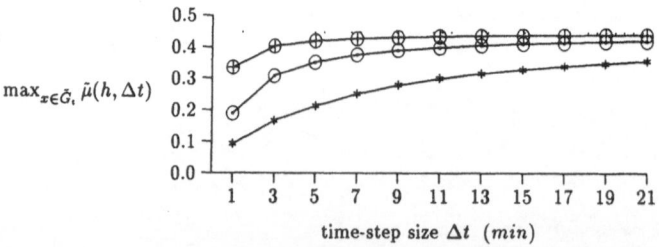

Fig. 4.7 Smoothing rates depending on Δt, CN, $T = 80$, $* \triangleq \tilde{G}_4$, CGSN, $\odot \triangleq \tilde{G}_5$, CGSN, $\oplus \triangleq \tilde{G}_6$, CGSN

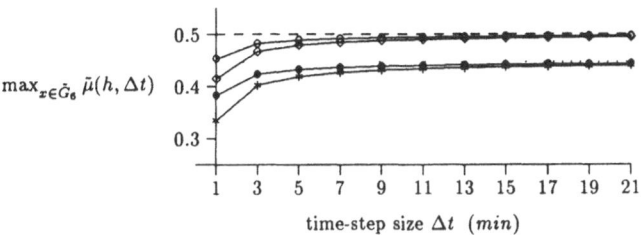

Fig. 4.8 Smoothing rates depending on Δt, different algorithms on $\tilde{G}_6$, $T80$, $\circ \hat{=} $ PGSN, BE, $\bullet \hat{=} $ CGSN, BE, $\diamond \hat{=} $ PGSN, CN, $* \hat{=} $ CGSN, CN

Conclusion 4.3. *The smoothing rates for PGSN and CGSN in combination with CN and BE are for large Δt very close to the reference values 0.5 and 0.4472 (Figure 4.8). This justifies the use of large time-steps without loosing the good smoothing capabilities. On fine grids, both PGSN and CGSN combined with either CN or BE asymptotically march towards $\mu^*_{PGS-lex,\Delta_h}$ and $\mu^*_{CGS-lex,\Delta_h}$. Numerical experiments on G_5 controlled by the time stepping procedure for an 80 minute simulation run into time-steps of twenty minutes and more without any convergence problems, thus verifying the local analysis by practical results.*

4.5 Experiments with the Crank-Nicolson Scheme and with the Fully Implicit Scheme

The main questions of interest concern the influence of the nonlinearity combined with the boundary movement. Numerical experiments verify the definitely smaller error of the CN scheme which therefore leads to considerably larger time-steps than BE. The second order approximation of CN, respectively the only first order approximation of BE is confirmed easily, too. This are the well-known properties of the implicit schemes "Fully implicit (backward) Euler (BE)" and "Crank-Nicolson (CN)" which are valid not only for linear model problems.

As explained in Section 2.8 CN may cause "oscillations" of the error, even for linear problems. To avoid this, it is recommended to use small time-steps within the initial phase to damp the high-frequent components or to use absolutely stable schemes like BE. With an initial Δt^1 due to the stability condition of the explicit Euler scheme, CN needs less time-steps for an eighty minute simulation than BE does.

For large times the low-frequent (nonoscillating) components start to dominate, and the solution approaches the stationary state. Several experiments with $T = 1000$ are performed to investigate the algorithm for this condition. CN needs considerably less time-steps than BE to finish the simulation. The

Table 4.5. *Number of time steps on G_5, $T = 1000$*

MGI V(2, 1) cycles 3 T = 1000	G_5 $\Omega = \Omega(t)$ $D = D(N)$		G_5 $\Omega = \Omega(t)$ $D \equiv$ const.		G_5 $\Omega = \Omega(0)$ $D = D(N)$		G_5 $\Omega = \Omega(0)$ $D \equiv$ const.	
PGS-lex	BE	CN	BE	CN	BE	CN	BE	CN
N_t time steps from $t \approx 100$ to T	44	37	38	19	39	28	28	15
	13	8	18	9	11	7	9	6

leftmost data part of Table 4.5 presents the results for the nonlinear problem on a time-dependent domain ($D = D(N)$ and $\Omega = \Omega(t)$). To separate the influence of the nonlinearity, the diffusion coefficient is kept constant ($D \equiv$ const.), reducing the problem to a linear one. Suppressing the oxidation leaves the computational domain unchanged ($\Omega = \Omega(0)$). The combination of these conditions now allows the interpretation of both the nonlinearity and the boundary movement.

Comparing the number of time-steps for the set of experiments shows the advantage of CN for large simulation times in any of the above mentioned situations. The advantage is not only true for large times but also for smaller ones ($T = 100$) and especially for the interval between $T = 100$ and $T = 1000$. The nonlinear problems with CN require approximately twice as much numerical work than the linear ones. For BE the difference between the linear and nonlinear cases is obvious, too, but not as large. Freezing the computational domain to $\Omega = \Omega(0)$ again reduces the required number of time-steps. This reduction is independent from the respective time discretization scheme. But the difference between "fixed" and "time-dependent" domain is smaller than the difference between "linearity" and "nonlinearity".

Conclusion 4.4. *Although BE is unconditionally stable the only first order approximation makes this scheme inferior to the CN scheme which is second order in time but which has to satisfy the already mentioned stability condition. With an increasing simulation time the advantage of the CN scheme becomes evident. This is true in spite of the fact that CN reacts very sensitive on the nonlinearity of the problem.*

These results separate the influence of both phenomena: nonlinearity and boundary movement. The negative effect of the nonlinearity is more important than the time-dependent domain, especially in combination with CN. The nonlinearity is a property of global influence, while the moving boundary typically is of local character. Therefore the above experiments will, in a practical way, corroborate the theorem of multigrid philosophy that global effects will not dominate over properly treated local effects. And, implicitly these results say that the boundary treatment seems to be appropriate.

4.6 The Prolongation of Grid Functions

When using Full-Multi-Grid (FMG) for the numerical solution of partial differential equations, theory prescribes using a higher-order interpolation when interpolating the current solution to a new finest grid if one wants to obtain a solution which is exact up to the discretization error after one FMG cycle. This interpolation is typically a fourth-order cubic interpolant. A straight-forward usage of these interpolation procedures may cause severe problems because of the typical oscillations, "standing waves", which may appear in the transition region where the concentration changes abruptly from a high value to essentially zero within a few grid-points. The oscillations such a standard cubic interpolation can cause are illustrated by the following example and in Figure 4.13.

Example 4.2. *Regarding a part of a one-dimensional doping profile (after a certain scaling) at* $x_0, x_1 = x_0 + H, x_2 = x_0 + 2H$ *and* $x_3 = x_0 + 3H$ *with values* $N_0 = 0.2D8$, $N_1 = 0.2D7$, $N_2 = 0.4D4$ *and* $N_3 = 0.6D1$, *respectively, results in an interpolated concentration of*

$$N(\hat{x}) \approx -0.1D6$$

at the mid-point $\hat{x} = x_1 + \frac{1}{2}H$ *when using the standard cubic polynomial interpolation. This is without any physical meaning and far away from the requirements of interpolating positive quantities. Thus the introduction of positivity—or monotonicity—preserving interpolations is imperative.*

High order interpolations are used in three places in the algorithm. The first place is in FMG when the solution on the current finest grid is transferred to the next finer grid. The second place where higher-order interpolations are used is on interior boundaries of grid refinements. In the multilevel adaptive technique (MLAT) used, the equations on grid refinements are posed with Dirichlet boundary conditions on the interior boundaries. The last situation in which higher order interpolations are needed is when the region covered by a local refinement becomes larger. Again, the values on the new refined regions are obtained from the coarser grid by higher-order interpolations.

For these purposes very simple positive, monotone and convex interpolations of fourth-order are applied. One of the reasons for their simplicity is, that they are tailored exactly to the needs of multigrid with standard coarsening, yielding interpolated values only at the mid-points between the given values. The methods will, after minor modifications [67], provide with values not only at the midpoint between data points, but also at any other point, maintaining the desired properties. On the other hand they are independent from the underlying problem and therefore suitable for other applications in which higher-order interpolations are required to transport grid functions from a given grid to a grid with a different structure.

4.6.1 Monotone Cubic Hermite-Interpolation

Let $a = x_1 < x_2 < \cdots < x_n = b$ be a partition of the interval $I = [a,b]$ and let $\{N_i\}_{i=1,n}$ be a set of monotone values at the x_i (either monotonous increasing or decreasing), then a piecewise cubic function $N(x)$ is wanted which is continuously differentiable, monotone and which satisfies $N(x_i) = N_i$ for $i = 1,\ldots,n$. In each subinterval $I_i = [x_i, x_{i+1}]$, $i = 1,\ldots,n-1$ and $x \in I_i$ the approach

$$N(x) = N_i \mathscr{H}_{1i}(x) + N_{i+1} \mathscr{H}_{2i}(x) + d_i \mathscr{H}_{3i}(x) + d_{i+1} \mathscr{H}_{4i}(x)$$

with $d_j = N'(x_j)$, $j = i, i+1$ is chosen. The $\mathscr{H}_{ki}(x)$ stand for the usual cubic Hermite-functions of I_i.

Using equidistant knots $x_{i+1} - x_i = H_i = H$ and the respective mid-point $\hat{x}_i = x_i + \frac{1}{2}H = x_i + h$ to be the position of interpolation the above formulation simplifies to

$$N(\hat{x}_i) = \tfrac{1}{2}(N_i + N_{i+1}) + \tfrac{1}{8}H(d_i - d_{i+1}).$$

The algorithm then consists in the construction of values $d_1, d_2, \ldots, d_n$ to guarantee for $N(x)$ the desired properties. Using the denotations

$$\Delta_i = \frac{N_{i+1} - N_i}{H}, \quad d_i = N'(N_i), \quad \alpha_i = \frac{d_i}{\Delta_i} \quad \text{and} \quad \beta_i = \frac{d_{i+1}}{\Delta_i} \qquad (4.32)$$

the set $\mathscr{S} = \{(\alpha_i, \beta_i) \,|\, \alpha_i^2 + \beta_i^2 \le 9, \alpha_i, \beta_i > 0\}$ is introduced for the following algorithm.

Algorithm 4.1. (*Fritsch-Carlson*)

Step 1 *Initialize $d_i, i = 1,\ldots,n$ in such a way that*
        $sign(d_i) = sign(d_{i+1}) = sign(\Delta_i)$. *If $\Delta_i = 0$ set $d_i = d_{i+1} = 0$.*
Step 2 *For each subinterval I_i with $(\alpha_i, \beta_i) \notin \mathscr{S}$ replace d_i by d_i^* with*
        $0 \le \alpha_i^* \le \alpha_i$ *and d_{i+1} by d_{*i+1} with $0 \le \beta_i^* \le \beta_i$ in order to have*
        $(\alpha_i^*, \beta_i^*) \in \mathscr{S}$ *(take care of $\alpha_i^* = d_i^*/\Delta_i$ and $\beta_i^* = d_{i+1}^*/\Delta_i$).*

Example 4.3. *The monotone cubic Hermite-Interpolation will, with the data of Example 4.2 result in $N(\hat{x}) \approx 0.9D6$ at the position $\hat{x} = x_1 + \frac{1}{2}H$. This is without any doubt the more reasonable result.*
The previously described method due to Fritsch and Carlson gives monotone values on monotone data. If the (α_i, β_i) don't belong to $\mathscr{S}$, that means that they don't satisfy the conditions for monotonicity, these are enforced by the projection of (α_i, β_i) onto the quarter circle around the origin with $r = 3$. The projected values (α_i^*, β_i^*) are used to calculate the interpolant. A modification due to Pollul [115] avoids this transformation. The resulting algorithm [63] then provides monotonicity and the additional property of smallest L_2-norm of N'' for the function N within the Sobolev space $H^2([a,b])$.

4.6.2 Monotone and Shape Preserving MG Interpolations

Let now $N(x)$ be a sufficiently smooth real-valued function which is either positive, monotone or convex. From its values $N_i = N(x_i)$, new data $T_{i+1/2}(N)$ at $\hat{x}_i$ have to be found such that the original data complemented with these interpolated values have the same properties as the original data and such that $|N(\hat{x}_i) - T_{i+1/2}(N)| = O(h^4)$.

The naive approach would be to let $T_{i+1/2}(N) = P(\hat{x}_i)$, where $P(x)$ is the cubic polynomial interpolating the data N_{i-1}, N_i, N_{i+1} and N_{i+2}. This is an $O(h^4)$ interpolation to $N(\hat{x}_i)$. However this interpolant preserves neither positivity, monotonicity nor convexity. The approach taken here to construct new multigrid interpolants, is to start with this cubic interpolant P, which interpolates the four values N_{i-1}, N_i, N_{i+1} and N_{i+2} of N at x_{i-1}, x_i, x_{i+1} and x_{i+2}. Each algorithm will yield a value $T_{i+1/2}$ at the mid-point $\hat{x}_i$ of the interval $[x_i, x_{i+1}]$. It is then modified to obtain the desired shape-preserving property without destroying its approximation order ("fit and modify" technique by Beatson and Wolkowicz, [4]).

Algorithm 4.2. (*A positive interpolant*) Let $N(x) \geq 0$.
Then $T_{i+1/2}(N) = \max(P(\hat{x}_i), 0)$.
The interpolant $T_{i+1/2}(N)$ is non-negative and is $O(h^4)$ if N is non-negative.

Algorithm 4.3. (*A monotone interpolant*) Let $N'(x) \geq 0$. Then

$$T_{i+1/2}(N) = \begin{cases} N_i & \text{if } P(\hat{x}_i) \leq N_i \\ N_{i+1} & \text{if } P(\hat{x}_i) \geq N_{i+1} \\ P(\hat{x}_i) & \text{otherwise.} \end{cases} \tag{4.33}$$

Let $N'(x) \geq 0$ in $I_i = [x_i, x_{i+1}]$. If $T_{i+1/2}(N)$ is defined as in (4.33), then it is a monotonely increasing $O(h^4)$ interpolant to N at $\hat{x}_i$.

The main idea of the proofs to the propositions concerning the properties of the previous algorithms and of those to follow is that if $P(\hat{x}_i)$ does not satisfy

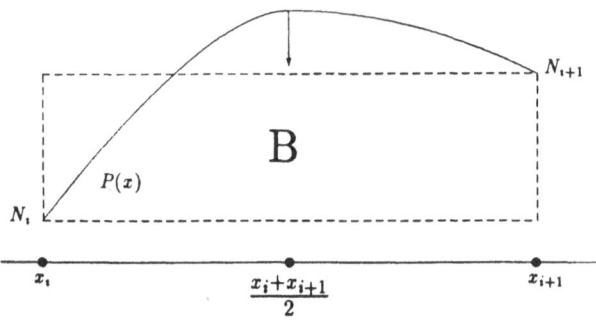

Fig. 4.9 Monotone interpolation

the same constraints as N does, then the approximation is improved by changing it to the closest value which does satisfy the constraints. This may be reformulated to saying that the value of the interpolant due to algorithm 4.3 must lie in the box B given in Figure 4.9.

Of course, a monotonely decreasing interpolant may be constructed in the same way. A combination of them can be used to obtain a comonotone interpolant in the sense that the interpolant is monotonely increasing where the function is increasing and monotonely decreasing where the function is monotonely decreasing.

Algorithm 4.4. (*A convex interpolant*) *Assume that $N'' \geq 0$. Let*

$$m_{i,i+1} = \max\left\{\frac{3}{2}N_i - \frac{1}{2}N_{i-1}, \frac{3}{2}N_{i+1} - \frac{1}{2}N_{i+2}\right\}. \tag{4.34}$$

Then

$$T_{i+1/2}(N) = \begin{cases} \dfrac{N_i + N_{i+1}}{2} & \text{if } P(\hat{x}_i) \geq \dfrac{N_i + N_{i+1}}{2} \\ m_{i,i+1} & \text{if } P(\hat{x}_i) \leq m_{i,i+1} \\ P(\hat{x}_i) & \text{otherwise.} \end{cases} \tag{4.35}$$

Let $N''(x) \geq 0$ for $x \in [x_{i-1}, x_{i+1}]$. Then the interpolant given in Algorithm 4.4 is $O(h^4)$ and convex in the sense that

$$N_i - N_{i-1} \leq \tfrac{1}{2}(T_{i+1/2}(N) - N_i) \leq \tfrac{1}{2}(N_{i+1} - T_{i+1/2}(N)) \leq N_{i+2} - N_{i+1}. \tag{4.36}$$

In the case of uniform grid spacing, the definition of the convex projection $m_{i,i+1}$ is equivalent to saying that the interpolant must lie in the triangle T formed by the three lines passing through (x_{i-1}, N_{i-1}) and (x_i, N_i), (x_i, N_i) and (x_{i+1}, N_{i+1}), respectively (x_{i+1}, N_{i+1}) and (x_{i+2}, N_{i+2}) (see Figure 4.10).

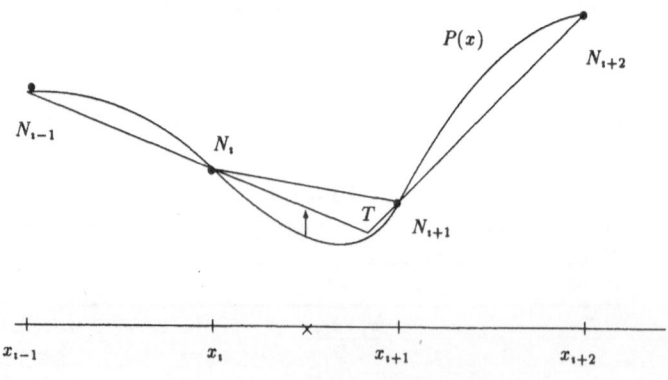

Fig. 4.10 Convex interpolation

In the Figure, P is the cubic interpolant which needs to be corrected at $\hat{x}_i$. A monotone two-dimensional multigrid interpolation can be constructed from any monotone one-dimensional interpolation. The starting point is the set of values of the function $N(x, y)$ at the grid-points of a coarse grid (see Figure 4.11). The solid lines are the coarse grid lines, while the dashed lines are fine grid lines which are no coarse grid lines. The values of $N(x, y)$ are given at the solid dots ($\bullet$), while the values to be interpolated are at the points given by the circles ($\circ$) and at the triangle ($\triangle$).

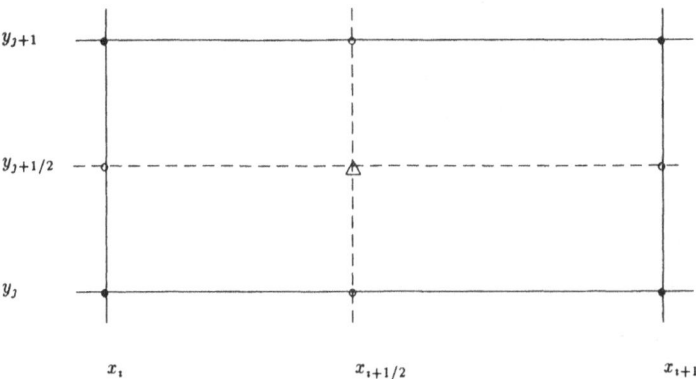

Fig. 4.11 Two-dimensional monotone interpolation

Algorithm 4.5. (*A monotone 2D interpolation*) *Let* $\partial N(x, y)/\partial x \geq 0$ *and* $\partial N(x, y)/\partial y \geq 0$ *on* $[x_i, x_{i+1}] \times [y_j, y_{j+1}]$. *The interpolant B will be calculated in several steps. It is assumed that a monotone univariate interpolation has been given.*

Step 1 Using values on the lines $x = x_i$ and $x = x_{i+1}$, respectively on the lines $y = y_j$ and $y = y_{j+1}$, monotone interpolations at the points $(x_i, \hat{y}_j)$, $(x_{i+1}, \hat{y}_j)$, respectively $(\hat{x}_i, y_j)$ and $(\hat{x}_i, y_{j+1})$ are calculated. These values are $B_0(x_i, \hat{y}_j)$, $B_0(x_{i+1}, \hat{y}_j)$, respectively $B_0(\hat{x}_i, y_j)$ and $B_0(\hat{x}_i, y_{j+1})$.

Step 2 If $B_0(x_{i+1}, \hat{y}_j) < B_0(x_i, \hat{y}_j)$, then
set $B(x_{i+1}, \hat{y}_j) = B_0(x_i, \hat{y}_j)$.
If not, then set $B(x_{i+1}, \hat{y}_j) = B_0(x_{i+1}, \hat{y}_j)$.
If $B_0(\hat{x}_i, y_{j+1}) < B_0(\hat{x}_i, y_j)$, then
set $B(\hat{x}_i, y_{j+1}) = B_0(\hat{x}_i, y_{j+1})$.
If not, then take $B(\hat{x}_i, y_{j+1}) = B_0(\hat{x}_i, y_{j+1})$. Leave the other values unchanged; i.e., $B(x_i, \hat{y}_j) = B_0(x_i, \hat{y}_j)$ and $B(\hat{x}_i, y_j) = B_0(\hat{x}_i, y_j)$.

Step 3 Use cubic interpolation to calculate a value $B_0(\hat{x}_i, \hat{y}_j)$ at $(\hat{x}_i, \hat{y}_j)$.

Step 4 Let
$$m_1 = \max\{B(x_i, \hat{y}_j), B(\hat{x}_i, y_j)\},$$
$$m_2 = \min\{B(\hat{x}_i, y_{j+1}), B(x_{i+1}, \hat{y}_j)\}.$$

Then

$$B(\hat{x}_i, \hat{y}_j) = \begin{cases} m_2 & \text{if } B_0(\hat{x}_i, \hat{y}_j) > m_2 \\ m_1 & \text{if } B_0(\hat{x}_i, \hat{y}_j) < m_1 \\ B_0(\hat{x}_i, \hat{y}_j) & \text{otherwise.} \end{cases}$$

Let $\partial N(x, y)/\partial x \geq 0$ *and* $\partial N(x, y)/\partial y \geq 0$ *in* $[x_i, x_{i+1}] \times [y_j, y_{j+1}]$. *If the univariant interpolant used above is monotone and* $O(h^4)$, *then the resulting interpolant B is monotone on the lines* $x = x_i, x = \hat{x}_i, x = x_{i+1}, y = y_j, y = \hat{y}_j$ *and* $y = y_{j+1}$. *Moreover it is an* $O(h^4)$ *approximation to* $N(x, y)$.

In the following, different interpolations are compared with regard to their approximation of functions and as the FMG interpolation of the algorithm for the model problem. The interpolations are:

1. The "linear" interpolation.
2. Cubic interpolation, which is called "standard cubic".
3. An "adaptive monotone" interpolation which works in the following way: If cubic interpolation is monotone, it is taken. If not, a quadratic interpolant is tried. If it is monotone, then it is taken. Otherwise, linear interpolation is performed.
4. "Fritsch-Carlson" monotone interpolation (Algorithm 4.1).
5. An algorithm due to Pollul [115], which we denote by "C_1-Pollul". This algorithm, which minimizes the L_2-norm of the second derivative of the spline interpolant, yields a convex C_1 spline interpolant by inserting additional nodes if necessary.
6. The "monotone cubic cut" interpolant of Algorithm 4.3.
7. A generalization of an algorithm of Irvine, Marin and Smith, [61], which yields a convex/concave interpolant when the data is convex/concave. The algorithm, which delivers a C_2-convex/concave cubic spline is due to Pollul [115], too. It is called the "C_2-Pollul" interpolation.
8. The "convex cubic cut" interpolation of Algorithm 4.4 based on the univariate cubic cut interpolation.
9. The "loglinear" interpolant, which is obtained by interpolating the logarithm of the data and then raising e to the result of this interpolation.
10. The "logcubic" interpolant, which is obtained as above, except that a cubic interpolant is used.

The original motivation for using monotone interpolations is to suppress the oscillations in the solution. It is also hoped that the quality of the approximation is improved by taking the additional properties of the function into account. The univariate test-function is a Gaussian distribution of the form $N(x) = c \cdot \exp(-(x - x_0)^2/2\sigma^2)$. This is a cross-section of a reasonable initial doping profile. Five different grids are used, G_3–G_7, from the coarsest to the finest one. The finest grid (G_7) has approximately ten grid points lying within a distance of one standard deviation from the maximum of N, while the coarsest grid (G_3) has one grid-point on each of the flanks of the curve. For cubic interpolation, the theoretical asymptotical rate of decrease of the error (which is 2^{-4}) was first attained when going from G_6 to G_7. For linear interpolation, the rate of decrease which is 2^{-2} asymptotically, is first obtained when going from G_5 to G_6. For a better representation and evaluation the concentration is normalized to have maximum value one. This is possible because most of the interpolations are quasi-linear in the sense that if T is the interpolant to the function N, then λT is the interpolant to the function λN for any $\lambda > 0$.

The conclusion to be made from the calculations with regard to the absolute error is that all of the monotone cubic interpolations (3–6), are the best and, of course, they suppress the oscillations. Near the profile, the standard cubic interpolation oscillates on grids G_3, G_4 and G_5. The linear interpolation is the worst except on the coarsest grids. All of the interpolations attain their maximum error at or near the maximum of the concentration. On finer grids, it is to be expected that the three $O(h^4)$ algorithms, standard cubic, cubic cut and convex cut become the best. Using convex interpolations does not pay. The logcubic is dramatically the best, because the logarithm of the function is a quadratic polynomial.

When looking at the relative error, things change quite a bit. The relative error is better suited to analyzing the behavior of the error in the transition area, where the concentration $N(x)$ changes from high values to values which are essentially zero, $0.001 \leq N \leq 0.05$ for a normalized concentration. The most striking difference between the error and the relative error is that the latter increases as $x \to \infty$ while the absolute error is largest at the maximum of the concentration. Figure 4.12 gives typical examples of such a behavior.

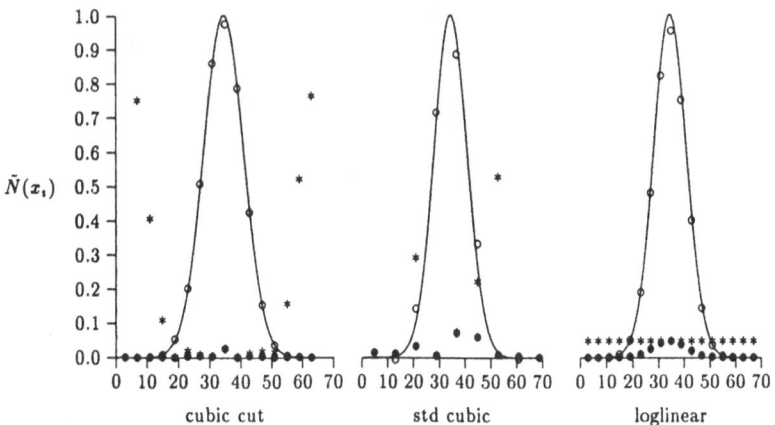

Fig. 4.12 Different interpolation methods, interpolated values and errors, $\circ \hat{=}$ values, $\bullet \hat{=}$ err, $* \hat{=}$ rel err

For most interpolations, the relative error tends to infinity. Nevertheless, some meaningful comparisons can be made by looking at the relative error in the transition area. There, the convex interpolants are the best, although the difference to monotone cubic interpolants is not large. Due to its oscillations on the coarser grids, the standard cubic interpolation is the worst of all.

The behavior of the relative errors of the logarithmic interpolants is remarkable. If ε is the machine round-off error, then the relative error is

$$\text{err}_{rel} = \frac{e^{P(x)} - N(x)}{N(x)} = e^{P(x) - \ln N(x) + \varepsilon} - 1,$$

where $P(x)$ is the interpolating polynomial. For small errors, this is approximately

$$P(x) - \ln N(x) + \varepsilon.$$

For moderate errors, this is dominated by the error of the interpolant. Thus, the loglinear interpolant is $O(h^2)$ and the logcubic interpolant is $O(h^4)$. If the interpolant is exact, then the round-off error dominates, and $\text{err}_{rel} = \varepsilon$, explains the constant relative error curve in Figure 4.12.

Since the underlying model problem is a two-dimensional one, the originally one-dimensional interpolations scan the points in the following way: first all fine-grid points on coarse-grid lines in the y-direction are interpolated. Then the remaining fine-grid points are interpolated in the x-direction. The result is a 2D-interpolation which is monotone on all lines in the x-direction and on coarse-grid lines in the y-direction but which is not necessarily monotone on fine-grid lines in the y-direction. The same holds true for the convex interpolations.

In addition, one version of the 2D-monotone interpolation of Algorithm 4.5 is tried. It uses the univariate monotone cubic cut interpolation. This 2D interpolation is called the "2D monotone cubic" interpolation.

The computational cost of the interpolations is measured during the experiments and given in per cent of the total CPU-time needed. The cost of linear interpolation is, of course, the lowest, being approximately 0.30 per cent of the total cost. All of the monotone cubic interpolations are more-or-less equally expensive, ranging from 3 to 4 times the cost of the linear interpolation. Of these, monotone cubic cut is the cheapest, being only ten per cent more expensive than the standard cubic interpolation on which it is based. The convex interpolations are the most expensive. The exact numbers are given in Table 4.6 for an 80 minute run. The 2D monotone cubic interpolation needs four times the CPU-time of the linear interpolation. Thus the overhead of a 2D interpolation compared to the use of the univariate interpolation on which it is based is only twenty per cent. This is quite remarkable in two regards. First, it is very difficult to obtain truly two-dimensional interpolations. To see this, one need only recall the additional difficulties occurred

Table 4.6. *CPU-time for different MG-interpolations*

Interpolation	% CPU time	Interpolation	% CPU time
linear	0.28	standard cubic	0.79
adapt. monotone	0.98	Fritsch.-Carl.	0.92
C_1-Pollul	1.00	cubic cut	0.88
C_2-Pollul	3.26	convex cut	1.40
2D-monotone	1.04	loglinear	0.92
logcubic	1.25		

in extending the Fritsch-Carlson algorithm to two dimensions [35]. The second point is that this implementation is possible in practice with such a small overhead.

It is surprising that the additional cost of the monotone cubic spline interpolations over the standard cubic interpolation is so small. The reason for this is that the standard cubic interpolation is monotone most of the time anyway. Thus the relatively complex changes required by some of the interpolations need not be carried out. The logarithmic interpolants are more expensive than their normal counterparts because the process of taking logarithms is relatively expensive. Nevertheless, the difference between the logcubic interpolant and the other cubic interpolants is negligible.

The question of computational cost is overshadowed by another factor. When using the standard cubic interpolation, it is not possible to start FMG on the coarsest grid because the algorithm then sometimes diverges. As a consequence, one FMG-cycle may not reach the desired accuracy on the finest grid. It is then necessary to make an additional V- or W-cycle to attain the prescribed precision (FMG + V, respectively FMG + W). When using monotone interpolations, on the other hand, one can always start on the coarsest grid and the prescribed accuracy is attained after one FMG-cycle. The savings obtained in this way range from thirty to fifty per cent (see Section 4.9). This is a dramatic reduction of computational cost when introducing monotone interpolation.

For example, in Table 4.7 the computational work for an 80 minute run is given in terms of relaxation units. The simulations are carried out using five

Table 4.7. *Computational cost in relaxation units*

V(2, 1)-cycle $G_H = G_1$	G_5		G_6	
	FMG	FMG + V	FGM	FMG + V
standard cubic	div	div	div	div
adapt. monotone	330	544	343	597
$G_H = G_3$	FMG	FMG + V	FMG	FMG + V
standard cubic	438	746	*div*	781
adapt. monotone	389	653	406	652

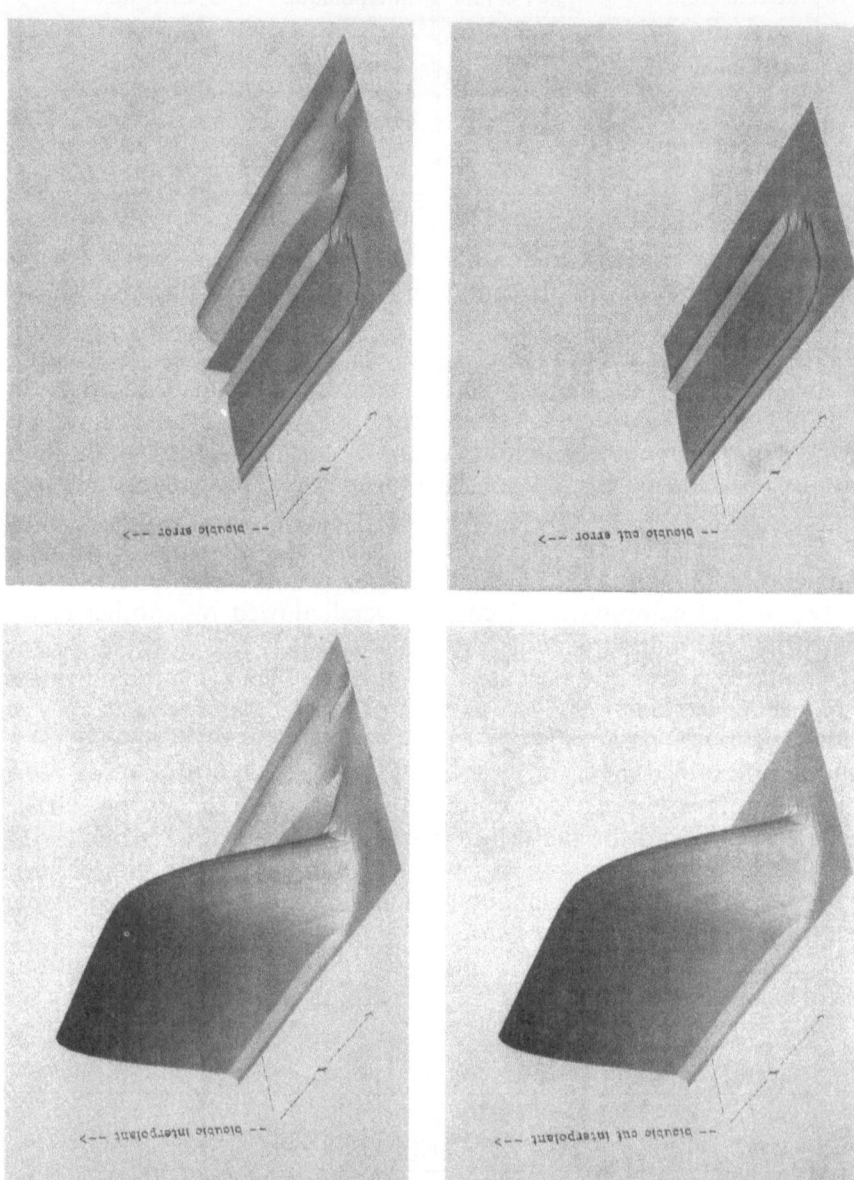

Fig. 4.13 Results after five time steps using cubic and cubic cut interpolation (function and error)

and six grids without local refinements. Note that the standard cubic interpolation sometimes diverges when starting from the coarsest grid (G_1).

The comparisons of the interpolations are made after one time-step, after five time-steps and after $T = 80$ minutes. The results show that monotone interpolations are definitely better than non-monotone interpolations. This difference is not noticeable after one time-step, but the error of the monotone cubic interpolations is only one third of that of the standard cubic interpolations after five time-steps. After 80 minutes, the errors of all of the interpolations are about the same because the dispersion has decreased the steepness of the concentration. All of the monotone cubic interpolations yield roughly the same results at all times even though their approximation order differs. The two convex interpolations yield the best results, although they are only slightly better than the monotone interpolations. The 2D monotone cubic interpolation ends with the same results as the other monotone cubic interpolations. After five time-steps, the maximum error of the standard cubic interpolation is attained at the boundary of the regions into which the doping agent had diffused and somewhat beyond. There the "standing waves" occur. The question is where the maximum error is achieved. It turns out that the maximum error is really attained on these "standing waves" and thus suppressing them improves the accuracy of the solution. After 80 minutes, the standard cubic interpolation has a maximum error of about five per cent. The region of highest error is where the standing waves have been. The monotone interpolations have their maximum errors at the maximum of the concentration.

Conclusion 4.5. *Positive, monotone, respectively convex fourth-order multigrid interpolants are very easy to calculate. Monotone interpolations result in a dramatic reduction of numerical work. They typically decrease the CPU time by a factor of 2 when using global grids. Not only are they more efficient, but they are more accurate in the small time range (0–10 minutes). Since the monotone cubic cut interpolation is slightly better than the other monotone cubic interpolations, since it is computationally cheaper to calculate and since it is a truly $O(h^4)$ interpolation, it can be recommended to use for the problem considered here. Linear interpolation is surprisingly good. This is due to the coarseness of the grids so that neither the asymptotic theory of FMG convergence nor the asymptotic behavior of interpolation errors do hold fully. As the grids become finer, it can be expected that the difference between the accuracy of linear and cubic interpolation will increase. When the question of accuracy is not of prime importance, linear interpolation is a viable option. Further details of the experiments and the proofs of the propositions concerning the properties of the Algorithms 4.2–4.5 are given in [67].*

4.7 The Restriction of Approximations and of Residuals

Within the FAS-scheme (2.74) it is necessary to restrict approximations to the solution $(\hat{I}_h^H)$. Because there is no indication to do it in another way, the simple injection is used to do this transfer from a fine grid to the next coarser one.

The proper choice of the residual transfer is important for the multigrid efficiency and robustness. Due to the chosen approach the boundary condition at the curved boundary is treated separately. The residuals of the corresponding equations are transferred by injection. For many applications the robust but expensive full weighting is the recommended restriction of residuals within the domain. In the present case, this operator is simply determined whenever the difference operator does not use boundary points. At the segments $\partial\Omega_1$, $\partial\Omega_2$ and $\partial\Omega_3$ where the respective boundary condition is used to eliminate the "ghost points", the standard modification of the operator comes to fruition (see Section 2.4). Near the curved boundary segment the standard approach is adapted due to the shape of the domain [6].

The general approach aims to conserve $\displaystyle\int_U r_h dV = \int_U I_h^H r_h dV$, where $U := \Omega \cap [x - h, x + h] \times [y - h, y + h]$ is an Ω-surrounding (shaded area in Figure 4.14) for a $z = (x, y) \in \mathcal{G}_h \cap \mathcal{G}_H \cap \Omega$ (marked by $\otimes$ in Figure 4.14). The idea is to have an operator which is represented by a stencil

$$I_h^H = \frac{1}{\chi}\begin{bmatrix} \xi_{-1,1} & \xi_{0,1} & \xi_{1,1} \\ \xi_{-1,0} & \xi_{0,0} & \xi_{1,0} \\ \xi_{-1,-1} & \xi_{0,-1} & \xi_{1,-1} \end{bmatrix} \quad \text{with} \quad \chi = \sum_{i,j=-1}^{1} \xi_{i,j}, \qquad (4.37)$$

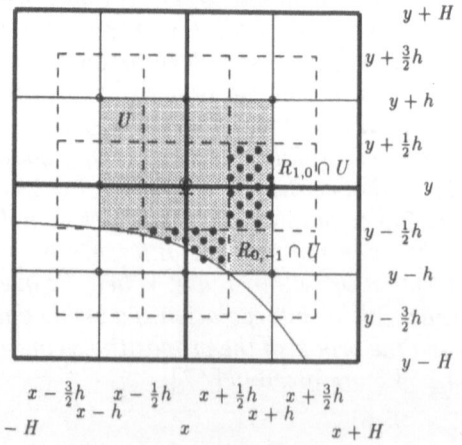

Fig. 4.14 Modified full weighting

and which coincides with the usual full weighting away from the boundary.
To determine

$$\xi_{i,j} = \begin{cases} \int\int_{R_{i,j} \cap U} dV & \text{if } (x+ih, y+jh) \in \bar{\Omega} \cap (\mathscr{G}_h - \partial\Omega_{4h}) \\ 0 & \text{otherwise} \end{cases}$$

the rectangles $R_{i,j} := [x+(i-\frac{1}{2})h, x+(i+\frac{1}{2})h] \times [y+(i-\frac{1}{2})h, y+(i+\frac{1}{2})h]$ are used. The computation of the above $\xi_{i,j}$ requires an integration over domains $R_{i,j} \cap U$ (examples are marked by the bullets in Figure 4.14). To avoid technical complications the integration over $R_{i,j} \cap U$ is approximated by the integration over $R_{i,j} \cap [x-h, x+h] \times [y-h, y+h]$. This simplifys the use of the modified full weighting (FW-mod).
FW-mod improves the algorithm considerably as comparisons with simple injection (INJ) or the combination of full weighting for inner points and injection for points near the boundary (FW-INJ) prove. This type of restriction provides convenient convergence rates (Table 4.8). Because this behavior is independent from the relaxation scheme used, the modified full weighting recommends itself to be the standard residual transfer for this model problem with curved boundary.

Example 4.4. *The geometry of Figure 4.14 where $(x-h, y-h)$ and $(x, y-h)$ belong to $\partial\Omega_{4h}$ leads to a representation of FW-mod as given below. Away from the curved boundary it coincides with the well-known full weighting.*

$$\text{FW-mod "near" the boundary: } I_h^H = \frac{1}{13} \begin{bmatrix} 1 & 2 & 1 \\ 2 & 4 & 2 \\ 0 & 0 & 1 \end{bmatrix}$$

The model problem analysis for linear equations recommends the half weighting of residuals for the red-black relaxation because the multigrid algorithm may diverge for the combination of chequerboarded Gauss-Seidel relaxation and residual injection, even for model problems. Therefore the "div." for PGS-ch with INJ in Table 4.8 is not a surprising result.

Table 4.8. *Convergence rates depending on the residual transfer*

cycle : $V(2, 1)$ time discr. : CN		G_5	
		$T = 80$	$\Delta t = 15$
scheme	INJ	FW-INJ	FW-mod
PGS-ch	div.	0.0670	0.0514
CGS-lex	0.2033	0.0748	0.0303
CGS-eo	0.4555	0.0533	0.0480

4.8 Technical Aspects of Time-Dependent Refinements

The doping material initially concentrates in small regions within the simulation area $\Omega(t)$ and diffuses into the domain. In those regions with then electrically important concentrations the accuracy requirements become more demanding than in "silent" parts of the domain where the concentrations are approximately constant or below values of physical relevance. The careful computation of the concentrations in those areas automatically leads to refined grids, which every serious approach should supply. This, in general, requires several disjoint refinement patches. In contrast to the more complex situation of Chapter 5, the model problem of this chapter is solved on only one connected refinement area. The only reason for this simplification is a technical one, which allows an easy implementation of refined grids within an algorithm which originates from a "global" multigrid approach. The advantage is, that there is no need for an optimizing algorithm to connect several small neighbouring patches to a larger one.

Remark 4.6. *This approach does not lead to a minimum number of points on the different grid levels. But the strictly use of uniform meshes will automatically provide efficient and clearly structured algorithms. For parallel algorithms the use of refinements which consist of rectangular subpatches seems to be a promising approach, because this may allow a good "load balancing" (that means, keeping processors busy).*

Starting with a refined grid for the time $t = t^n$ (upper part of Figure 4.15) the next time-step may allow a smaller refinement patch. There is no problem to determine the new right hand side of the actual problem due to (2.85). This holds for an initial approximation to the new solution on the smaller refinement, too. The reduction of both the computational domain and of the refinement area turns out to be of second importance, from the numerical point of view.

If new refinement areas become necessary due to the redistribution of impurities, a careful calculation of values on these new refinements is required. If the refinement has to cover a larger subdomain for the new time-step, there exist grid points which have no corresponding fine grid point in previous time-steps. Then there is no direct way for taking over the right hand side and an initial approximation for the discrete fine grid problem of the $n + 1$-th time-step (compare the dotted area in the lower part of Figure 4.15).

In principle, this problem to extend a grid function to new regions is solved in a standard way: the nested and regular grids, where fine grid boundaries coincide with coarse grid lines, make interpolation methods easy to use. It is natural to take advantage of the FMG-interpolation which is necessary, anyhow. Oscillating start approximations may lead to convergence problems

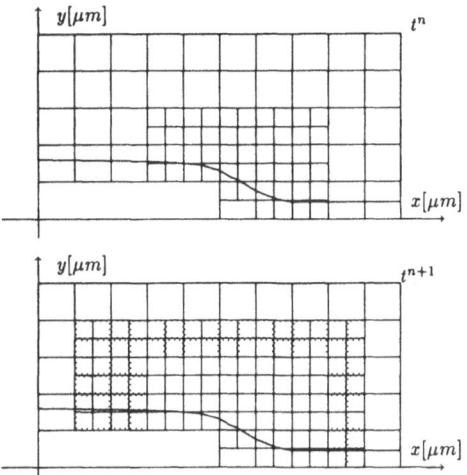

Fig. 4.15 Changing of refinements from time t^n to t^{n+1}

of the nonlinear relaxation method. The problem of oscillations is increased if right hand side values are interpolated. They influence the quality of the solution dramatically, because the oscillations of the right hand side are not smoothed. This again leads to the idea of monotone and shape preserving interpolation methods.

The decision where and when to adapt the grids for the next time-step has to be based on a sufficiently good approximation to the solution, respectively on a satisfactory estimation of the discretization error. Therefore it is recommended to determine the next grid refinement only after a certain amount of numerical work, which means after some cycles, because the invested work is necessary to obtain an appropriate estimation of the discretization error (see Section 4.9). This situation is naturally given when the solution for the actual time-step is accepted.

4.9 Numerical Results for the Model Problem

All results of the following sections refer to the grid constellation which was described at the beginning of this chapter. The standard time discretization scheme is CN. If BE is used, then it is explicitly mentioned. T is always equal to 80.

By theoretical results concerning linear model problems it is known that the convergence rate of the cycle used, should be less than 0.1 to guarantee convergence upto the level of truncation error within one FMG-cycle. Algorithms with less than three smoothing sweeps are not adequate because of their poor convergence properties ($\rho_{av} > 0.1$). The W-cycles do not show such an advantage that justifies this more expensive cycle type. Therefore the following results are presented only for V-cycles.

4.9.1 Experiments on Global Grids—Standard Approach

The expression "standard approach" points to the fact that the "standard" bicubic polynomial interpolation method serves as FMG-interpolation.
Using the coarsest grid $G_H = G_1$ to start the FMG iteration immediately runs into problems because the extreme variation of the grid functions results in oscillations and even in negative concentrations. One reasonable remedy to these ugly phenomena is the use of asymmetric interpolation formula. Additionally,. the FMG-algorithm has to start on a finer grid: $G_H = G_3$. These modifications together with additional cycles mostly avoid the bad oscillations, but they do not help in any case.
Knowing about the problems with FMG, a natural idea is to increase the number of cycles on each level during the FMG-iteration before inter- polation and performing additional cycles on the finest grid. Corresponding experiments show, that it is sufficient to cycle only once on each level between the coarsest and the finest grid and to do one additional cycle on the finest one. This means that the standard FMG-iteration followed by one addi- tional cycle (FMG + V) on the finest level is a good choice.
Table 4.9 shows characteristic values of experiments using FMG-, respec- tively MG-iteration. The given values of relaxation units contain the relax- ation effort as well for the Δt-step as for the two $\Delta t/2$-steps. The initial time- step size Δt^1 is chosen to meet the stability condition (4.19) of the explicit scheme on grid $G_H = G_3$. Varying the number of post-smoothing steps shows that the additional work of a V(2,2)-cycle will not pay out compared

Table 4.9. *Experiments on global grids, MGI/FMG, V(2,1)/V(2,2)*

FMG + V, $G_H = G_3$		G_5				G_6			
scheme	CN	N_t	Δt_{max}	ΣRU	$\dfrac{\Sigma RU}{N_t}$	N_t	Δt_{max}	ΣRU	$\dfrac{\Sigma RU}{N_t}$
CGS-eo	$V(2,1)$	28	14.4	829	29.6	28	17.5	781	27.9
PGS-ch	$V(2,2)$	26	14.9	997	38.4	26	22.4	1005	38.7
MGI	3 cycles		G_5				G_6		
scheme	CN	N_t	Δt_{max}	ΣRU	$\dfrac{\Sigma RU}{N_t}$	N_t	Δt_{max}	ΣRU	$\dfrac{\Sigma RU}{N_t}$
CGS-eo	$V(2,1)$	26	15.3	970	37.3	28	20.1	1079	38.6
PGS-ch	$V(2,2)$	26	15.0	1295	49.8	26	20.5	1295	49.8

to the cheaper V(2,1)-cycle. The relaxation scheme is not of main importance with respect to the overall behavior of the algorithm.

4.9.2 Experiments on Global Grids—Improved Algorithm

Still executing on global grids, it is intended to point out the improvement by the use of modified interpolation techniques. The results complete the basis for an estimation of the efficiency gain when refined grids are introduced. As already pointed out, interpolations which preserve special properties of the data improve the algorithm for several aspects: FMG may start on the coarsest grid $G_H = G_1$ and produces no oscillations. Table 4.10 shows the influence of the interpolation method on the algorithm and extends the information of Table 4.7. The first line again says that FMG using standard

Table 4.10. *Number of time-steps using different interpolation methods*

cycle : $V(2,1)$	G_5		G_6	
$G_H = G_1$	FMG	FMG + V	FMG	FMG + V
bicubic	div.	div.	div.	div.
adapt. bicub.	20	19	20	20
monot. Hermite	20	19	21	20
$G_H = G_3$	FMG	FMG + V	FMG	FMG + V
bicubic	27	26	div.	28
adapt. bicub	24	23	24	22

Table 4.11. *Experiments using different FMG-interpolations*

cycle : $V(2,1)$	G_5				G_6			
algorithm	N_t	Δt_{max}	ΣRU	$\dfrac{\Sigma RU}{N_t}$	N_t	Δt_{max}	ΣRU	$\dfrac{\Sigma RU}{N_t}$
interpolation	bicubic							
$G_H = G_3, FMG + V$	26	15.3	749	28.8	26	13.5	754	29.0
interpolation	adaptive bicubic							
$G_H = G_3, FMG + V$	23	13.2	653	28.4	22	14.5	652	29.7
$G_H = G_1, FMG + V$	19	17.0	544	28.7	20	18.4	597	29.9
$G_H = G_3, FMG$	24	11.7	389	16.3	24	21.5	406	17.0
$G_H = G_1, FMG$	20	15.1	330	16.5	20	16.7	343	17.2
interpolation	monot. Hermite							
$G_H = G_1, FMG + V$	19	17.2	545	28.6	20	18.4	597	29.8
$G_H = G_1, FMG$	20	15.1	330	16.4	21	22.0	359	17.0

bicubic interpolation diverges when starting on the coarsest grid. An additional cycle cannot avoid this. Even a finer start grid $G_H = G_3$ does not always guarantee convergence.

Conclusion 4.6. *The benefit of using monotone interpolation methods can be summarized by the following: FMG algorithms may start the iteration on the coarsest grid. In case of convergence the number of time-steps used is smaller than for algorithms using the bicubic interpolation. With monotone interpolation methods the number of time-steps is approximately constant for the same experiments and the extra cycle after one FMG-iteration (FMG + V instead of FMG) is not absolutely necessary with respect to time stepping.*

Algorithms using the more sophisticated interpolation methods are about twice as fast as the standard approach what proves both the influence and the superiority of these interpolation formula. On the other hand the numerical experiments don't exhibit a significant dependence on a special monotone interpolation.

The time stepping for FMG-methods using a finest global grid G_6 and the adaptive bicubic FMG-interpolation is presented by Figure 4.16. The different growth of Δt after about ten time-steps is obvious. The time-step size of the algorithms with $G_H = G_1$ grows more rapidly than that of algorithms which start on grid $G_H = G_3$ ($\circ$ and $\bullet$ are above $*$ and $\diamond$). An additional cycle (FMG + V) makes the Δt grow faster, although the acceleration is in no adequate proportion to the extra work ($\bullet$ lies above $\circ$ and $*$ above $\diamond$, but not as much as desired). For comparison, the dotted curve shows the time-step size development for the standard experiment with FMG in combination with BE ($G_H = G_1$) using the monotone Hermite-interpolation (compare the corresponding CN-development, marked by $\circ$).

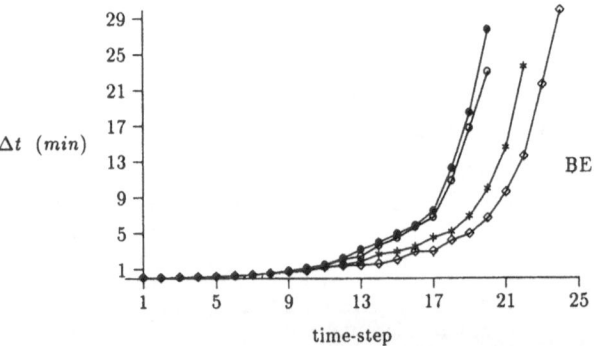

Fig. 4.16 Time steping on global grids, adaptive bicubic interpolation, $\circ \hat{=}$ FMG, $\bullet \hat{=}$ FMG + V, $G_H = G_1$, $\diamond \hat{=}$ FNG, $* \hat{=}$ FMG + V, $G_H = G_3$

4.9.3 Experiments Using Locally Refined Grids

The theoretical background of different refinement strategies has been discussed earlier. In principle, the refinement area is determined with the aid of

$$\tau^{sup} := \sup_{(x,y)\in G_H} \| \tau_H^h(x, y) \|_\infty. \tag{4.38}$$

The question is to find positions where the condition

$$\| \tau_H^h(x, y) \|_\infty \le Tol_{ref} * \tau^{sup} \tag{4.39}$$

with a certain Tol_{ref} is violated. If it is violated, the actual position should belong to the refinement patch. This very cheap analysis on G_H provides information which allows an immediate recalculation into $G_{h/2}$ information for the next grid adaptation. A somewhat crucial point is the proper choice of the Tol_{ref}. Discretizing with a second order formula and using standard coarsening causes the discretization error to decrease locally by a factor of 4 proceeding from the $2h$-grid to the h-grid. Therefore the choice of $Tol_{ref} = 4$ is a reasonable and theoretically motivated one.

The τ-values depend on time. Consequently the condition (4.39) does so, too. Keeping the τ-values fixed for several time-steps induces the use of elder and usually smaller refinement patches. Such a "freezing" may offer CPU-time advantages compared to the time accurate creation of patches. The higher the accuracy requirements are the more appropriate it is to evaluate the τ_H^h-information within each time-step and to determine the refinements corresponding to the actual state of the solution process.

The following experiments use a certain number of levels as global grids, denoted by "m_g". The number of refinement levels then is $\kappa = j - m_g$ where j denotes the level of the finest mesh used. The monotone Hermite-interpolation (Fritsch-Carlson) is applied everywhere within the algorithm where interpolation is required. The situation which is already observed in Section 4.9.2 for global grids is confirmed for refined grids: with the monotone interpolations the extra cycle on the finest level does not result in a real

Table 4.12. *Experiments on refined grids, MGI/FMG, PGS-ch*

cycle	: V(2, 1)	G_5				G_6			
method	$m_g = 3$	N_t	Δt_{max}	ΣRU	$\dfrac{\Sigma RU}{N_t}$	N_t	Δt_{max}	ΣRU	$\dfrac{\Sigma RU}{N_t}$
FMG+V	$G_H = G_1$	20	16.3	286	14.3	20	16.4	233	11.6
FMG	$G_H = G_1$	19	15.8	172	8.6	20	16.7	139	6.9
MGI	3 cycles	23	13.8	407	17.7	22	21.3	320	14.6
MGI	2 cycles	23	13.8	271	11.8	23	22.1	228	9.9
MGI	1 cycles	25	11.8	150	6.0	28	12.0	145	5.2

improvement compared to the standard FMG-iteration. With respect to the total amount of work and considering the numerical cost per time-step the standard FMG-cycle is recommended.

Experiments performing multigrid iterations with only one cycle per time-step require the lowest number of ΣRU. On the other hand these algorithms show a poor time stepping, resulting in a large number of time-steps. This comes from the weak approximation to the solution after only one cycle which is not yet good enough to provide small local discretization errors in time to allow a satisfactory time stepping. On the other hand, multigrid cycling with three cycles for each time-step does not show that improved behavior to justify the considerably increased numerical effort.

4.9.4 The Relative Local Discretization Error on Refined Grids

The numerical effort to solve the model problem is directly proportional to the size of the refinements. Therefore it is natural to ask whether the refinement control strategy may provide smaller refinement areas if an additional solution effort supplies with an improved approximation τ_H^h to the discretization error. This should produce refinement areas with less points per level and thereby reduce the total amount of numerical work. But if τ_H^h were a good approximation to the discretization error only after a large amount of additional iterative work just this additional effort might destroy the gain of the possibly smaller refinement area.

The development of τ_H^h as observed in Table 4.13 guides to the choice of an appropriate algorithm. There the relative deviation of τ_H^h from that reference value which is adopted in case of convergence, that means after approximately ten cycles for calculations on a grid sequence upto G_7 is given in per cent. The underlying investigations have been performed on seven levels of discretization. $\tau_H^h(x, y)$ has been evaluated on the two finest grids where it is defined, consequently on grids G_5 and G_6. The reference values are approximated upto a factor of 10^{-5} within four to five cycles. The τ_H^h-approximation

Table 4.13. *Relative deviations of τ_H^h-values, finest grid G_7*

finest grid: G_7 $m_g = 2$		on G_5		on G_6	
method		$\sup \tau_H^h$	$\|\tau_H^h\|_{L_2}$	$\sup \tau_H^h$	$\|\tau_H^h\|_{L_2}$
MGI	1 cycle	43.877	2.668	4.116	6.832
MGI	2 cycles	1.112	0.029	0.058	0.093
MGI	3 cycles	0.025	0.005	0.003	0.004
MGI	4 cycles	0.004	0.0	0.0	0.0
FMG	$G_H = G_1$	5.587	0.321	0.714	0.026
FMG + V	$G_H = G_1$	0.224	0.023	0.075	0.002

of τ_h after one standard FMG-cycle is not yet satisfying although it is considerably better than for MGI with only one cycle.

Conclusion 4.7. *From the refinement control point of view it is sufficient to use either standard FMG or MGI with two cycles, because in spite of the additional iterative work a significant reduction of the refinement area is not observed. The same recommendation has been given from the time stepping and defect reduction point of view. MGI with two cycles within every time-step provides excellent values for the refinement control, but this is not immediately converted into numerical advantages. So the standard FMG may be recommended even for this nonlinear problem.*
The improved interpolation allows FMG to start on the coarse grid $G_H = G_1$ and the additional cycle on the respective finest grid is no longer needed. This already reduces the numerical effort compared to the standard approach by a factor of two. Performing the same experiments using refined grids again yields approximately the same enhancement factor compared to the already improved FMG-algorithm (see Table 4.14). The overall acceleration factors for FMG range between four and five.
The experiments point to the fact that refined grids will lead to speed-up factors which are larger than ten compared to algorithms exclusively using global grids. This statement is valid for this model application although the refinement control allows a degeneration of refinement patches to global grids and although no point minimizing technique for the simply structured refinements is applied. Refinement strategies which are designed to minimize the number of points will automatically come to further improvements (see Chapter 5).

Table 4.14. *Comparison of multigrid methods for global and refined grids*

cycle $V(2, 1)$		G_5				G_6			
method	N_t	Δt_{max}	ΣRU	$\dfrac{\Sigma RU}{N_t}$	N_t	Δt_{max}	ΣRU	$\dfrac{\Sigma RU}{N_t}$
FMG+V $G_H = G_3$	bicubic							
no refinement	26	14.1	746	28.7	28	17.5	781	27.9
FMG $G_H = G_1$	monotone Hermite							
$m_g = 2$	20	15.8	159	7.9	22	21.0	162	7.3
$m_g = 3$	19	15.8	172	8.6	20	16.7	139	6.9
no refinement	20	15.1	330	16.4	21	22.0	359	17.0
MGI 2 cycles								
$m_g = 2$	21	13.4	234	11.1	21	16.9	195	9.3
$m_g = 3$	23	13.8	271	11.8	23	22.1	228	9.9
no refinement	26	14.0	656	25.3	28	14.8	726	26.0

4.9.5 CPU-Time Requirements

CPU-time measurements point to those parts of the algorithm which are the most time consuming ones and which are candidates to be optimized. The FAS-cycle (program FASFIX) participates with 95%. This is an expected magnitude because this routine is the core of the multigrid solver. Additionally, it shows that all the other tasks like grid generation, creating the right hand side due to the implicit time discretization and the defect calculation for iteration control purposes are of less importance.

The defect calculations of the main program (3.9%) determine termination criteria for iterations and statistical information. The effort to generate the time-dependent grids and other tasks which are due to the time-dependent problem (computing the new right hand side, calculating the information for the time stepping, storing information for restart) is remarkably small. Within the FAS-cycle the different multigrid components contribute to the total CPU-time in an expected range.

The time for the relaxation routines contains the update of coefficients immediately after computing a new value. Reasonable alternatives with the aim to reduce CPU-time can be the "freezing" of coefficients and updating them only after complete cycles respectively before or after grid switching or just keeping them fixed for one relaxation sweep. The Newton-method then degenerates to a Picard-like iteration. On the other hand, such a freezing may damage convergence properties and increases the storage requirements.

Table 4.15. *CPU-time requirements of different components, MGI*

G_5 $V(2,1)$ CN PGS-lex			
main program	100%	FASFIX	100%
grid generation	0.8	defect calculation	10.1
time stepping	0.5	relaxation (total)	88.2
defect calculation	3.9	thereof boundary relax.	(10.2)
FASFIX	94.9	restriction	1.5
		prolongation (corrections)	0.2

Procedures for Adaptive Multigrid Simulation of Evolution Processes 5

This chapter presents procedures for the adaptive multigrid simulation of evolution processes. In order to simulate evolution processes within processing sequences including intermediate auxiliary processes, the procedures are designed for

- the construction of the adaptive discrete approximation of initial states at an arbitrary stage of the processing sequence according to the results of previous processing steps, given in compact or discrete formulations,
- the simulation of evolution problems, which implies an adaptive solution of the discrete incremental elliptic problems at each temporal grid node and an adaptive time-step size selection.

The algorithmic procedures are primarily introduced with an emphasis on local principles which make the multilevel adaptive and multigrid solution approaches advantageous over classical simulation techniques. They can be also used as guidance for the practical programming of other adaptive multilevel strategies for process simulation tools.

The procedures represent a practical implementation of the adaptive multilevel grid selection strategies, described in Chapter 3, with the principal goal to control the level of discretization errors using problem-independent error control parameters. This adaptivity goal is extended to the multigrid solution procedure to supply it with an iteration stopping criterion based on the extrapolation of the discretization error.

The procedures concerning the spatial local grid refinement are based on strictly regular grids. The adaptive grid structures are defined in terms of simple and manageable data structures which naturally fit into the solution procedure without impeding its efficiency. The multigrid solver and the components it consists of are not discussed in detail here (see Chapters 2 and 4). They are assumed to be tailored for the particular process simulation problem. The basic algorithms are easily extended in order to meet the particular needs of process simulation problems. A grid-decoupling strategy is proposed to reduce the numerical effort which has to be invested whenever only one grid structure is used for all solution components of the multiparticle evolution problems.

An application of the proposed algorithmic procedures is illustrated by examples of two-dimensional ion implantation and diffusion processes. Finally, a general processing sequence consisting of critical ion implantation and diffusion processes for the fabrication of BiCMOS integrated circuits is presented. All calculations have been done with MUSIC [103, 104], a multigrid simulator for integrated circuits processes.

5.1 Practical Algorithmic and Programming Techniques for Multilevel Local Grid Refinement

The adaptive grid construction is simplified by using only uniform grids, both global and local (on subsequent discretization levels), enabling the formulation of efficient techniques for the local grid refinement. This section presents such a grid refinement procedure. It is based on the concept of elementary refinement patches and uses a simple data structure for the description of local uniform grids which allow an efficient handling of the refinement.

5.1.1 Elementary Refinement Patches and a Basic Two-Level Local Grid Refinement Procedure

Suppose that a currently finest local grid G_l is given and that an elementary discretization cell $\Omega_l^{\vec{p}}$ surrounding an interior grid node $\vec{p} = (x_p, y_p) \in G_l$ has to be refined. The typical refinement strategy in singlegrid adaptive approaches is to introduce a new partition of the elementary discretization cell, which also modifies neighboring discretization cells. For instance, in case of nonuniform finite difference grids, the local modification of the grid partition implies the modification of all discretization cells lying along grid lines introduced for the local grid refinement. According to the multilevel adaptive concept, the refining of a single elementary discretization cell $\Omega_l^{\vec{p}}$ can be performed by simply introducing a grid patch

$$E_{l+1}(\vec{p}) = \{\vec{q} \mid \vec{q} = (x_p + \alpha h_{l+1}, y_p + \beta h_{l+1}); |\alpha| \leq k, |\beta| \leq k; \alpha, \beta \in \mathbb{Z}\}$$

on the $l+1$-st level. In the sequel this is referred to as the elementary refinement patch, which is, for the given grid node $\vec{p}$, defined by two parameters: the discretization level $l+1$ and the patch width $k \geq 1$.

For example, in Figure 5.1 a simple two-grid adaptive structure is shown. The coarse grid G_l covers a global square domain Ω and therefore contains only global boundary grid nodes. Suppose that the elementary discretization cell $\Omega_l^{\vec{p}}$ surrounding the grid node $\vec{p}$ requires a better grid resolution. The

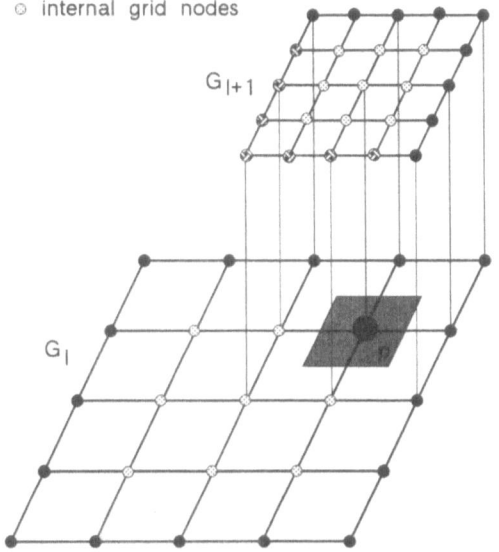

Fig. 5.1 Elementary refinement patch

local grid G_{l+1} on the subsequent level $l+1$ is then represented by an elementary refinement patch $E_{l+1}(\vec{p})$ with width parameter $k=2$. The discrete boundary of the local grid G_{l+1} includes both global boundary grid nodes and internal boundary grid nodes.

In addition to the goal to improve the grid resolution of the elementary discretization cell $\Omega_l^{\vec{p}}$, an elementary refinement patch should also provide a safe overlap with the area $\Omega\backslash\Omega_l^{\vec{p}}$. This overlap is necessary to obtain internal boundary conditions from those coarse grid nodes whose elementary discretization cells do not require a better grid resolution. However, the size of this overlap area should not introduce unnecessarily refined regions. In that sense, the elementary refinement patch with the width parameter $k=2$ turns out to be a reasonable choice. If an elementary refinement patch has to refine elementary discretization cells corresponding to global discrete boundary nodes, its basic structure is modified to include grid nodes lying inside Ω and only those grid nodes lying outside Ω which participate in the discrete approximation of boundary conditions.

The complete local grid G_{l+1} is obtained by the union of all elementary refinement patches introduced over all G_l grid nodes, whose elementary discretization cells require a better grid resolution. The basic two-level grid refinement procedure is described in Algorithm 5.1.

Algorithm 5.1. (Construct a new level)
Input parameters: $(\varepsilon_l, u_l)[G_l]$, Rel, Abs
Output parameters: G_{l+1}
The discrete variables ε_l and u_l are the discrete error estimator and the solution defined on the uniform grid G_l, corresponding to the currently finest level. These grid functions, along with the relative and absolute error parameters Rel and Abs, are required for the local refinement criterion. This algorithm produces a local uniform grid G_{l+1}: the local refinement of G_l.

Step 1. [Initialize] Set $G_{l+1} := \varnothing$
Step 2. Repeat Step 3 for all $\vec{p} \in G_l$
Step 3.   If $(\varepsilon_l > Rel \cdot u_l + Abs)$
                Set $G_{l+1} := G_{l+1} \cup E_{l+1}$
                [End of If structure]
          [End of Step 2 loop]
Step 4. Exit.

Step 1 of the Algorithm 5.1 implies that the grid refinement is the complete construction of a grid structure on the next finer level rather than the reconstruction of the already existing grid by adding new or deleting old grid nodes. There are two reasons for this. First, with reliable and accurate refinement criteria based on the extrapolated local or global error estimators, the coupled refinement-solution procedure is of approximately a *priori* character. This means that the local grid reconstruction is very rarely needed. The second reason is that even if the local grid reconstruction is inevitable, the computation of a completely new grid structure using a static data structure (see Section 5.1.2) is more efficient than a reconstruction of the existing grids using dynamic data structures. In spite of the fact that local grids are always considered as newly constructed ones, the interpolation of grid functions from the old grid structure is needed only in those new grid structure parts which do not overlap the old ones.

Revisiting the qualitative comparison of the multilevel and singlegrid adaptive concepts, it is important to note that, at first sight, the number of grid points introduced for the local refinement of a single coarse grid elementary discretization cell (Figure 5.1) seems to be larger than the corresponding number in case of some singlegrid adaptive concepts. This fact obviously holds for anisotropic problems which require local grid refinement in specific directions. However, anisotropic problems can be usually treated as problems of lower dimension or cut out by symmetry lines from the purely two-dimensional parts of the problem. Moreover, even if a single elementary refinement patch introduces grid node excess in the isotropic problem, it is obvious that this excess lies only along the internal grid bound aries. They are again of lower dimension and therefore the grid node excess is negligible compared with the total number of grid nodes. Strictly uniform local grids in the multilevel adaptive concept provide an environment for higher approximation orders, thus leading to a faster decrease of the discretization error in the refinement process than the approaches based on irregular adaptive grids. Consequently, the adaptive multigrid concept may require less refinement steps to obtain an error comparable to those obtained by nonuniform singlegrid approaches.

5.1.2 Efficient Data Structures for Spatial Local Refinement

The efficiency of the adaptive simulation strongly depends on the data structures for the description and the manipulation of the multilevel adaptive grid structures. Generally, the data structures should provide an easy and inexpensive way to modify grid structures and mechanisms to handle various discrete operations on the grid structure.
In many singlegrid adaptive approaches a grid node (or an associated elementary discretization cell) is typically used as a basic data structure element [116]. It is characterized by the absolute position in space and additionally supplied with a set of pointer variables to describe its relative position to the neighboring grid nodes. Such data structures vary from very simple to handle, like those for finite difference grids given in matrix and vector forms, to complex data structures, like those for finite element grids whose representation is characterized by an extensive bookkeeping of information. In order to restrict the interpolation of discrete functions only to those grid parts which are modified by a refinement step, the data structures are typically designed as dynamic. This additionally increases their complexity.
The multilevel adaptive concept based on the use of uniform global and local grids leads to significant simplifications in the data structure design. It is possible to choose among various particular approaches which exploit the regularity of the elementary refinement patches. Some efforts are also made towards unification and standardization of the multilevel grid structure programming by introducing economic data structures and fully portable, problem-independent routines and macro statements for the programming of general grid operations. Two examples are the collections of subroutines GRIDPACK [13] and BASIS [52]. The former is oriented towards typical finite difference grids, while the latter one is specialized to finite element approaches.
In the following some basic ideas for the description and manipulation of local uniform grid structures constructed from elementary refinement patches are demonstrated. Instead of the grid node, the basic constitutive element to describe a uniform grid structure here is a *string* of grid nodes which on level l is defined by

$$s_l^{\alpha, \beta}(n_p) = \{(x, y) | x = \alpha \cdot h_l, y = j \cdot h_l; \beta \leq j \leq \beta + n_p - 1; \alpha, \beta \in \mathbb{Z}\}$$

which represents a lexicographically ordered set of consecutive grid nodes on the same grid line α. β and n_p denote the position of the first string node in the lattice and the string length, respectively. The strings can also be defined in the x-direction. A set of strings along the same lattice line α makes a column

$$c_l^\alpha(n_s) = \{s_l^{\alpha,\beta_j}(n_{pj})| 1 \leq j \leq n_s\} \tag{5.1}$$

where n_s is the number of strings within a column which satisfy the condition $\beta_j > \beta_{j-1} + n_{j-1}$ if $n_s \geq 2$. The column definition also includes the trivial case $c_l^\alpha \equiv \varnothing$ if $n_s = 0$.

The local uniform subgrid on level l can be defined as a set of columns:

$$G_l = \{c_l^\alpha(n_{s\alpha})|\alpha_l^1 \leq \alpha \leq \alpha_l^{N_l}\}$$

where $N_l = 2^{l-1} \cdot (N_1 - 1) + 1$ is the total number of columns on the l-th level. α_l^1 and $\alpha_l^{N_l}$ denote the absolute position of the first and the last grid line in the lattice.

The principal advantage of using a string-based data structure compared to the point-based data structures is the significant reduction of information required for the local grid structure. The collection of grid nodes which constitutes a string is defined by 3 integers, while the relative position of grid nodes inside node strings is defined by their lexicographical ordering. The amount of logical information describing a uniform grid is proportional to the number of strings which is usually much smaller than that describing the total number of grid nodes. The string-based data structure easily allows local parts of uniform grids to be made accessible to various local grid operations.

The string-based data structures simplify the procedure of elementary refinement patch couplings (see Algorithm 5.1). It can be also performed on the string level. The basic grid operation for that purpose is the coupling of two strings $s_l^{\alpha_1,\beta_1}(n_{p_1})$ and $s_l^{\alpha_2,\beta_2}(n_{p_2})$ with $\alpha_1 - \alpha_2$ an even number on level l. The string coupling operation for this situation is defined by

$$s_l^{\alpha_1,\beta_1}(n_{p_1}) \cup s_l^{\alpha_2,\beta_2}(n_{p_2})$$

$$= \begin{cases} s_l^{(\alpha_1+\alpha_2)/2,\min(\beta_1,\beta_2)}(n_{pM}) & \text{if } n_{pM} \leq n_{p_1} + n_{p_2} \\ s_l^{(\alpha_1+\alpha_2)/2,\beta_1}(n_{p_1}); s_l^{(\alpha_1+\alpha_2)/2,\beta_2}(n_{p_2}) & \text{if } n_{pM} > n_{p_1} + n_{p_2} \end{cases}$$

where $n_{pM} := \max(\beta_1 + n_{p_1}, \beta_2 + n_{p_2}) - \min(\beta_1, \beta_2)$. The result are one, respectively two strings within the column placed in the middle of the corresponding columns of the initial strings. Similarly, a column coupling operation

$$c_l^{\alpha_1}(n_{s_1}) \cup c_l^{\alpha_2}(n_{s_2}) = c_l^{(\alpha_1+\alpha_2)/2}(n_{s_{12}}) \tag{5.2}$$

can be composed of the sequential applications of the string coupling operations to the strings belonging to these two columns.

Figure 5.2 schematically shows an example of the local refinement procedure with the string-based data structure. The first step is to introduce the refined

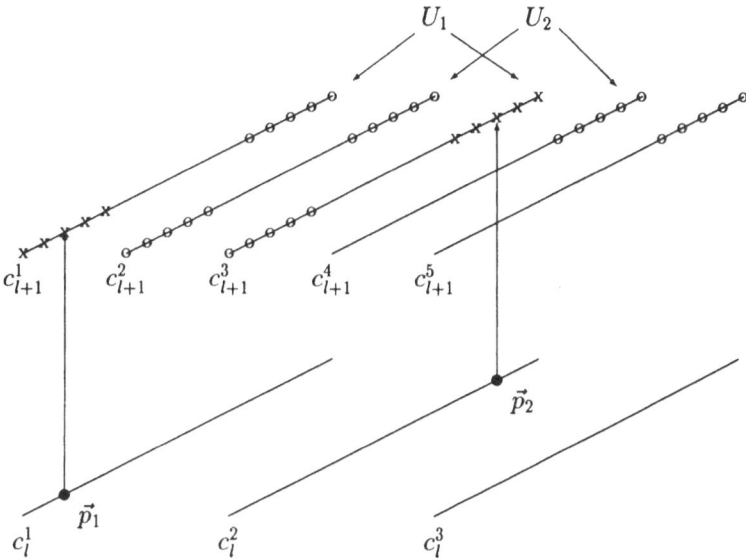

Fig. 5.2 Local refinement with a string-based data structure

string structures over all coarse grid nodes whose elementary discretization cells require better grid resolution (grid nodes p_1 and p_2 in Figure 5.2). This step also implies the coupling of the overlapping strings inside the fine columns which corresponds to the coarse one. This step can be performed independently for each coarse grid column. The length of the refined string introduced for a single coarse grid node is determined by the width of the elementary refinement patch (which in Figure 5.2 is $k = 2$). The resulting intermediate "one-dimensional" refined grid structure in Figure 5.2 consists of the fine grid columns c_{l+1}^1, c_{l+1}^3 and c_{l+1}^5 obtained by locally refining the coarse grid columns c_l^1, c_l^2 and c_l^3.

The final two-dimensional refined local grid is obtained after the first k column couplings in a sequence $(U_1, U_2, U_1, U_2 \cdots)$ where the column coupling operators are defined by

$$U_1 : c_{l+1}^{2i} = c_{l+1}^{2i-1} \cup c_{l+1}^{2i+1}, \quad i = \alpha_l^1, \dots, \alpha_l^{N_l - 1}$$
$$U_2 : c_{l+1}^{2i-1} = c_{l+1}^{2i-2} \cup c_{l+1}^{2i}, \quad i = \alpha_l^1, \dots, \alpha_l^{N_l}$$

The column coupling operation at the first and the last grid columns ($i = \alpha_l^1$ and $i = \alpha_l^{N_l}$) assumes a symmetric column structure. The obvious advantage of the string-based data structure is the possibility to split the global coupling steps U_1 and U_2 into independent two column coupling operations which can be performed in parallel.

5.2 The Adaptive Construction of an Initial State Discrete Approximation

An adaptive transient simulation of the discrete evolution problem (3.8) is characterized by both the evolution of the discrete solution $(u_l^n)_{l=1,M^n}$ and its adaptive multilevel space grid structure $G^n \equiv (G_l^n)_{l=1,M^n}$ for $0 \le n \le N_t$. Generally, this implies the existence of a multilevel space-grid structure G^0 for the adaptive discrete approximation of an initial state distribution. The principal role of G^0 is to supply the discrete operators of the first time-step with an appropriate discrete approximation as it is analogously done for the transition from G^{n-1} to G^n during transient simulation.

This section gives the basic procedures to construct spatial multilevel adaptive grid structures for the discrete approximation of compact initial state distributions. Such distributions can be both results of auxiliary processes and results of previously performed processing steps.

5.2.1 A Procedure for the Adaptive Discrete Approximation of a Compact Initial State Distribution

The initial state distribution for the evolution problem (3.1) may be given in compact form as a result of auxiliary processes either in terms of

1. an analytical expression, or
2. as a sum of a huge number of analytical expressions, or
3. as an interpolation rule for a certain grid structure.

Examples for all three descriptions are found in the two-dimensional compact modeling of ion implantation processes. The first compact form is typically used to model ion implantation profile selectively introduced into the semiconductor substrate through the planar surface with vertical mask edges. The resulting impurity profiles are commonly described by compact two-dimensional analytical expressions based on a single probability distribution function in the vertical direction, modified by an error function in the lateral direction [19]. The second type is typically given by the modeling of ion implantation processes through nonplanar mask structures. In that case the impurity distributions are commonly defined by a large sum of compact point-response functions with vertical and lateral distribution functions for a single ion [44]. The third type is inevitable if the ion implantation processes are simulated using more complex numerical models (e.g. Monte-Carlo method [59]). In order to attach these externally produced profiles as an initial state for the following simulation it is necessary to define an appropriate interpolation formula between possibly incompatible grid structures.

Even for simple and cheaply evaluated compact distributions of the first type, the natural question arises whether it is necessary to construct an initial state discrete approximation. In particular, if this compact initial state distribution is sufficiently smooth, the discrete approximation can be avoided by evaluating the analytical expression directly whenever its value is required at a certain grid node. Initial state distributions which are characterized by steep profiles (this typically appears in process simulation problems) could be wrongly represented on the very coarse discretization levels, especially when the accuracy of the local dose conservation is considered. An obvious remedy is to apply the averaging projection $F^h u^0$ instead of $(u^0)_h$ at the level with meshsize h. In general, however, the averaging projection operator (3.10) is not possible to be evaluated in the closed form. With an adaptive multi-level discrete approximation of the compact initial state distribution, the continuous averaging projection can be easily replaced by a discrete averaging procedure. Such a procedure applies the corresponding relative restriction operator F_h^{2h} sequentially, starting from the finest grid towards the coarsest one. This is the main reason which makes the construction of an adaptive multilevel grid structure for the compact initial state distribution attractive, even if it is given in a simple analytical form.

The construction of an adaptive discrete approximation for the second type of an initial state in compact form is further motivated by the time consuming summation of a huge number of analytical expressions whenever it is needed at a certain grid node. The construction of the multilevel discrete approximation for such complex models of auxiliary processes may be considered as a part of their own simulation procedures, and also represents the most convenient way to use them for further processing, especially to use them as an initial state for discrete evolution problems.

A procedure to construct an adaptive multilevel grid structure G^0 as a discrete approximation of the compact initial state distribution u^0 is given in the Algorithm 5.2.

Algorithm 5.2. (Adaptive discrete approximation of a compact initial state distribution)

Input parameters: $u^0, (G_l^0)_{l=1,2}, Rel, Abs$

Output parameters: $(u_l^0[G_l^0])_{l=1,M^0}$

u^0 *is the compact initial state distribution.* $(G_l^0)_{l=1,2}$ *are the first two global levels of* G^0, *as a core for its further construction which is controlled by the relative and absolute error parameters Rel and Abs. This algorithm produces the complete adaptive multilevel discrete approximation* $(u_l^0[G_l^0])_{l=1,M^0}$.

Step 1. [*Initialize*]
   (a) *Set* $M := 1$
   (b) *Repeat for all* $\vec{p} \in G_M^0$
               Set $u_M^0 := (u^0)_M$
   [*End of loop*]

Step 2. Repeat Steps 3 and 4 while $G^0_{M+1} \neq \emptyset$

Step 3.   *[Put the compact distribution on the currently finest level]*
          (a) Set $M := M + 1$
          (b) Repeat for all $\vec{p} \in G^0_M \cap G^0_{M-1}$
               Set $u^0_M := u^0_{M-1}$
          [End of loop]
          (c) Repeat for all $\vec{p} \in G^0_M \setminus G^0_{M-1}$
               Set $u^0_M := (u^0)_M$
          [End of loop]

Step 4.   *[Local grid refinement]*
          (a) **Evaluate generalized local error estimator**
          $(u^0_{M-1}[G^0_{M-1}], u^0_M[G^0_M]; \varepsilon_M[G^0_M])$
          (b) **Construct a new level** $((\varepsilon_M, u^0_M)[G^0_M], Rel, Abs; G^0_{M+1})$
     [End of Step 2 loop]

Step 5. Exit.

The construction of the multilevel adaptive grid structure is a sequentially performed creation of local grids, beginning with the two global grids which serve as a core grid structure for the initialization of the extrapolation procedure to evaluate the local error estimator. In order to avoid the introduction of an approximation error which might be different from the discretization error of the discrete approximations the initial state approximation should be generated with a criterion compatible to those used in the further simulation. As demonstrated in Section 3.2.2, the most convenient choice for an error estimator is in this case the generalized local error (3.50) because it is fully compatible to the global error estimator obtained by the averaging projection operator. In order to allow the extrapolation of the generalized local error by the Richardson extrapolation technique the initial state compact distribution u^0 is projected into grids by injection. For all levels greater than one, the evaluation of the compact distribution is performed only in those grid nodes which do not coincide with nodes on coarser grids.

The procedure to evaluate the generalized local error estimator is described in Algorithm 5.3 [106].

Algorithm 5.3. (Evaluate generalized local error estimator)
Input parameters: $u^0_{l-1}[G^0_{l-1}], u^0_l[G^0_l]$
Output parameters: $\varepsilon_l[G^0_l]$

$u^0_{l-1}[G^0_{l-1}], u^0_l[G^0_l]$ *are values of the compact distribution u^0 injected onto two subsequent levels. This algorithm produces the error estimator $\varepsilon_l[G^0_l]$, based on the generalized local error (3.50) evaluated using the Richardson extrapolation.*

Step 1. Repeat for all $\vec{p} \in G_{l-1} \cap G_l$

$$Set\ (\varepsilon_l)_{l-1} := \frac{1}{2^p - 1} |u^0_{l-1} - F^{l-1}_l u^0_l|$$

$$\text{Set } \varepsilon_l := (\varepsilon_l)_{l-1}$$
[*End of loop*]
Step 2. Repeat for all $\vec{p} \in G_l \backslash G_{l-1}$
$$\text{Set } \varepsilon_l := I^l_{l-1}(\varepsilon_l)_{l-1}$$
[*End of loop*]
Step 3. *Exit.*

The extrapolated generalized local error is directly evaluated in nodes of the l-th discretization level which also belong to the $(l-1)$-st discretization level. It is therefore denoted by $(\varepsilon_l)_{l-1}$. In order to represent the error estimate on the whole discretization level l where the most recent information for the next grid adaptation is available, it is interpolated from level $(l-1)$ grid nodes.

5.2.2 A Procedure for Adaptive Superposition of Discrete and Compact Initial State Distributions

When the transient simulation of the evolution process is interrupted for some reasons (for example, the observation of intermediate results) and restarted again without auxiliary processes, an initial state for the further evolution processing is trivially given on the last grid structure used before the interrupt. In general, any initial multilevel space grid structure should be properly based on the complete previously performed processing, including the simulation results of both the evolution processes and auxiliary processes. This superposition, resulting in an initial state approximation for the following evolution process, has to be performed adaptively.
Suppose that an intermediate multilevel space grid $G^a \equiv (G^a_l)_{l=1,M^a}$ structure is generated to approximate the compact distribution u^a, while $G^{old} \equiv (G^{old}_l)_{l=1,M^{old}}$ is the space grid structure for the same physical quantity, but resulting from the preceding processing step. Simply combining

$$G^0_l \leftarrow G^{old}_l \cup G^a_l$$

for all existing levels is not the appropriate choice. G^0_l has to reflect the error in the discrete approximation of $u^{old}_l + (u^a)_l$ instead of the sum of the individual errors of the discrete approximations of u^{old}_l and $(u^a)_l$. In order to assure the compatibility of the discrete and injected compact distribution, u^{old}_l should be, prior to the adaptive superposition procedure, sequentially injected from the finest towards the coarsest level. This is especially required for the discrete solutions of the previous evolution problems which are typically characterized by the coarse discrete solution corresponding to the averaged values of the finer ones.
A procedure how to construct an adaptive multilevel space-grid structure by an adaptive superposition of a compact and a discrete initial state distribution is given in Algorithm 5.4.

Algorithm 5.4. (Adaptive superposition of compact and discrete initial state distributions)

Input parameters: u^a, $(u_l^{old}[G_l^{old}])_{l=1,M^{old}}$, $(G_l^0)_{l=1,2}$, Rel, Abs

Output parameters: $(u_l^0[G_l^0])_{l=1,M^0}$

u^a *is the compact distribution.* $(u_l^{old}[G_l^{old}])_{l=1,M^{old}}$ *is the discrete distribution of the same quantity resulting from previous processing.* $(G_l^0)_{l=1,2}$ *are two global discretization levels as a core for the further construction of the initial multilevel space-grid structure, controlled by the relative and absolute error parameters Rel and Abs. This algorithm produces the complete initial multi-level discrete approximation* $(u_l^0[G_l^0])_{l=1,M^0}$.

Step 1. *[Initialize]*
  (a) *Set* $M := 1$
  (b) *Repeat for all* $\vec{p} \in G_M^0$
      Set $u_M^0 := u_M^{old} + (u^a)_M$
  [End of loop]

Step 2. *Repeat Steps 3 and 4 while* $G_{M+1}^0 \neq \emptyset$

Step 3.   *[Superposition on the currently finest level]*
  (a) *Set* $M := M + 1$
  (b) *Repeat for all* $\vec{p} \in G_M^0 \cap G_{M-1}^0$
      Set $u_M^0 := u_{M-1}^0$
  [End of loop]
  (c) *Repeat for all* $\vec{p} \in (G_M^0 \cap G_M^{old}) \setminus G_{M-1}^0$
      Set $u_M^0 := u_M^{old} + (u^a)_M$
  [End of loop]
  (d) *Repeat for all* $\vec{p} \in (G_M^0 \setminus G_M^{old}) \setminus G_{M-1}^0$
      Set $u_M^0 := I_{M-1}^M(u_{M-1}^0 - (u^a)_{M-1}) + (u^a)_M$
  [End of loop]

Step 4.   *[Local grid refinement]*
  (a) **Evaluate generalized local error estimator**
      $(u_{M-1}^0[G_{M-1}], u_M^0[G_M]; \varepsilon_M[G_M])$;
  (b) **Construct a new level** $((\varepsilon_M, u_M^0)[G_M^0], Rel, Abs; G_{M+1}^0)$;
  [End of Step 2 loop]

Step 5. *Exit.*

In general $G_M^{old} \neq G_M^0$. Therefore the compact space distribution u^a is injected and directly superimposed to the discrete distribution u_M^{old} in all grid nodes of the coarser level and in selected grid nodes of $G_M^0 \setminus G_{M-1}^0$ which coincide with G_M^{old} grid nodes. At remaining nodes $(G_M^0 \setminus G_M^{old}) \setminus G_{M-1}^0$ the compact distribution $(u^a)_M$ is superimposed to $I_{M-1}^M u_{M-1}^{old} = I_{M-1}^M(u_{M-1}^0 - (u^a)_{M-1})$.

5.2.3 Examples of Ion-Implantation Process Modeling

In modern semiconductor technology, ion implantation is an indispensable process for the selective introduction of impurity profiles. From the simulation point of view, implanted impurity profiles represent an initial state for the subsequent diffusion and oxidation processes, which govern their evolution towards the final doping profile of the semiconductor device. To demonstrate the proposed procedures the construction of multilevel adaptive structures for compact two-dimensional models of ion implantation processes is presented.

For that purpose the rectangular simulation domain $\Omega := \{(x, y)|0 < x < l_x; 0 < y < l_y\}$ represents the semiconductor substrate. l_x and l_y denote its length and depth, respectively. The planar semiconductor surface is the physical boundary $\partial\Omega_p := \{(x, y)|(0 \leq x \leq l_x; y = 0\}$ which is covered by a nonplanar masking layer $(- d_m(x) \leq y \leq 0; 0 \leq x \leq l_x)$, where $d_m \geq 0$ defines the depth of the mask. It is assumed to be the target for the ion implantation beam perpendicular to the semiconductor surface. The boundaries $\partial\Omega_{s1} := \{(x, y)|x = 0; 0 \leq y \leq l_y\}$ and $\partial\Omega_{s2} := \{(x, y)|x = l_x; 0 \leq y \leq l_y\}$ are the symmetry lines.

The compact modeling of the ion implantation profiles is performed by the superposition method [43] which belongs to the second type of compact initial state representations. The total response to a homogeneous beam is assumed to be equal to the sum of all incremental responses to punctiform beams of finite width, which are equidistributed over $\partial\Omega_p$. For that purpose the punctiform beams are associated with the nodes of an auxiliary one-dimensional grid

$$\partial\Omega_{p\Delta x} = \{x_i | x_i = (i - 1/2)\cdot\Delta x; i = 1, \dots, n_p\}$$

constructed along $\partial\Omega_p$. The step size $\Delta x = l_x/n_p$ actually represents the width of the punctiform beams. The selection of the Δx is independent of the mesh-sizes used in the multilevel adaptive grid structure. It strongly depends on the lateral statistical moment for the implanted impurity and on the complexity of the various mask structures on the top of the semiconductor surface and can be nonuniformly distributed over $\partial\Omega_p$. Inside each of the punctiform beam segments $(x_i - \Delta x/2 \leq x \leq x_i + \Delta x/2)$ the mask structure is approximated by locally planar mask elements, positioned at $y_{mi} = - d_m(x_i)$. The implanted impurity profile is described by

$$N(x, y) = N_d \cdot \Delta x \cdot \sum_{i=1}^{n_p} (f_i(x, y) + f_i(- x, y) + f_i(2l_x - x, y)). \qquad (5.3)$$

N_d represents the implantation dose. $f_i(x, y)$ is the statistical distribution function for a single ion entering the target at $x = x_i$. Usually it is modeled as a product of the corresponding vertical and lateral statistical distributions. The implemented model due to [58]

$$f_i(x, y) = f_i^{vert}(y) \cdot f_i^{lat}(x, y) \qquad (5.4)$$

takes into account the very important dependence of the lateral statistical moments on the vertical coordinate.

Different statistical distribution functions for both the masking layers and the semiconductor substrate are handled by the numerical range scaling model [121]. The vertical distribution function f_i^{vert} is expressed by the separate distribution functions f_m^{vert} and f_s^{vert} corresponding to the independent ion implantation into the masking layer with infinite depth and bare semiconductor substrate, respectively:

$$f_i^{vert}(y) = \begin{cases} f_m^{vert}(y - y_{mi}) & \text{for } y_{mi} \leq y \leq 0 \\ \alpha_i \cdot f_s^{vert}\left(y - \dfrac{R_{ps}}{R_{pm}} \cdot y_{mi}\right) & \text{for } y \geq 0 \end{cases} \tag{5.5}$$

In order to account for the different stopping powers of the semiconductor and mask materials, the incremental depth of the masking layer used for the coordinate transformation in the semiconductor region are scaled by the ratio of the projection ranges R_{ps} and R_{pm}, corresponding to the semiconductor and the mask, respectively. The parameter α_i has to be chosen to satisfy the condition

$$\int_{y_{mi}}^{\infty} f_i^{vert}(y) dy = 1.$$

Similarly, the lateral distribution is expressed by

$$f_i^{lat}(x, y) = \begin{cases} f_m^{lat}(x - x_i, y - y_{mi}) & \text{for } y_{mi} \leq y \leq 0 \\ f_s^{lat}\left(x - x_i, y - \dfrac{R_{ps}}{R_{pm}} \cdot y_{mi}\right) & \text{for } y \geq 0 \end{cases} \tag{5.6}$$

Pearson IV [121, 126] and modified Gaussian distribution functions [58] are used for the vertical and lateral distribution functions both in the masking layer and in the semiconductor substrate. The parameters for the statistical moments are due to the LSS theory [83] and are evaluated by low order polynomial representations in terms of the implantation energy [125].

To illustrate the construction of a multilevel adaptive grid structure for a single compact distribution, it is considered for a boron implantation through a nonplanar oxide mask structure which is composed of three segments: two planar-layered segments and a unit slope segment as it might be produced by chemical etching. The simulation parameters and the mask shape description are given in Table 5.1

The multilevel adaptive grid structures for the implanted boron profile are constructed with the error parameters $Rel = 10^{-1}$ and $Rel = 10^{-2}$. The absolute error parameter is chosen to be $Abs = 10^{14}\,cm^{-3}$ in both cases. The surface plots of the resulting boron profiles are represented on the corresponding composite grid structures in Figures 5.3 and 5.4.

In both cases the local grid refinement process stops at the maximum level $l = 6$ starting from the coarsest grid $l = 1$ with 5×5 grid nodes. With the

Table 5.1. *Parameter set for the boron implanation*

Processing	Parameter	Selection
0. Substrate	length (l_x) depth (l_y)	$1\,\mu m$ $1\,\mu m$
1. Implantation	energy dose mask position	$80\,keV$ $5 \cdot 10^{13}\,cm^{-2}$ $(x \geq 0.5\,\mu m; d_m = 0.3\,\mu m)$ $(x \leq 0.2\,\mu m; d_m = 0)$ $(0.2\,\mu m < x < 0.5\,\mu m; d_m = (x - 0.2)\,\mu m)$

refinement criterion based on the local dose conservation, the finest local grid occurs where the steepest profile slopes appear. In the case of boron implantation this typically occurs at the profile edge directed towards the semiconductor substrate. However, the nonplanar mask modulates the boron profile so that only areas below the planar mask segments require the

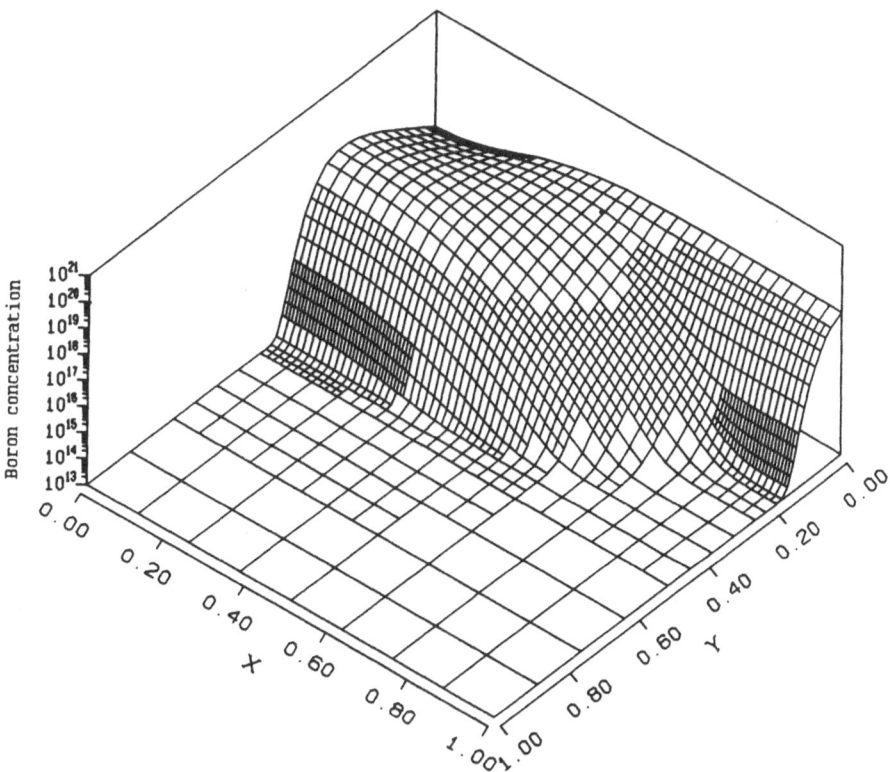

Fig. 5.3 Implanted boron profile $(Rel = 10^{-1})$

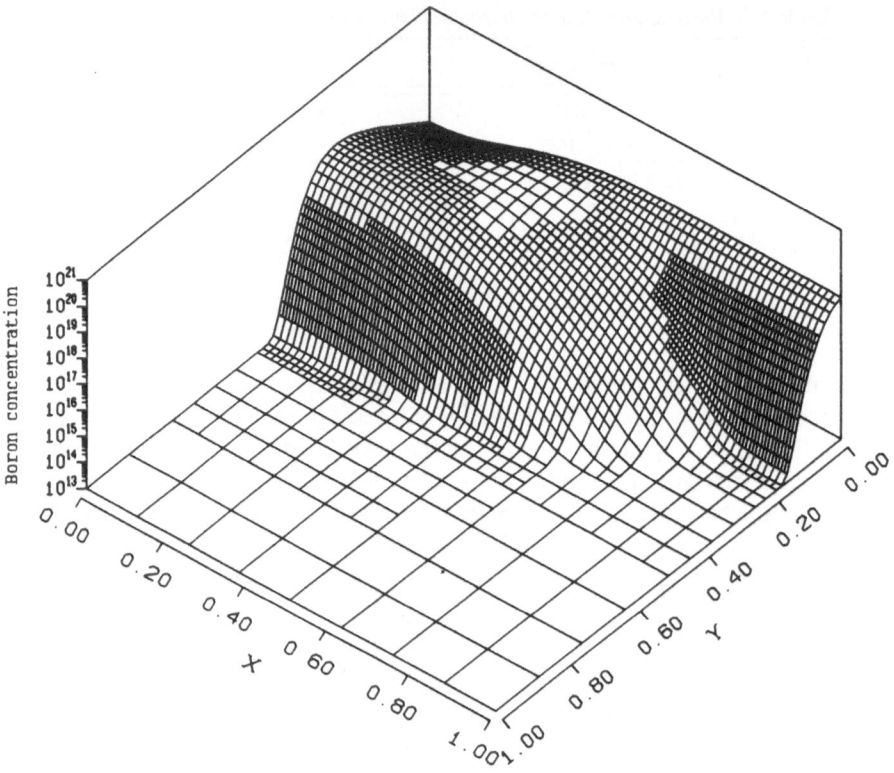

Fig. 5.4 Implanted boron profile $(Rel = 10^{-2})$

Table 5.2. *Parameter set for the double phosphorus implanation*

Processing	Parameter	Selection
0. Substrate	length (l_x) depth (l_y)	$0.6\,\mu m$ $0.6\,\mu m$
1. Implantation	energy dose mask position	$30\,keV$ $10^{13}\,cm^{-2}$ $(x \geq 0.4\,\mu m; d_m \to \infty)$ $(x < 0.4\,\mu m; d_m = 0.2\,\mu m)$
2. Implantation	energy dose mask position	$80\,keV$ $5 \cdot 10^{15}\,cm^{-2}$ $(x \geq 0.1\,\mu m; d_m \to \infty)$ $(x < 0.1\,\mu m; d_m = 0)$

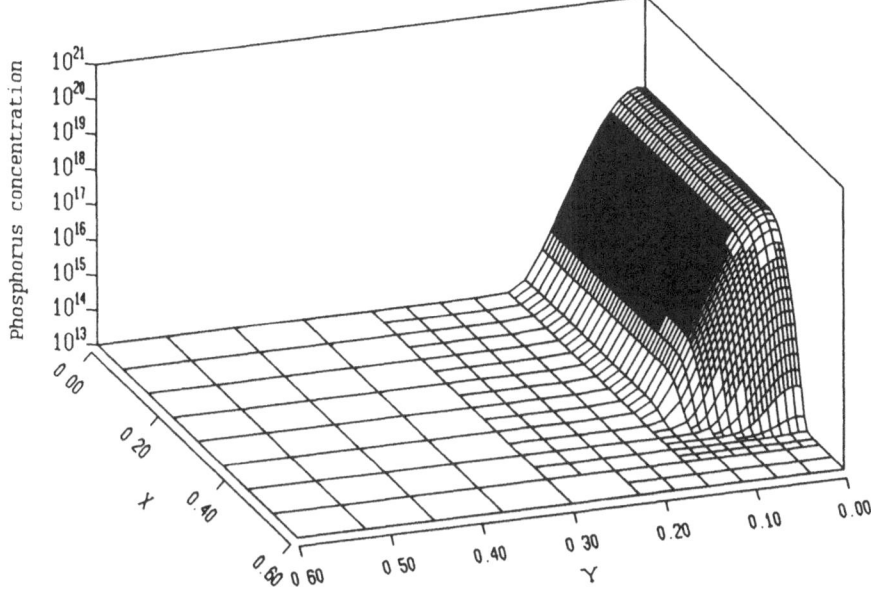

Fig. 5.5 Phosphorus profile after the first implantation only

finest local grid resolution. On the other hand, comparing Figures 5.3 and 5.4 the influence of the relative error parameter on the size of the areas covered by local grids on different levels is obvious.

In order to illustrate the adaptive superposition of both the discrete and the compact distributions of the same physical quantity the second simulation example considers the double phosphorus implantation. Each of them is performed through planar layered mask segments but with different positions of the steep mask edge. The simulation uses the processing parameters given in Table 5.2. The control parameters are $Rel = 5 \cdot 10^{-2}$ and $Abs = 10^{14}\,\mathrm{cm}^{-3}$.

The composite grid surface plots of the profiles after the first and the second phosphorus implantation, performed separately, are shown in Figure 5.5 and Figure 5.6, respectively.

In both cases the finest local grid structures ($M = 6$) follow the steepest front of the phosphorus profiles. Figure 5.7 shows the phosphorus profile obtained by the subsequent application of these two implantation processes due to the superposition Algorithm 5.4. The adaptive superposition of the two implantation profiles, Figure 5.7, results in a grid structure which significantly differs from that grid structure which results from the straight union of those grid structures used for the independently implanted profiles.

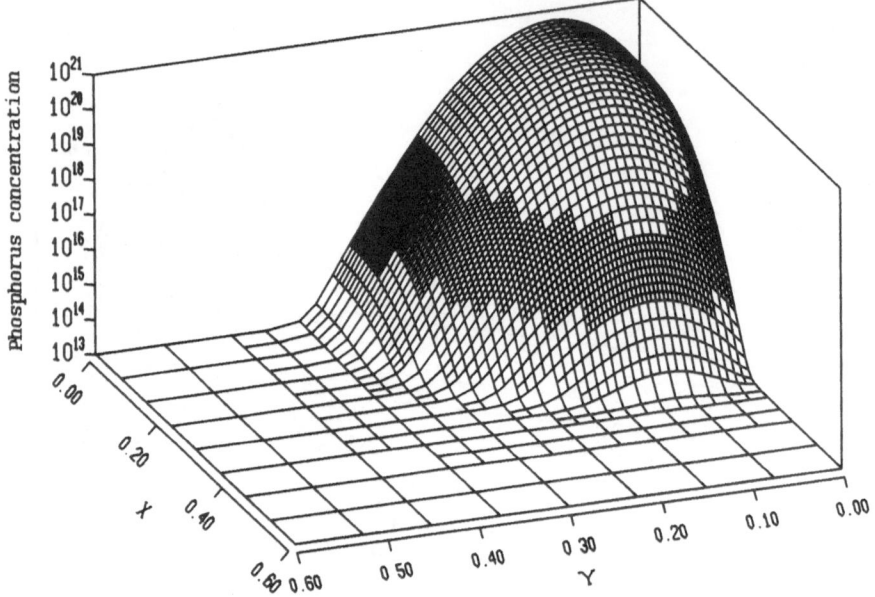

Fig. 5.6 Phosphorus profile after the second implantation only

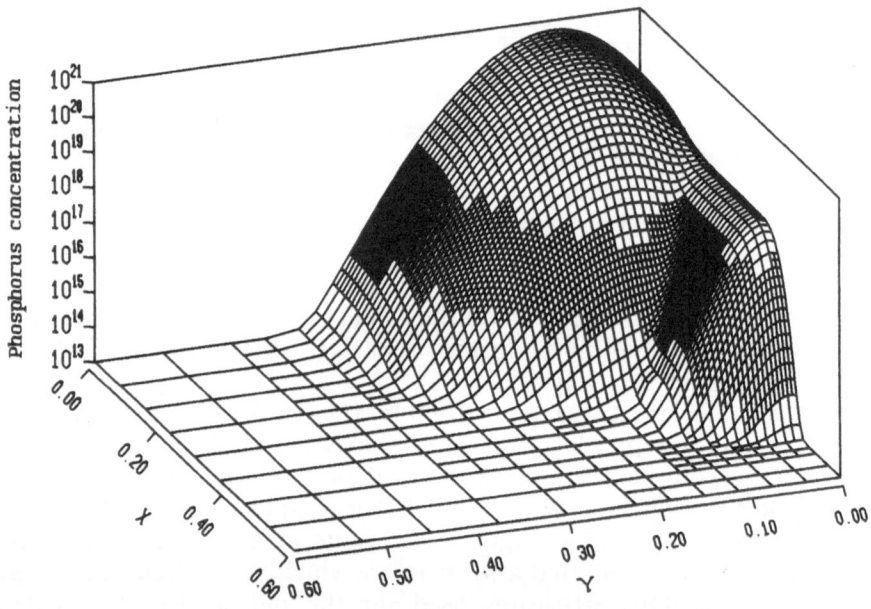

Fig. 5.7 Phosphorus profile after subsequent implantations

5.3 Basic Procedures for the Adaptive Transient Simulation of Evolution Problems

One of the most important advantages of the adaptive multilevel simulation approach is the possibility to achieve a tight coupling between the solution algorithm and the grid generation steps. The presented procedures for the adaptive treatment of the discrete incremental elliptic problem for each time-step are based on the MLAT technique and on the full multigrid algorithm. An adaptive iteration stopping criterion stops the multigrid cycling within the FAS-scheme. Finally, the basic refinement and solution procedures are combined with automatic two-level time-step selection procedures.

5.3.1 Cycling Multigrid Solution Procedure with an Adaptive Iteration Stopping Criterion

The discrete incremental elliptic problem

$$L_{l,\Delta t}(u_l) = f_l \quad (l = 1, \dots, M) \tag{5.7}$$

with the nonlinear discrete operator $L_{l,\Delta t}$ on a multilevel grid structure $G = (G_l)_{l=1,M}$ with the implicit Euler time-discretization is considered. For convenience it is assumed that the discrete operator $L_{l,\Delta t}$ includes the global and internal boundary conditions.

Besides the already known properties of the FAS multigrid approach—we recall here only that FAS supplies, in a natural way, a full discrete solution both on parts of the grids which are the currently finest and on parts which are covered by even finer grids—it is very promising with respect to the definition of iteration stopping criteria. The fact that the discrete solution should be obtained on an adaptive grid structure created with the goal to keep the discrete approximation error below some predetermined tolerance, implies that the solution procedure should also operate on such a grid structure. The final algebraic error has to be at the level of the discretization error. In other words: its magnitude reached by the solution procedure has to be controlled by the same predetermined tolerance used for the adaptive grid construction. Any computational work involved in a reduction of the algebraic accuracy below the discretization error is a pure waste and may in many cases represent a serious source of inefficient simulations. Quite convenient and compatible quantities which are available within the iterative solution procedure are the residual of the discrete problem and the local error. The general form of the stopping criterion for the iterative solution procedure

$$\|r_l\| \le \|\tau_l\| \quad (l = 1, \dots, M) \tag{5.8}$$

combines both of them and is incorporated into the FAS scheme in a natural and inexpensive way by the τ-extrapolation technique. The evaluation of the relative local error is a constituent part of the FAS restriction phase. The stopping criterion (5.8) can be rewritten as [99]

$$\|r_l\| \leq \begin{cases} \dfrac{1}{2^p - 1} \cdot \|I_{l-1}^l \tau_{l-1}^l\| & (l = 2, \ldots, M) \\[3mm] \dfrac{2^p}{2^p - 1} \cdot \|\tau_l^{l+1}\| & (l = 1) \end{cases}$$

(5.9)

The computational work for this stopping criterion (5.9) is negligible. The basic procedure for the corresponding multigrid procedure which exploits the adaptive stopping criterion (5.9) is given in Algorithm 5.5 [102].

Algorithm 5.5. (Cycling multigrid solution procedure)
Input parameters: $((u_l, L_{l,\Delta t}, f_l)[G_l])_{l=1,M}$
Output parameters: $(u_l[G_l])_{l=1,M}$
$((u_l, L_{l,\Delta t}, f_l)[G_l])_{l=1,M}$ *represents an initial solution, the incremental discrete elliptic operator and its discrete right hand side defined on the fixed local grid G_l for all levels of the adaptive grid structure G. This algorithm produces the discrete solution $(u_l[G_l])_{l=1,M}$ of the discrete incremental elliptic problem. The local error and the residual for the iteration stopping criterion are monitored with variables LEN[1..M] and REN, while the status of the stopping criterion is given in the logical variable Convergence[1..M].*

Step 1. [*Initialize*]
         Repeat for $l = 1, \ldots, M$
                 Set Convergence[l]:= false
         [*End of loop*]
Step 2. Repeat Steps 3, 7 and 8 until Convergence[l] = true for each
         $l \in (1, \ldots, M)$
Step 3.   [*Restriction phase*]
                 Repeat Steps 4 to 6 for $l = M, \ldots, 2$ by -1
Step 4.   Repeat for all $\vec{p} \in G_l$
                                 Set $u_l := \mathcal{S}_l^{(v_1)}(u_l)$
                         [*End of loop*]
Step 5.   [Repeat for all $\vec{p} \in G_l \cap G_{l-1}$
                                 (a) Set $u_{l-1} := \hat{I}_l^{l-1} u_l$
                                 (b) Set $e_{l-1} := L_{l-1,\Delta t}(u_{l-1}) - I_l^{l-1}(L_{l,\Delta t}(u_l))$
                                 (c) Set $f_{l-1} := I_l^{l-1} f_l + e_{l-1}$
                         [*End of loop*]
Step 6.   Set $LEN[l] := (2^p - 1)^{-1} \|I_{l-1}^l e_{l-1}\|$
                 [*End of Step 3 loop*]
Step 7.   [*Solution on the coarsest level*]
                 Set $LEN[1] := 2^p \cdot LEN[2]$

$\qquad$ *Repeat until* $REN < \varsigma \cdot LEN[1]$
$\qquad\qquad$ *(a) Set* $u_1 := \mathcal{S}_1^{(v_{sol})}(u_1)$
$\qquad\qquad$ *(b) Set* $REN := \| f_1 - L_{1,\Delta t}(u_1) \|$
$\qquad$ *[End of loop]*
$\qquad$ *Set Convergence*$[1] :=$ *true*

Step 8. $\qquad$ *[Prolongation phase]*
$\qquad$ *Repeat Steps 9 and 10 for* $l = 2, \ldots, M$

Step 9. $\qquad\qquad$ *Repeat for all* $\vec{p} \in G_l$
$\qquad\qquad\qquad$ *(a) Set* $u_l := u_l + I_{l-1}^l(u_{l-1} - \hat{I}_l^{l-1} u_l)$
$\qquad\qquad\qquad$ *(b) Set* $u_l := \mathcal{S}_l^{(v_2)}(u_l)$
$\qquad\qquad\qquad$ *(c) Set* $e_l := f_l - L_{l,\Delta t}(u_l)$
$\qquad\qquad$ *[End of loop]*

Step 10. $\qquad$ *[Check the convergence]*
$\qquad\qquad$ *(a) Set* $REN := \| e_l \|$
$\qquad\qquad$ *(b) If* $REN < \varsigma \cdot LEN[l]$ *then:*
$\qquad\qquad\qquad$ *Set Convergence*$[l] :=$ *true*
$\qquad\qquad$ *[End of If structure]*
$\qquad$ *[End of Step 8 loop]*
$\qquad$ *[End of Step 2 loop]*
Step 11. *Exit.*

The local error norm is evaluated on each level during the restriction phase of the cycling in order to be compared with the residual in the prolongation phase. The iteration is stopped if the residual norm is below the local error norm on all levels above the coarsest one. The coarse grid problem is solved by v_{sol} iterations of the smoothing scheme used. Optionally, this can be replaced by any reasonable direct solver. The control parameter $\varsigma < 1$ (a typical value is $\varsigma = 0.1$) is introduced for safety, since the local error norm is evaluated prior to the coarse grid correction, that is, using an insufficiently accurate discrete solution. Its usage can be avoided by a simultaneous evaluation of both the local error and the residual norms in the prolongation phase. However, this would require additional computational work for the evaluation of the relative local error which is not automatically provided within the prolongation phase. The principal components of the cycling multigrid algorithm: smoothing iteration $\mathcal{S}_l^{(v)}$, restriction and prolongation operators $\hat{I}_l^{l-1}$, I_l^{l-1}, I_{l-1}^l have to be properly tailored to the problem at hand (see Chapters 2 and 4).

5.3.2 *The Local Refinement-Solution Procedure*

The procedure which couples both the grid refinement and the solution algorithm for the incremental elliptic problems resembles the full multigrid algorithm and local refinement procedures for elliptic problems as they are presented in Section 2.7.3 or in [118]. Three basic steps are involved:

1. Formulate the discrete incremental elliptic problem $L_{M,\Delta t}(u_M) = f_M$ on the currently finest level.
2. Solve the set of global and local discrete incremental elliptic problems on all levels $l \leq M$ by the cycling multigrid algorithm.
3. Construct a new refinement over the currently finest level.

These steps are recursively repeated, in the same order and starting from the coarsest global level, until no further refinement is needed. As soon as a new local discrete problem is defined on the new and currently finest level a set of coupled local and global problems is solved by FAS cycling. With such an approach the number of refine-solve loops is fixed to the maximum number of levels involved in the local grid refinement procedure.
It is straightforward to set up the discrete incremental elliptic problems. However, the time discrete operators require the transfer of the solution of the previous time step u_l^{old} from G_l^{old} to G_l which are in general different from each other. The procedure to establish the time and space discrete operators is given in Algorithm 5.6 [102].

Algorithm 5.6. (Formulate discrete incremental elliptic problem)
Input parameters: $\Delta t, u_M^{old}[G_M^{old}], (u_l^{old}[G_l])_{l=1,M-1}, G_M$
Output parameters: $u_M^{old}[G_M], L_{M,\Delta t}, f_M$
$u_M^{old}[G_M^{old}]$ and $(u_l^{old}[G_l])_{l=1,M-1}$ *represent the discrete solutions of the previous temporal grid node defined on the multilevel grid structures G^{old} and G. This algorithm produces the discrete approximation of the "old" discrete solution on the "new" uniform grid G_M as well as discrete incremental analogues $L_{M,\Delta t}$ and f_M for the evolution problem $L(u) = f$.*

Step 1. [Restore the previous temporal node discrete solution]
      Repeat for all $\vec{p} \in G_M \cap G_M^{old}$
          Set $u_M^{old}[G] := u_M^{old}[G^{old}]$
      [End of loop]
      If $M \geq 2$ then:
          Repeat for all $\vec{p} \in G_M \setminus G_M^{old}$
              Set $u_M^{old}[G] := I_{M-1}^M u_{M-1}^{old}[G]$
          [End of loop]
      [End of If structure]
Step 2. [Formulate discrete analogues]
      (a) Set $L_{M,\Delta t} := I^{\Delta t}(I^M(L, h_M), \Delta t, u_M^{old})$
      (b) Set $f_M := I^M(f, h_M)$
Step 3. Exit.

Since the local grid structures G_M^{old} and G_M are strictly regular and defined on the same lattice, a transfer of grid functions onto the intersection $G_M \cap G_M^{old}$ is performed directly by injecting the "old" discrete solution into G_M grid nodes. An interpolation is required only for grid nodes $G_M \setminus G_M^{old}$. But even in such cases it exploits the regular coarse grid structure. Possible choices for this transfer have been presented in Section 4.6. The operators I^M and $I^{\Delta t}$ describe the discretization of $L(u) = f$ with respect to h_M and Δt, respectively.

Details of the procedure which couples both the solution algorithm and the grid refinement process are given in Algorithm 5.7 [102].

Algorithm 5.7. (Local grid refinement-solution procedure)
Input parameters: $\Delta t, (u_l^{old}[G_l^{old}])_{l=1,M^{old}}, (G_l)_{l=1,2}, Rel, Abs$
Output parameters: $(u_l[G_l])_{l=1,M}$
Δt *is the time-step size.* $(u_l^{old}[G_l^{old}])_{l=1,M^{old}}$ *represent the discrete solution of the previous temporal grid node,* $(G_l)_{l=1,2}$ *represent the first two global levels which are the core for the further adaptive construction controlled by the relative and absolute error parameters Rel and Abs. This algorithm produces the discrete solution of the incremental elliptic problem* $(u_l[G_l])_{l=1,M}$.

Step 1. [*Initialize*]
        Set $M := 1$
Step 2. *Repeat Steps 3 to 6 while* $G_M \neq \emptyset$
Step 3.   **Formulate discrete incremental elliptic problem**
                $(\Delta t, u_M^{old}[G_M^{old}], (u_l^{old}[G_l])_{l=1,M-1}, G_M; u_M^{old}[G_M], L_{M,\Delta t}, f_M)$
Step 4.   [*Put an initial solution*]
                If $M \geq 2$ then:
                        Repeat for all $\vec{p} \in G_M$
                                Set $u_M := I_{M-1}^M u_{M-1}$
                        [*End of loop*]
                Else:
                        Repeat for all $\vec{p} \in G_M$
                                Set $u_M := u_M^{old}[G_M]$
                        [*End of loop*]
                        [*End of If structure*]
Step 5.   **Cycling multigrid solution procedure**
                $(((u_l, L_{l,\Delta t}, f_l)[G_l])_{l=1,M}; (u_l[G_l])_{l=1,M})$
Step 6.   [*Local grid refinement*]
                If $M \geq 2$ then:
                        (*a*) **Evaluate global error estimator**
                                $(u_{M-1}[G_{M-1}], u_M[G_M]; \varepsilon_M[G_M])$ *or*
                                Evaluate local error estimator
                                $((L_{M-1,\Delta t}, f_{M-1})[G_{M-1}], u_M[G_M]; \varepsilon_M[G_M])$
                        (*b*) **Construct a new level**
                                $((\varepsilon_M, u_M)[G_M], Rel, Abs; G_{M+1})$
                        [*End of If structure*]
                        Set $M := M + 1$
                [*End of Step 2 loop*]
                Set $M := M - 1$
Step 7. *Exit.*

The discrete solution of the previous time-step is used as an initial guess on the coarsest level, while on all finer levels the start approximation is obtained by interpolation of the coarse grid solution. In the latter case, the initial solution automatically poses initial internal boundary conditions for the currently finest local incremental problem.

In order to permit the above approach, the local refinement criteria have to employ reliable error estimators. For this purpose the extrapolation of both global and local errors are applicable. Since the FAS-based multigrid approach naturally provides discrete solutions on different levels, this strategy offers a natural environment to estimate directly and cheaply the global error by Richardson extrapolation and the local error by τ-extrapolation. The practical implementation of these two procedures acting on two consecutive levels of the space grid structure is described in the Algorithms 5.8 and 5.9 [106, 102].

Algorithm 5.8. (Evaluate global error estimator)
Input parameters: $u_{l-1}[G_{l-1}]$, $u_l[G_l]$
Output parameter: $\varepsilon_l[G_l]$
$u_{l-1}[G_{l-1}]$ and $u_l[G_l]$ are discrete solutions on two subsequent discretization levels. This algorithm produces the error estimate $\varepsilon_l[G_l]$ for the local grid refinement process based on the Richardson extrapolation of the global error.

Step 1. *Repeat for all* $\vec{p} \in G_l \cap G_{l-1}$
         (a) *Set* $(e_l)_{l-1} := (2^p - 1)^{-1} \cdot |\hat{I}_l^{l-1} u_l - u_{l-1}|$
         (b) *Set* $\varepsilon_l := (e_l)_{l-1}$
     [End of loop]
Step 2. *Repeat for all* $\vec{p} \in G_l \backslash G_{l-1}$
         Set $\varepsilon_l := I_{l-1}^l (e_l)_{l-1}$
     [End of loop]
Step 3. *Exit.*

Algorithm 5.9. (Evaluate local error estimator)
Input parameters: $(L_{l-1,\Delta t}, f_{l-1})[G_{l-1}]$, $u_l[G_l]$
Output parameter: $\varepsilon_l[G_l]$
$(L_{l-1,\Delta t}, f_{l-1})[G_{l-1}]$ are the discrete operator and its right hand side on the coarse level. $u_l[G_l]$ is the discrete solution on the fine level. This algorithm produces the error estimate $\varepsilon_l[G_l]$ for the local grid refinement process based on the τ-extrapolation of the local error.

Step 1. *Repeat for all* $\vec{p} \in G_l \cap G_{l-1}$
         (a) *Set* $(\tau_l)_{l-1} := (2^p - 1)^{-1} \cdot |L_{l-1}(\hat{I}_l^{l-1} u_l) - f_{l-1}|$
         (b) *Set* $\varepsilon_l := (\tau_l)_{l-1}$
     [End of loop]

Step 2. Repeat for all $\vec{p} \in G_l \backslash G_{l-1}$
$$\text{Set } \varepsilon_l := I_{l-1}^l (\tau_l)_{l-1}$$
[*End of loop*]
Step 3. Exit.

In the case of the global error extrapolation it is very important to use the coarse grid discrete solution u_{l-1} obtained with the $l-1$-st level as the finest level instead of the corresponding solution u_{l-1} which is available during the cycling for the solution u_l with level l being the finest one (within the FAS scheme the fine-to-coarse grid correction updates the coarse grid approximation which makes the Richardson extrapolation useless).

5.3.3 *Automatic Time Stepping Procedures in the Framework of Multilevel Spatial Grid Structures*

In Chapter 3 the two-level time stepping strategy is applied to the temporal semi-discrete problem $L_{\Delta t} = f_{\Delta t}$. As long as the error splitting (3.18) and (3.19) is valid, the extrapolated local-global and local errors do not depend on the special discrete approximation. In other words: replacing the semi-discrete solution $u_{\Delta t}^n$ and the operator $L_{\Delta t}^n$ by its completely discretized analogues u_l^n and $L_{l,\Delta t}^n$ does not influence the final result of the two-level extrapolation of the temporal local-global and local errors. Nevertheless, an optimal evaluation of the temporal error estimators requires the proper algorithmic treatment of space discrete quantities on the auxiliary time discretization level. Also, it is important that only the discrete solution defined on the composite grid participates in the evaluation of the temporal error estimators and in the automatic time-step size control.

The evolution of the discrete solution and its multilevel spatial grid structure in a two-step integration sequence on the fine time level (which is the only level responsible for the time-step size control) looks like

$$\cdots \rightarrow (u^{n-2,2}, G^{n-2}) \xrightarrow{\Delta t^{n-1}} (u^{n-1,2}, G^{n-1}) \xrightarrow{\Delta t^n} (u^{n,2}, G^n) \rightarrow \cdots.$$

$G^n \equiv (G_l^n)_{l=1,M^n}$ is the multilevel spatial grid structure at the n-th temporal grid node and $u^{n,k} \equiv (u_l^{n,k})_{l=1,M^n}$ is the corresponding discrete solution of the k-th $(k = 1, 2)$ time level. In the two-level step-doubling strategy this sequence is repeated with $\Delta t^{n-1} = \Delta t^n$ for each even n. In addition to the original time level integration path a single coarse level integration step

$$\cdots \rightarrow (u^{n-2,2}, G^{n-2}) \xrightarrow{2 \cdot \Delta t^n} (u^{n,1}, G^n) \rightarrow \cdots$$

results in the coarse level solution $u^{n,1}$ which is required for the Richardson extrapolation of the local-global error. Due to the mutually independent temporal and spatial discretization errors, both the coarse and the fine integration steps use the same multilevel spatial grid structures G^{n-2} and G^n, respectively. Consequently, local grid refinement steps and the computational work involved in it at the coarse temporal grid level are not necessary. Such an integration step which uses the already known grid structure both at the beginning and at the end of the time-step and which only produces the new coarse time level solution $u^{n,1}$ is denoted by

$$\cdots \rightarrow (u^{n-2,2}, G^{n-2}, G^n) \xrightarrow{2 \cdot \Delta t^n} (u^{n,1}) \rightarrow \cdots.$$

The procedure for the transient simulation based on the two-level step-doubling strategy is given in Algorithm 5.10 [102].

Algorithm 5.10. (Time stepping procedure with two-level step-doubling strategy)
Input parameters: $(u_l^0[G_l^0])_{l=1, M^0}, \Delta t^1, T, Rel, Abs$
Output parameters: $(u_l^n[G_l^n])_{l=1, M^n} (n=1, \ldots, N_t)$
$(u_l^0[G_l^0])_{l=1, M^0}$ *is the multilevel adaptive discrete approximation of an initial state.* Δt^1 *is an initial time-step size. T is the simulation time. Rel and Abs are relative and absolute error parameters. This algorithm produces the incremental discrete solution* $(u_l^n[G_l^n])_{l=1, M^n}$ *at each temporal grid node* $t^n (n=1, \ldots, N_t$, *where* $t^{N_t} = T$).

Step 1. [Initialize]
       (a) Set $t:=0$;
       (b) Set $\Delta t:= \Delta t^1$
       (c) Set $\Delta T:= 2\Delta t$
       (d) Set $n:= 2$
Step 2. Repeat Steps 3 to 6 while $t < T$
Step 3.   [Integration on the fine level]
              Repeat for $i = n-1, n$
                     Local grid refinement-solution procedure
                     $(\Delta t, (u_l^{i-1,2}[G_l^{i-1}])_{l=1, M^{n-1}}, (G_l^i)_{l=1,2}, Rel, Abs;$
                     $(u_l^{i,2}[G_l^i])_{l=1, M^i})$
              [End of loop]
Step 4.   [Integration on the coarse level]
              (a) Repeat for $l = 1, \ldots, M^n$
                     Formulate discrete incremental elliptic problem
                     $(\Delta T, u_l^{n-2,2}[G_l^{n-2}], (u_k^{n-2,2}[G_k^n])_{k=1, l-1}, G_l; u_l^{n-2,1}[G_l^n],$
                     $L_{l,\Delta T}^{n,1}, f_l^{n,1})$
              [End of loop]
              (b) **Cycling multigrid solution procedure**
                     $(((u^{n,1}, L_{l,\Delta T}^{n,1}, f_l^{n,1})[G_l^n])_{l=1, M^n}; (u_l^{n,1}, [G_l^n])_{l=1, M^n})$
Step 5.   [Evaluate the error estimator]
              Repeat for $l = 2, \ldots, M^n$
                     Repeat for all $\vec{p} \in G_l^n \backslash G_l^n \cap G_{l+1}^n$

$$Set\ \varepsilon_l^n := \frac{1}{2(2^q - 1)} \cdot |u_l^{n,2} - u_l^{n,1}|$$

> [End of loop]
> [End of loop]

Step 6.   [Local refinement]
> (a) **New time-step** $(\Delta t, (\varepsilon_{*}^n, u_{*}^n)[G_{*}^n], T, t, Rel, Abs; \Delta t,$
>   Rejected)
> (b) Set $\Delta T := 2\Delta t$
> (c) If not Rejected then:
>   Set $t := t + 2 \cdot \Delta t$
>   Set $n := n + 2$
> [End of If Structure]
> [End of Step 2 loop]

Step 7. Exit.

In case of the τ-extrapolation based time stepping scheme, the principal integration on the fine time level is open for a flexible step-by-step modification of the time-step size, that is, Δt^n may be different from Δt^{n-1} (the only exception are the first two time-steps which have to be selected as equal in size). The solution step on the coarse time integration path is eliminated. Beginning with the second time grid node, the coarse time level is visited only to compute the coarse level defect using the fine time level solution. In spite of the fact that this fictitious coarse level integration step does not require any solution step, it has to be correctly posed. The corresponding procedure does this on the grid structure G^n starting with an accepted discrete solution defined on the structure G^{n-2} and using the coarse level time-step size $\Delta t^n + \Delta t^{n-1}$. The procedure for the transient simulation based on the two-level τ-extrapolation time stepping strategy is given in Algorithm 5.11.

Algorithm 5.11. (Time stepping procedure with two-level τ-extrapolation based strategy)
Input parameters: $(u_l^0 [G_l^0])_{l=1,M^0}, \Delta t^1, T, Rel, Abs$
Output parameters: $(u_l^n [G_l^n])_{l=1,M^n}$ $(n = 1, \dots, N_t)$
$(u_l^0 [G_l^0])_{l=1,M^0}$ *is the multilevel adaptive discrete approximation of an initial state. Δt^1 is an initial time-step size, T is the simulation time. Rel and Abs are relative and absolute error parameters. This algorithm produces the incremental discrete solution $(u_l^n [G_l^n])_{l=1,M^n}$ at each temporal grid node $t^n (n = 1, \dots, N_t,$ with $t^{N_t} = T$).*

Step 1. [Initialize]
> (a) Set $t := 0$;
> (b) Set $\Delta t := \Delta t^1$
> (c) Set $\Delta T := 2\Delta t$
> (d) Set $n := 1$

Step 2. Repeat Steps 3 to 6 while $t < T$

Step 3.   [*Initial integration step on the fine level*]
                 If $n = 1$ then:
                    (*a*) **Local grid refinement-solution procedure**
                    $(\Delta t, (u_l^{n-1,2}[G_l^{n-1}])_{l=1,M^{n-1}}, (G_l^n)_{l=1,2}, Rel, Abs;$
                    $(u_l^{n,2}[G_l^n])_{l=1,M^n})$
                    (*b*) *Set $n := 2$*
                 [*End of If structure*]

Step 4.   [*An ordinary integration step*]
                 Local grid refinement-solution procedure
                 $(\Delta t, (u_l^{n-1,2}[G_l^{n-1}])_{l=1,M^{n-1}}, (G_l^n)_{l=1,2}, Rel, Abs; (u_l^{n,2}[G_l^n])_{l=1,M^n})$

Step 5.   [*Evaluate the error estimator*]
                 (*a*) *Repeat for $l = 1, \ldots, M^n$*
                           Formulate discrete incremental elliptic problem
                           $(\Delta T, u_l^{n-2,2}[G_l^{n-2}], (u_k^{n-2}[G_k^n])_{k=1,l-1},$
                           $G_l^n; u_l^{n-2,2}[G_l^n], L_{l,\Delta T}^1, f_l^{n,1})$
                 [*End of loop*]
                 (*b*) *Repeat for $l = 2, \ldots, M^n$*
                           Repeat for all $\vec{p} \in G_l^n \setminus G_l^n \cap G_{l+1}^n$

                           $$Set \ \varepsilon_l^n := \frac{1}{2^q - 1} \cdot |L_{l,\Delta T}^{n,1}(u_l^{n,2}) - f_l^{n,1}|$$

                           [*End of loop*]
                 [*End of loop*]

Step 6.   [*Local refinement*]
                 (*a*) **New time-step** $(\Delta t, (\varepsilon_{\mathscr{H}}^n, u_{\mathscr{H}}^n)[G_{\mathscr{H}}^n], T, t, Rel, Abs; \Delta t^{new}, Rejected)$
                 (*b*) *If not Rejected then:*
                           Set $\Delta T := \Delta t + \Delta t^{new}$
                           Set $t := t + \Delta t^{new}$
                           Set $n := n + 1$
                      Else If $n = 2$ then:
                           Set $\Delta T := 2\Delta t^{new}$
                           Set $t := 0$
                           Set $n := n - 1$
                      Else
                           Set $\Delta T := \Delta T - \Delta t + \Delta t^{new}$
                      [*End of If Structure*]
                 (*c*) *Set $\Delta t := \Delta t^{new}$*
              [*End of Step 2 loop*]
Step 7. Exit.

The algorithm 5.12 [102] gives the basic details of the automatic time-step selection strategy used in the framework of the composite grid structure $G_{\mathscr{H}}$.

Algorithm 5.12. (New time-step)
Input parameters: $\Delta t, (\varepsilon_{\mathscr{H}}, u_{\mathscr{H}})[G_{\mathscr{H}}], T, t, Rel, Abs$

Output parameters: Δt^{new}, Rejected

Δt is the current time-step size. $(\varepsilon_{\mathscr{H}}, u_{\mathscr{H}})[G_{\mathscr{H}}]$ are the local error estimator and the discrete solution. Rel and Abs are relative and absolute error parameters. This algorithm produces a new guess for the time-step size Δt^{new} and returns the decision on rejection of the current time-step or not by the logical variable Rejected.

Step 1. [Initialize]
        Set Facmin := 1/2
        Set Facmax := 2
        Set Safety := 0.9
Step 2. [Integral form of the error estimator and the tolerance]
        (a) Repeat for all $\vec{p} \in G_{\mathscr{H}}$

$$Set\ r_{\mathscr{H}} := \frac{\varepsilon_{\mathscr{H}}}{Rel \cdot u_{\mathscr{H}} + Abs}$$

        [End of loop]
        (b) Set $\|r_{\mathscr{H}}\| := \max_{\vec{p} \in G_{\mathscr{H}}} r_{\mathscr{H}}[\vec{p}]$
Step 3. [A new time-step size]
        (a) Set $\Delta t^{new} := \Delta t \cdot \|r_{\mathscr{H}}\|^{-\frac{1}{q+1}}$
        (b) Set $\Delta t^{new} := \Delta t \cdot \min(Facmax, \max(Facmin, Safety \cdot \Delta t^{new}/\Delta t))$
        (c) Set $\Delta t^{new} := \dfrac{T - t}{\lceil (T - t)/\Delta t^{new} - \epsilon \rceil}$
Step 4. [Local refinement criterion]
        If $\|r_{\mathscr{H}}\| < 1$ then:
            Set Rejected := false
        Else
            Set Rejected := true
        [End of if structure]
Step 5. Exit.

Analogously to the spatial grid refinement, in the case of discrete evolution systems the time integration procedure proceeds with the worst case time-step size obtained scanning all discrete solution components.

5.3.4 An Example of Impurity Diffusion Simulation

An adaptive multigrid simulation of an impurity diffusion process using an initially implanted profile illustrates the proposed simulation strategies. The redistribution of the j-th impurity is modeled by the evolution problem [102]:

$$\frac{\partial N_j}{\partial t} - \frac{\partial}{\partial x}\left(D_j(N) \cdot \frac{\partial N_j}{\partial x}\right) - \frac{\partial}{\partial y}\left(D_j(N) \cdot \frac{\partial N_j}{\partial y}\right) = 0 \text{ in } \Omega \times \Omega_t$$

$$(5.10)$$

$$\frac{\partial N_j}{\partial \vec{n}} = 0 \text{ at } \partial\Omega_p, \partial\Omega_{s1} \text{ and } \partial\Omega_{s2} \tag{5.11}$$

$$N_j = N_{j,\min} \text{ at } \partial\Omega_d$$
$$N_j = N_j^0 \quad \text{ at } \quad t = 0.$$

The diffusion equation (5.10) is commonly used in two-dimensional process simulation programs. The Neumann boundary conditions (5.11) on the corresponding boundary segments confine the diffusion process analysis to the impurity diffusion in an inert ambient. The equations are coupled by the diffusion coefficients which depend on all impurities present. This is indicated by omitting the index j for N in $D_j(N)$. Ω is the rectangular simulation domain $(0 \leq x \leq l_x) \times (0 \leq y \leq l_y)$. The physical boundary $\partial\Omega_p = \{(x, y)|0 \leq x \leq l_x, y=0\}$ coincides with the planar semiconductor surface. The distant boundary $\partial\Omega_d = \{(x, y)|0 \leq x \leq l_x, y=l_y\}$ lies deep in the semiconductor substrate and far from the diffusion scene. The boundaries

$$\partial\Omega_{s1} = \{(x, y)|x = 0, 0 \leq y \leq l_y\}$$
and
$$\partial\Omega_{s2} = \{(x, y)|x = l_x, 0 \leq y \leq l_y\}$$

are the symmetry lines. $N_{\min} = 10^{13} \text{ cm}^{-3}$ is the impurity concentration for all species at the bottom of the simulation domain. The minimum concentration of interest for the adaptive simulation which actually defines the absolute error parameter is $Abs = 10^{14} \text{ cm}^{-3}$.

It is well known that the point defects (vacancies and interstitials) represent the principal diffusion vehicles for impurities. Assuming that they are at their equilibrium concentrations immediately adjusted to the changing impurity concentrations [30, 31], the diffusion coefficient $D_j(N)$ is modeled by the vacancy diffusion model [57]

$$D_j = D_j^0 + D_j^- \cdot \frac{n}{n_i} + D_j^+ \cdot \frac{n_i}{n} + D_j^= \cdot \left(\frac{n}{n_i}\right)^2 \tag{5.12}$$

where each of the terms on the right hand side accounts for the j-th impurity interaction with different states of semiconductor vacancies. D_j^0, D_j^-, D_j^+ and $D_j^=$ are the intrinsic diffusivities corresponding to neutral, single negative, positive, and double negative vacancies, respectively. The intrinsic diffusivities are modeled as temperature dependent in Arrhenius-like form [112, 126]. n and n_i are the electron and intrinsic carrier concentrations at the processing temperature. The electron concentration is expressed approximately from the charge neutrality condition as

$$n = \frac{N_{\text{net}} + \sqrt{N_{\text{net}}^2 + 4n_i^2}}{2}.$$

The quantity N_{net} represents the total net concentration of all ionized impurities defined by

$$N_{net} = -\sum_{j=1}^{n} Z_j \cdot N_j$$

where Z_j determines the charge state and degree of ionization of the particles. The time discretization is performed by the implicit backward Euler scheme while spatial derivatives are discretized by 5-point central differences. For the discretization of the Neumann boundary conditions the so-called "mirror imaging" method has been used.

The numerical experiments are performed for a high-concentration arsenic diffusion [101] which is typically needed for VLSI NMOS source-drain areas. The processing and simulation parameters together with the chosen multigrid components are given in Table 5.3 and Table 5.4, respectively.

Figure 5.8 shows the perspective isoconcentration plot and the composite grid structure (obtained with the relative error parameter $Rel = 10^{-2}$) for

Table 5.3. *Parameter set for arsenic implantation and diffusion*

Processing	Parameter	Selection
0. Substrate	length (l_x)	$0.4\ \mu m$
	depth (l_y)	$0.4\ \mu m$
	type	P
	concentration	$10^{15}\ cm^{-3}$
1. Implantation	energy	$100\ keV$
	dose	$6 \cdot 10^{15}\ cm^{-2}$
	mask position	$(x < 0.1\ \mu m; d_m = 0)$
		$(x \geq 0.1\ \mu m; d_m \to \infty)$
2. Diffusion	temperature	$1000\ ^\circ C$
	time (T)	$20\ min$

Table 5.4. *The basic multigrid components*

Multigrid component	Parameter	Selection
relaxation	type	Gauss-Seidel
	ordering	red-black
	ν_1	2
	ν_2	1
	ν_{sol}	5
restriction	type	full weighting
prolongation	type	linear interpolation
refinement in space	error estimator	global error
	grid size ($l = 1$)	5×5
refinement in time	error estimator	local error
	first time-step size	$10\ s$

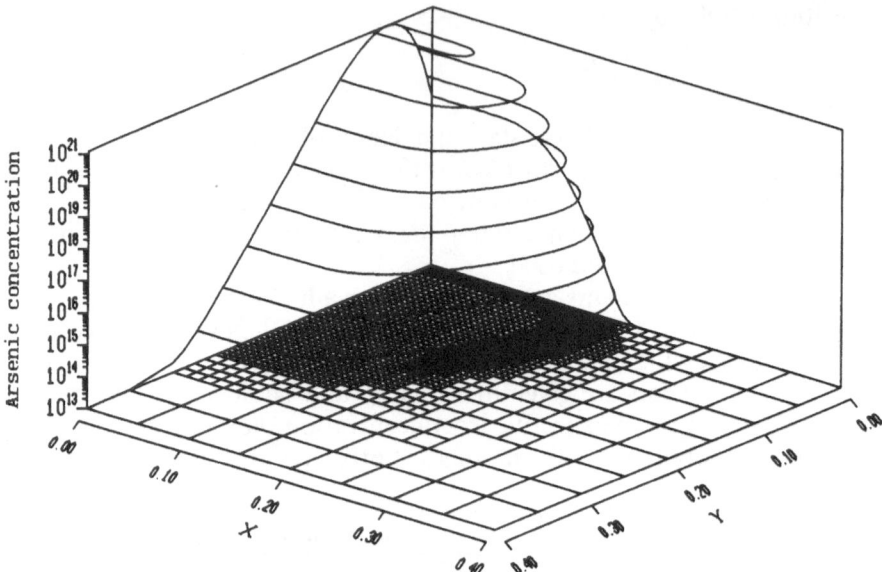

Fig. 5.8 A perspective isoconcentration plot over the composite grid structure of the implanted arsenic profile ($Rel = 10^{-2}$)

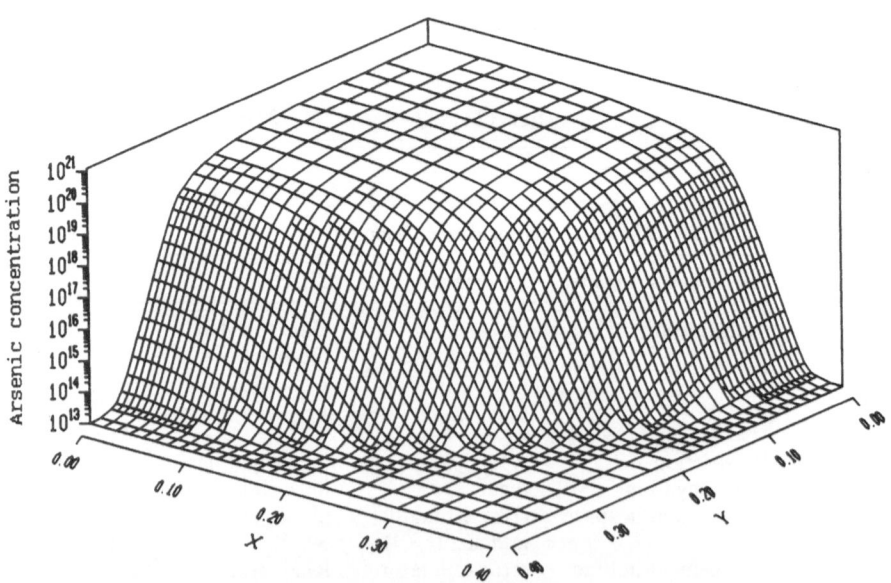

Fig. 5.9 Arsenic profile after the diffusion process ($Rel = 10^{-1}$)

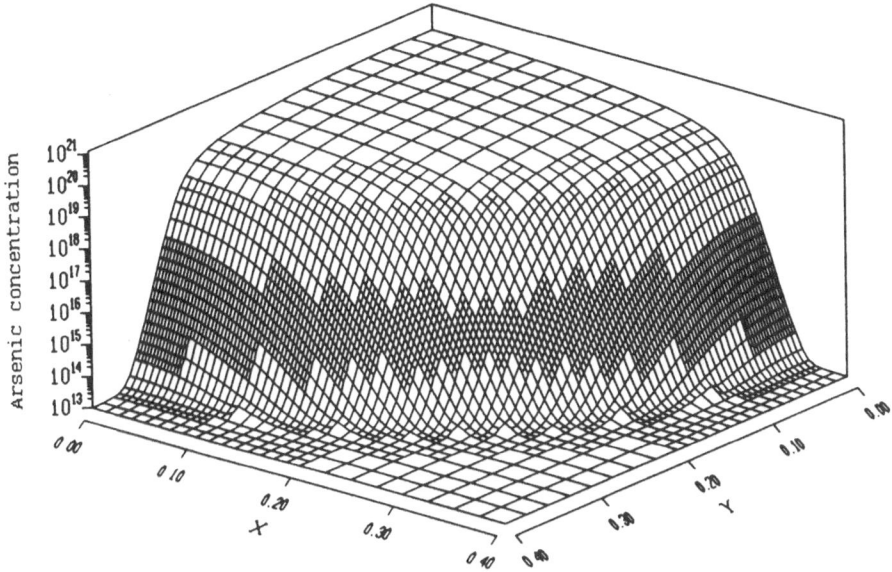

Fig. 5.10 Arsenic profile after the diffusion process $(Rel = 5 \cdot 10^{-2})$

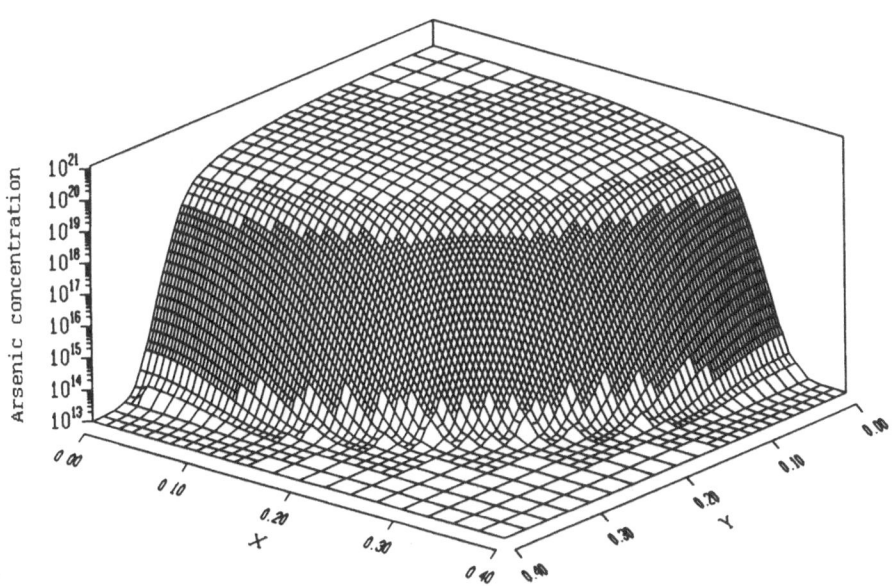

Fig. 5.11 Arsenic profile after the diffusion process $(Rel = 10^{-2})$

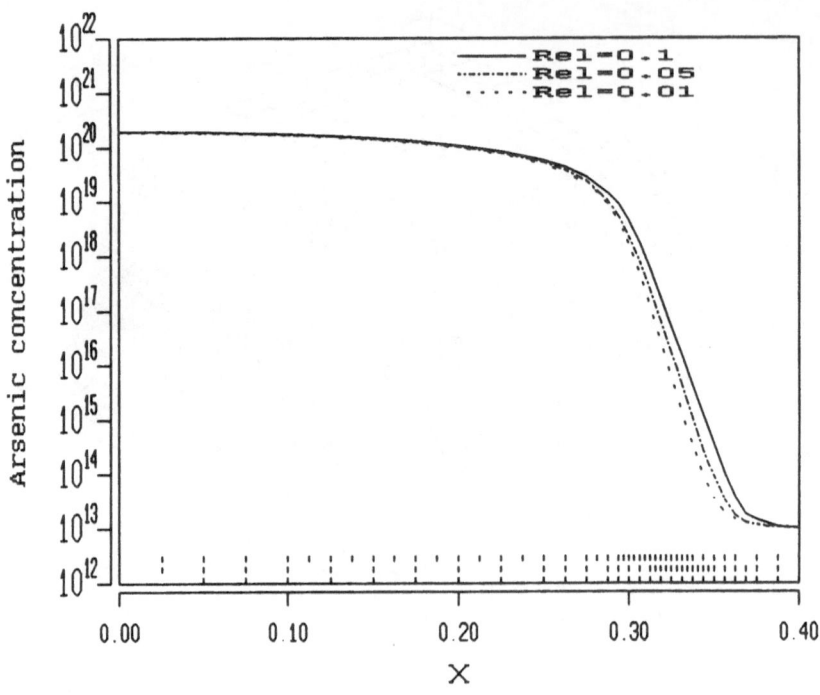

Fig. 5.12 A comparison of arsenic profiles with different relative error parameters at the boundary $\partial \Omega_p$

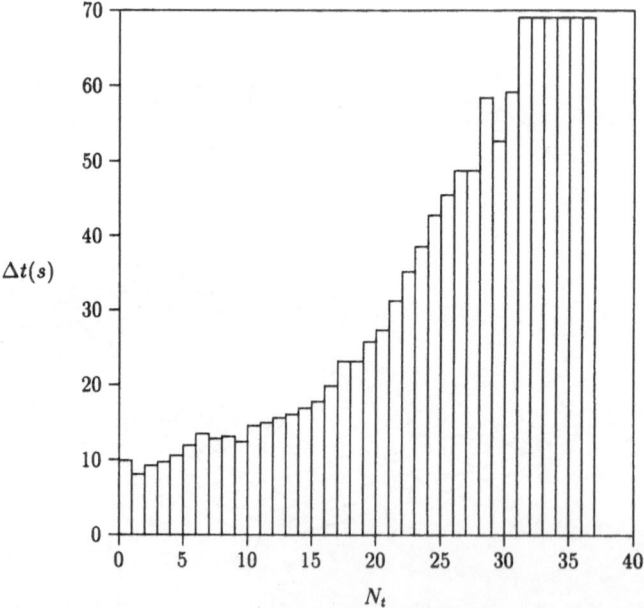

Fig. 5.13 The evolution of the time-step size during the diffusion process $(Rel = 10^{-1})$

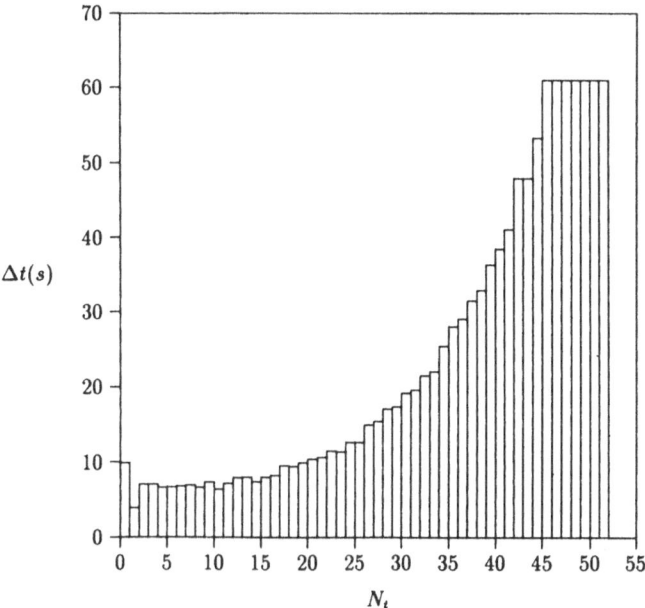

Fig. 5.14 The evolution of the time-step size during the diffusion process $(Rel = 5 \cdot 10^{-2})$

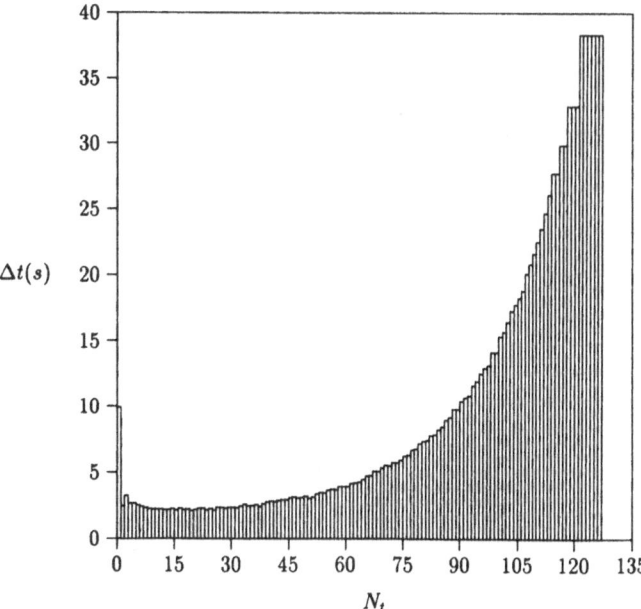

Fig. 5.15 The evolution of the time-step size during the diffusion process $(Rel = 10^{-2})$

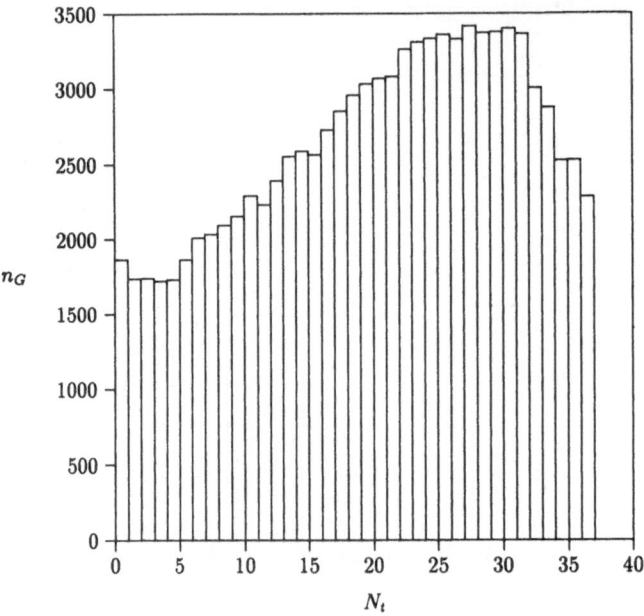

Fig. 5.16 The evolution of the total number of grid points during the diffusion process
$(Rel = 10^{-1})$

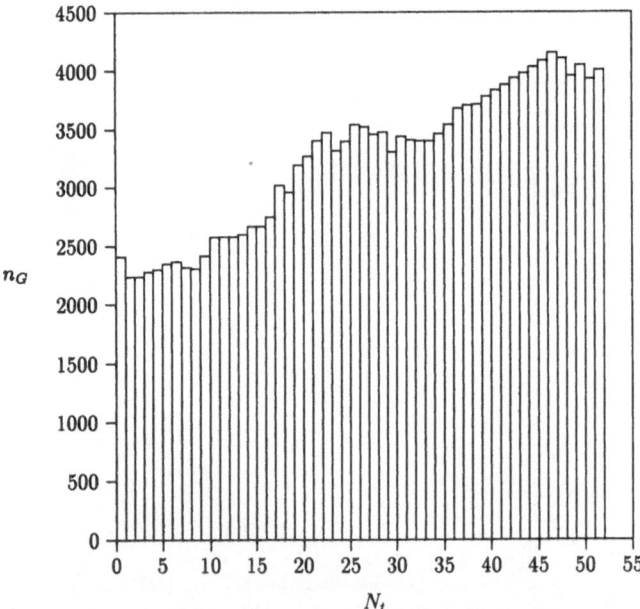

Fig. 5.17 The evolution of the total number of grid points during the diffusion process
$(Rel = 5 \cdot 10^{-2})$

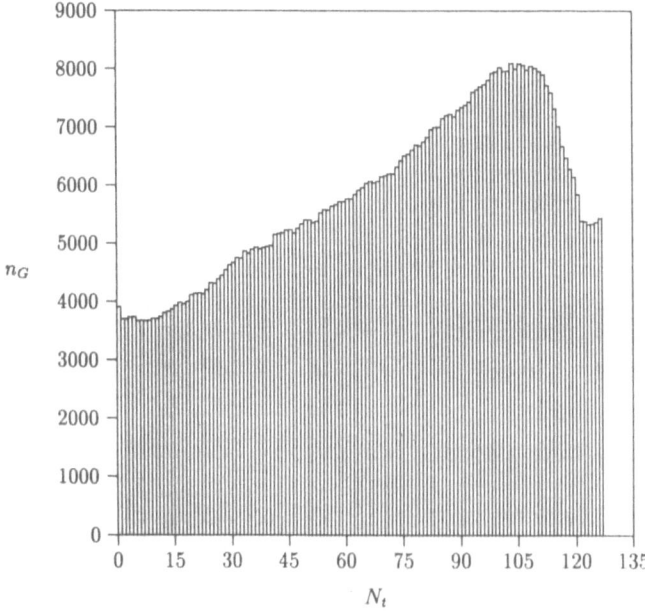

Fig. 5.18 The evolution of the total number of grid points during the diffusion process
$(Rel = 10^{-2})$

the implanted arsenic profile. This profile is used as initial state for the sub-sequent diffusion process.

The composite grid surface plots of the final arsenic profile after the diffusion process obtained with the relative error parameters $Rel = 10^{-1}$, $Rel = 5 \cdot 10^{-2}$ and $Rel = 10^{-2}$ are shown in Figures 5.9, 5.10 and 5.11.

It is obvious that the grid evolution during the simulation follows the steep front of the impurity concentration profile. The arsenic profiles in the lateral direction at the semiconductor surface defined by the boundary $\partial\Omega_p$, obtained with different relative error parameters are compared in Figure 5.12. The adaptive transient simulation has been monitored with respect to the time-step size and the total number of grid points. Figures 5.13–5.15 and 5.16–5.18 show the evolution of these two characteristic quantities for the relative error parameters $Rel = 10^{-1}$, $Rel = 5 \cdot 10^{-2}$ and $Rel = 10^{-2}$, respectively.

5.4 Local Grid-Decomposition Algorithmic Strategy for Simulation of Multiparticle Evolution Processes

The emphasis in today's process modeling as well as the obvious direction for its further development is put towards kinetic models governing the evolution of multiparticle complexes [108]. The local grid refinement, the multigrid solution and the automatic time stepping procedures, of the previous section, can be directly generalized for an application to multiparticle evolution processes. In this case the solution of the multiparticle system may be considered as vector-valued functions which have to be approximated on a multilevel adaptive grid structure (see Section 2.9). However, the straightforward application of these procedures faces a serious problem in achieving an efficient simulation if the same adaptive grid structure is used for all components of the multiparticle system with vastly different characteristic lengths. This problem is present in general and independent of the chosen approach—either singlegrid or multigrid. A great advantage of the multigrid approach is the natural extension to a local grid-decomposition strategy which may effectively eliminate this deficiency.

5.4.1 Multiparticle Process Simulation Challenges

The multiparticle diffusion with N_{eq} different types of particles (impurities and defects) is described here by a system of evolution problems where the redistribution of the j-th particle is governed by [105]

$$\frac{\partial u_j}{\partial t} - \nabla(D_j \cdot \nabla u_j) + R_j = 0 \text{ in } \Omega \times \Omega_t$$

$$D_j \cdot \nabla u_j \cdot \vec{n} + R_{pj} - G_{pj} = 0 \text{ at } \partial \Omega_p$$

$$\nabla u_j \cdot \vec{n} = 0 \text{ at } \partial \Omega_s \tag{5.13}$$

$$u_j = 0 \text{ at } \partial \Omega_d$$

$$u_j = u_j^0 \text{ in } \bar{\Omega} \text{ at } t = 0$$

Note that the coupling of the different problems (5.13) generally is achieved by the two basic coupling mechanisms:

- reaction coupling described by the reaction terms R_j and R_{pj}, in the substrate and at physical boundaries, and
- diffusion coupling described by the diffusion coefficients D_j.

According to the different reaction coupling terms, the multiparticle system can be split into subsystems concerning those particles which participate in the same reaction [139]. Inside the s-th subsystem which involves k particles $j_{s1} \le j \le j_{sk}$ for the corresponding generation-recombination rate

$$R_j = \hat{\imath}_j \cdot R_s(u_{j_{s1}}, \ldots, u_{j_{sk}})$$

holds, where $R_s(u_{j_{s1}}, \ldots, u_{j_{sk}})$ describes the governing generation-recombination mechanism for the s-th subsystem. A reference sign $\hat{\imath}_j$ is defined for each particle j which is either $+1$ or -1 and denotes whether u_j is reagent or product of R_s. The generation term G_{pj} accounts for the externally induced particle generation like interstitial injection at physical boundaries.

In general, the diffusion coefficient for the j-th particle is

$$D_j = D_j(u_1, \ldots, u_{N_{eq}}),$$

that is, for the given processing temperature, it is a nonlinear function of all particles present in the diffusion system. For simplicity the other physical coupling mechanisms like those caused by the internal electric field or those like artificially introduced couplings by cross diffusivities are neglected.

The diffusion-reaction coupling mechanisms have qualitatively different roles in impurity and point defect kinetics. The coupling through the diffusion coefficient D_j is the most important coupling mechanism for impurity diffusion equations, while in the case of point defect equations, diffusion coefficients are commonly assumed to be independent of other particle concentrations and constant for the given processing temperature. On the other hand, the reaction coupling is essential for point defect kinetics, while the impurity kinetics is usually treated without recombination-generation processes. In particular, for the limited impurity types at high concentrations, the reaction coupling can be used to model the dynamic clustering effects. The most convenient way to incorporate the clustering effects into an impurity diffusion model is to separate total impurity concentration into clustered and electrically active parts which are considered as separate particles governed by the separate transport and continuity equations and coupled through the corresponding clustering/declustering reaction terms.

Severe difficulties in the adaptive simulation of multiparticle diffusion processes arise from the large discrepancy in the impurities and defect transport coefficients. Although a considerable disagreement exists about the values of point defects diffusivities, they appear to be several orders of magnitude greater then impurity diffusivities. From the numerical simulation point of view the presence of the vastly different impurity and point defect streams causes the problem to be stiff, that is, the solution components involve a wide range of temporal and spatial characteristic lengths. The stiffness can be also associated with the point defect streams alone because of the reaction coupling. For example, the point defect diffusion length, which is typically greater than $40\,\mu m$ in the vertical direction towards the semiconductor substrate, is only a couple of microns in the lateral direction due to the rapid surface recombination rate [72]. It is well known that the proper treatment of such problems largely depends on the discretization schemes [23].

It is obvious that approaches with only a single adaptive grid structure for all particle profiles are not able to adequately resolve particle profiles with significantly different characteristic lengths. Such particle profiles typically require grid refinements at different subdomain areas with small overlap among them and a different number of local refinement steps. Consequently, the use of only one grid structure for the adaptive approximation of all particle profiles present produces in each individual particle approximation unnecessary fine grid regions introduced to satisfy the approximation needs of other particle profiles. This implies a huge computational waste, as soon as the excess grid nodes participate in the solution procedure. Besides, unnecessary fine grid regions additionally increase the stiffness of the discrete problem.

To partly overcome this problem, it has already been suggested that the simulation of the defects and impurities is performed using separate spatial grid structures [72]. Because the point defect diffusivities are much larger than impurity diffusivity the grids required for the accurate simulation of these two species differ considerably. However, the vacancies and the interstitials as well as separate impurities may also behave completely different from the discrete approximation point of view. The origin of such behavior is the fact that in the general processing sequence both oxidation processes, which are responsible for the surface interstitial generation, and ion implantation processes, used for the introduction of impurity profiles into the semiconductor, are typically performed locally on surface areas with small overlap. Even if subsequent ion implantation processes use the same processing areas, they are typically performed with different processing parameters. As a consequence, the resulting particle profiles require a different number of local grid refinement steps in different subdomains. This naturally requires additional grid separations and finally leads to the full grid-decomposition of the discrete approximation where each component of the multiparticle system has its own adaptive grid structure. With the local framework of multigrid methods using the multilevel adaptive approach with strictly regular local grids on different levels, the grid-decomposition approach is simply and inexpensively implemented both with the adaptive approximation step and within the solution process. The grid-decomposition strategy naturally applies to the construction of an initial state discrete approximation using different adaptive grid structures for different implanted impurities.

5.4.2 An Adaptive Multigrid Algorithm with Local Grid Decomposition

If in the case of discrete evolution systems the error estimator and the tolerance can be considered as vector-valued functions $\varepsilon_l = (\varepsilon_{l,1}, \ldots, \varepsilon_{l,N_{eq}})^{\mathrm{T}}$ and $Tol_l = (Tol_{l,1}, \ldots, Tol_{l,N_{eq}})^{\mathrm{T}}$. The local refinement criterion on the l-th level is commonly performed on the worst case basis with local refinement criterion:

$$\max(\varepsilon_{l,1}, \ldots, \varepsilon_{l,N_{eq}})[\vec{p}] > \min(Tol_{l,1}, \ldots, Tol_{l,N_{eq}})[\vec{p}] \qquad (5.14)$$

resulting in a single multilevel adaptive grid structure for discrete approximation of all particles. On the other hand, if the ratio of the corresponding error estimators and tolerances is vastly different, an independent application of the local refinement criteria

$$\varepsilon_{l,j}[\vec{p}] > Tol_{l,j}[\vec{p}] \quad (j = 1, \ldots, N_{eq})$$

leads to the spatial discrete approximation based on N_{eq} separately constructed adaptive multigrid structures

$$G_1 = (G_{l,1})_{l=1,M^1}, \quad G_2 = (G_{l,2})_{l=1,M^2}, \ldots, G_{N_{eq}} = (G_{l,N_{eq}})_{l=1,M^{N_{eq}}}$$

for each particle profile. It is obvious that

$$G = (G_{l,1} \cup G_{l,2} \cup \cdots \cup G_{l,N_{eq}})_{l=1,\max_{1 \leq j \leq N_{eq}} M^j}$$

Due to the different properties of various particles, which imply different needs for the local grid refinement and the maximum number of levels, the number of nodes which belong to the grid structure G_j associated with the j-th particle can be expected to be considerably smaller than the number of nodes which belongs to the grid structure G. Thus, the grid-decomposition technique may significantly reduce the total number of grid nodes involved in the solution procedure eliminating the waste of computational work associated with the unnecessary grid nodes $G_l \backslash G_{l,j}$. Nevertheless, the solution procedure requires that the local grids of the multilevel adaptive structure corresponding to the j-th particle also contains the information concerning all the other particle concentrations:

$$u_l[G_{l,j}] = (u_{l,j,1}, \ldots, u_{l,j,i}, \ldots, u_{l,j,N_{eq}})[G_{l,j}].$$

$u_{l,j,i}$ is the i-th particle concentration on the l-th level of the j-th particle adaptive multigrid structure. In order to perform the coupled multiparticle diffusion simulation, with the local grid-decoupling technique, the basic multigrid procedure has to be extended by procedures for the exchange of the discrete solution components among different adaptive grid structures. The easily realized modifications of the multigrid procedure which are proposed here for the data exchange only concern both relaxation and prolongation [105].

The relaxation is the main multigrid component which couples different particle concentrations. Therefore, the exchange of the particle concentrations among different grid structures should be performed prior to each relaxation step but restricted to the mutual intersections of the local grids on the same level. This leads to the modified relaxation procedure for the coupled multiparticle diffusion system on the l-th level, which is described in Algorithm 5.13.

Algorithm 5.13. (Modified relaxation)

Step 1. Repeat Step 2 for $j = 1, \ldots, N_{eq} - 1$
Step 2.   Repeat Step 3 for $i = j + 1, \ldots, N_{eq}$

Step 3.   *Repeat for all* $\vec{p} \in G_{l,j} \cap G_{l,i}$
                        (a) *Set* $u_{l,j,i}[G_{l,j}] := u_{l,i,i}[G_{l,i}]$
                        (b) *Set* $u_{l,i,j}[G_{l,i}] := u_{l,j,j}[G_{i,j}]$
                    [*End of loop*]
                [*End of Step 2 loop*]
            [*End of Step 1 loop*]
Step 4. Repeat Step 5 for $j = 1, \ldots, N_{eq}$
Step 5.   *Repeat for all* $\vec{p} \in G_{l,j}$
                        Set $u_{l,j,j} := S_{l,j}^{(v)}(u_{l,j,1}, \ldots, u_{l,j,N_{eq}})$
                [*End of loop*]
            [*End of Step 4 loop*]
Step 6. Exit.

The update of the variables different from that variable actually relaxed is performed by an ordinary prolongation only in the grid parts of $G_{l,j}$ which do not overlap with other particles' local grids on the same level. The modified prolongation procedure is described in Algorithm 5.14

Algorithm 5.14. (Modified prolongation)

Step 1. Repeat for all $\vec{p} \in G_{l,j}$ *and* $j = 1, \ldots, N_{eq}$
            Set $u_{l,j,j} := u_{l,j,j} + I_{l-1}^{l}(u_{l-1,j,j} - \hat{I}_{l}^{l-1} u_{l,j,j})$
        [*End of loop*]
Step 2. Repeat Step 3 for $j = 1, \ldots, N_{eq}$
Step 3.   *Repeat Step 4 for* $i = 1, \ldots, N_{eq}$ *and* $i \neq j$
Step 4.   *Repeat for all* $\vec{p} \in G_{l,j} \setminus G_{l,i}$
                        Set $u_{l,j,i} := I_{l-1}^{l}(u_{l-1,j,i})$
                    [*End of loop*]
                [*End of Step 3 loop*]
            [*End of Step 2 loop*]
Step 5. Exit.

The most appropriate selection for the interpolation operator I_{l-1}^{l} in Step 4 is the operator which is used as full multigrid interpolation. Except for the exchange of particle concentrations among local grid structures before relaxation, all multigrid components, including the modified prolongation procedure as well as the multilevel local grid refinement, are performed on the separate multilevel adaptive structures and can be done in parallel.

5.4.3 An Example of Coupled Impurity Diffusion Simulation

A first practical application of the local grid-decomposition strategy is the simulation of coupled impurity diffusion. For that purpose the impurity diffusion model from Section 5.3.4 is used to simulate the coupled arsenic

Table 5.5. *Technology parameters*

Processing	Parameter	Selection
0. Substrate	length (l_x)	$0.6\,\mu m$
	depth (l_y)	$0.6\,\mu m$
	type	P
	concentration	$10^{15}\,cm^{-3}$
	pad oxide	$0.1\,\mu m$
1. Implantation	impurity type	arsenic
	energy	120 keV
	dose	$6 \cdot 10^{15}\,cm^{-2}$
	masked region	$x \geq 0.1\,\mu m$
2. Implantation	impurity type	phosphorus
	energy	110 keV
	dose	$5 \cdot 10^{13}\,cm^{-2}$
	masked region	$x \geq 0.1\,\mu m$
3. Diffusion	temperature	950°C
	time	1500 s
	ambient	inert

and phosphorus diffusion. The coupling is achieved by the equilibrium vacancy concentration dependent diffusion coefficients. Implanted arsenic and phosphorus profiles are used as an initial state. The values of the processing parameters are given in Table 5.5 while the numerical simulation is performed with the multigrid components which are given in Table 5.4 in combination with the parameters $Rel = 5 \cdot 10^{-2}$ and $Abs = 10^{14}\,cm^{-3}$.

The composite grid surface plots of the implanted arsenic and phosphorus profiles are shown in Figures 5.19 and 5.20. Figures 5.21 and 5.22 show the corresponding plots of arsenic and phosphorus profiles after coupled diffusion simulation with grid-decomposition. For comparison, the phosphorus profile which is obtained under the same processing conditions but completely decoupled (solved as an independent evolution problem) from the arsenic is shown in Figure 5.23.

Figures 5.22 and 5.23 show that the presence of the high arsenic concentration in areas with the low phosphorus concentration significantly increases the phosphorus diffusion coefficient and thus produces a flat phosphorus profile at this area. In case of the decoupled diffusion the phosphorus profile still has an implantation-like shape. The reverse effect is not observed since the phosphorus concentration is below the intrinsic carrier concentration at the processing temperature. The difference between the two grid structures of arsenic and phosphorus is obvious. The extremely steep arsenic profiles lead to the maximum $M = 6$ levels while for the phosphorus profile the local refinement procedure terminated with $M = 5$ levels. Moreover, local grid structures cover different domain areas.

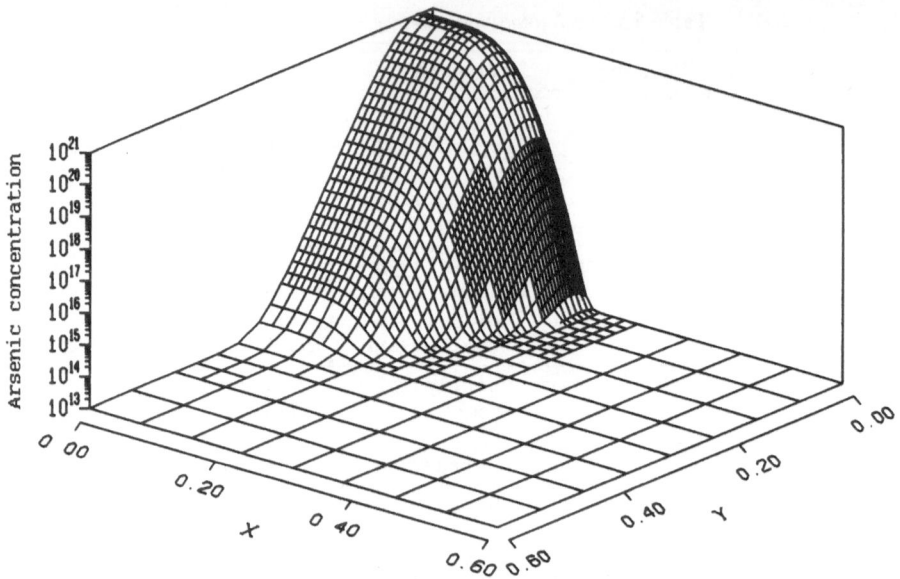

Fig. 5.19 Arsenic profile after implantation

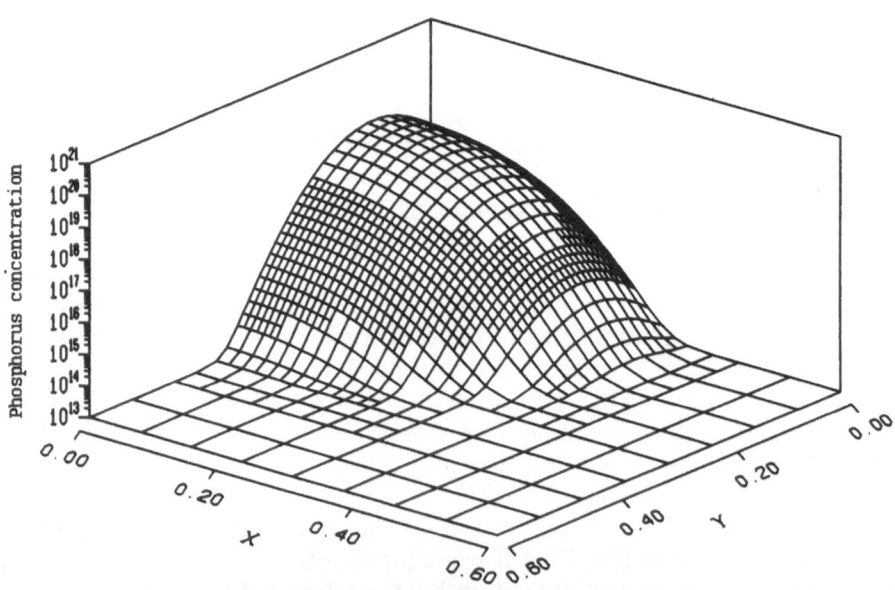

Fig. 5.20 Phosphorus profile after implantation

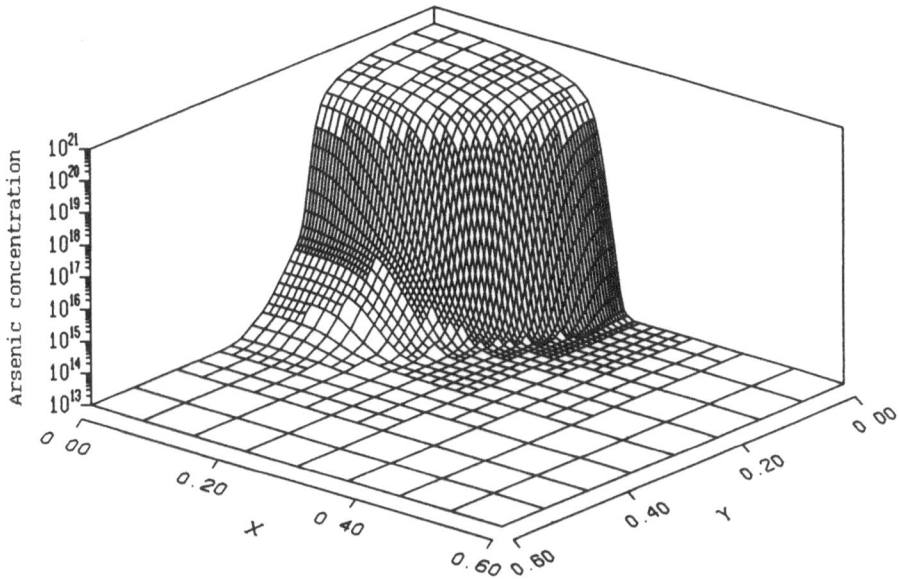

Fig. 5.21 Arsenic profile after coupled diffusion

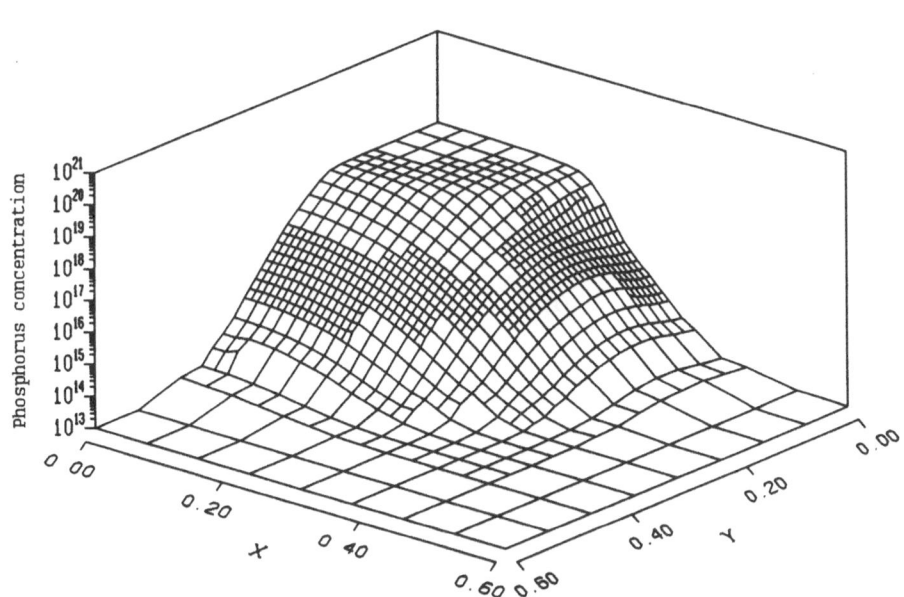

Fig. 5.22 Phosphorus profile after coupled diffusion

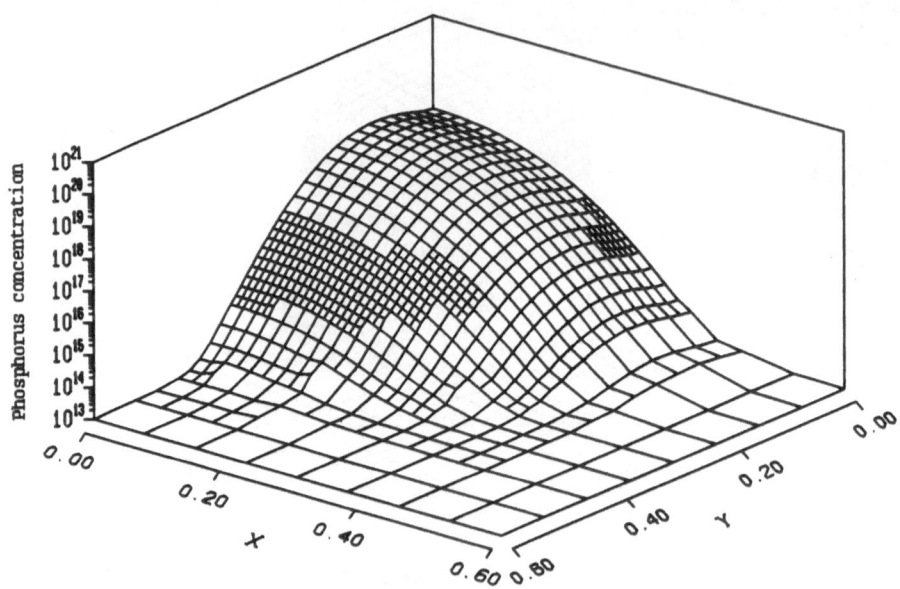

Fig. 5.23 Phosphorus profile after decoupled diffusion

5.4.4 An Example of Coupled Interstitial-Vacancy Diffusion Simulation During Local Oxidation

Another practical application of the grid-decomposition strategy to diffusion-reaction coupled systems is the coupled interstitial and vacancy diffusion during the local oxidation process. It is assumed that point defect kinetics is governed by the two-particle evolution problem [72]:

$$\frac{\partial N_{I/V}}{\partial t} - D_{I/V} \cdot \left(\frac{\partial^2 N_{I/V}}{\partial x^2} + \frac{\partial^2 N_{I/V}}{\partial y^2} \right) + k_R (N_I N_V - N_I^* N_V^*) = 0 \text{ in } \Omega \qquad (5.15)$$

$$D_I \cdot \frac{\partial N_I}{\partial y} - G_I + k_I (N_I - N_I^*) = 0 \text{ at } \partial\Omega_{p1}$$

$$D_V \cdot \frac{\partial N_V}{\partial y} + k_V (N_V - N_V^*) = 0 \text{ at } \partial\Omega_{p1}$$

$$D_{I/V} \cdot \frac{\partial N_{I/V}}{\partial y} + k_{I/V} (N_{I/V} - N_{I/V}^*) = 0 \text{ at } \partial\Omega_{p2}$$

$$\frac{\partial N_{I/V}}{\partial x} = 0 \text{ at } \partial\Omega_{s1}, \ \partial\Omega_{s2}$$

$$N_{I/V} = N_{I/V}^* \text{ in } \bar{\Omega} \text{ at } t = 0$$

The subscripts I and V refer to interstitials and vacancies, respectively. The superscript $*$ denotes their equilibrium concentrations. I/V in some of the above equations stands for two identical equations, one for I and one for V. The interstitial and vacancy diffusivities are assumed to be constant and the generation-recombination term $k_R(N_I N_V - N_I^* N_V^*)$ in the semiconductor substrate couples the interstitial and vacancy diffusion streams. The equations are symmetric with the only exception that the boundary condition for the interstitials includes the local generation term G_I which is introduced to model the interstitial injection due to the local oxidation.

Apart from the previously considered impurity diffusion problem, in case of point defects there are two physical boundaries

$$\partial\Omega_{p1} = \{(x, y)|0 \le x \le l_x; y = 0\}$$

and

$$\partial\Omega_{p2} = \{(x, y)|0 \le x \le l_x; y = l_y\}$$

associated with the top and the bottom planar semiconductor surfaces, respectively. The top semiconductor surface actually represents the $Si - SiO_2$ interface where the local oxidation takes place. The boundaries $\partial\Omega_{s1} = \{(x, y)|x = 0; 0 \le y \le l_y\}$ and $\partial\Omega_{s2} = \{(x, y)|x = l_x; 0 \le y \le l_y\}$ are the symmetry lines.

The interstitial generation term G_I at the top silicon surface exposed to the local oxidation process is modeled due to [72] by

$$G_I = n_{Si} \cdot \theta \cdot \dot{U}_{\bar{n}}$$

n_{Si} is the atomic density of silicon. θ is the fraction of consumed silicon injected into the substrate. $\dot{U}_{\bar{n}}$ is the velocity of the $Si - SiO_2$ interface in the direction of the normal vector, which is modeled according to the Deal-Grove formulation [22, 31] as a function of time and processing temperature and using (4.3) to express its dependence of the lateral coordinate along the oxidizing surface.

Table 5.6. *Technology parameters*

Processing	Parameter	Selection
0. Substrate	length (l_x)	15 μm
	depth (l_y)	15 μm
	type	intrinsic
	pad oxide	0.01 μm
2. Local oxidation	temperature	1000 °C
	time	60 s, 600 s
	ambient	wet
	masked region	$x \ge 3\,\mu$m

Table 5.7. *The model parameters for the coupled defect diffusion simulation*

D_I	$2.03 \cdot 10^{-10}$ cm^2/s	D_V	$2.03 \cdot 10^{-10}$ cm^2/s
C_I^*	$1.56 \cdot 10^{16}$ cm^{-3}	C_V^*	$1.56 \cdot 10^{16}$ cm^{-3}
k_I	$1.1 \cdot 10^{-6}$ cm/s	k_V	$1.1 \cdot 10^{-6}$ cm/s
k_R	$7.14 \cdot 10^{-21}$ cm^3/s	θ	0.005
n_{Si}	$5 \cdot 10^{22}$ cm^{-3}		

The technology parameters of the example are given in Table 5.6 while the physical parameters used for the formulation of the interstitial and the vacancy equations at the processing temperature have the values given in Table 5.7 [72].
Composite grid surface plots of both the interstitial and vacancy profiles after 60s of the local oxidation process are shown in Figures 5.24 and 5.25. The rapid initial oxidation in the wet ambient and the resulting interstitial injection significantly increases the interstitial concentration near the oxidizing surface reaching supersaturation after an extremely short time. In the region covered by an interstitial profile whose concentration is

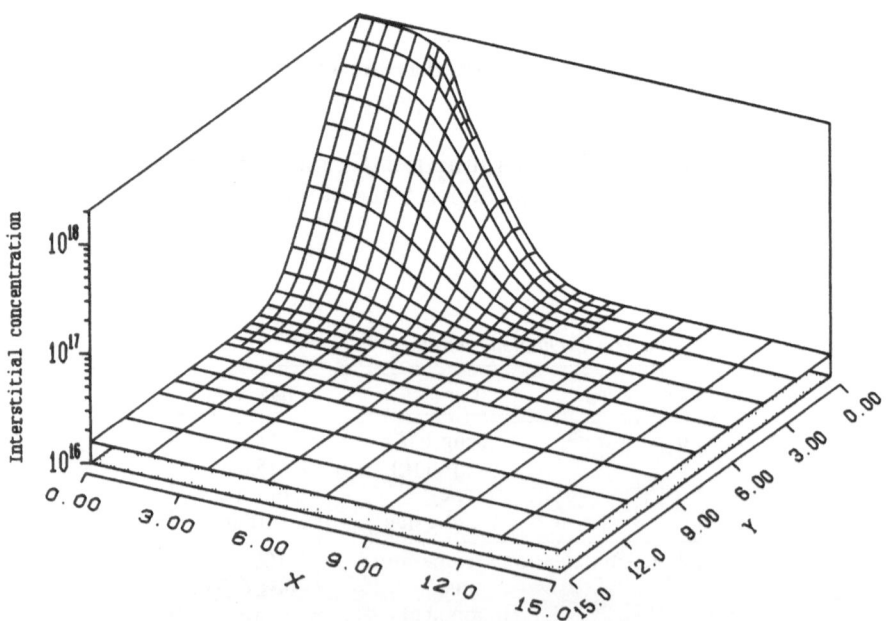

Fig. 5.24 Interstitial profile after 60s of the local oxidation process

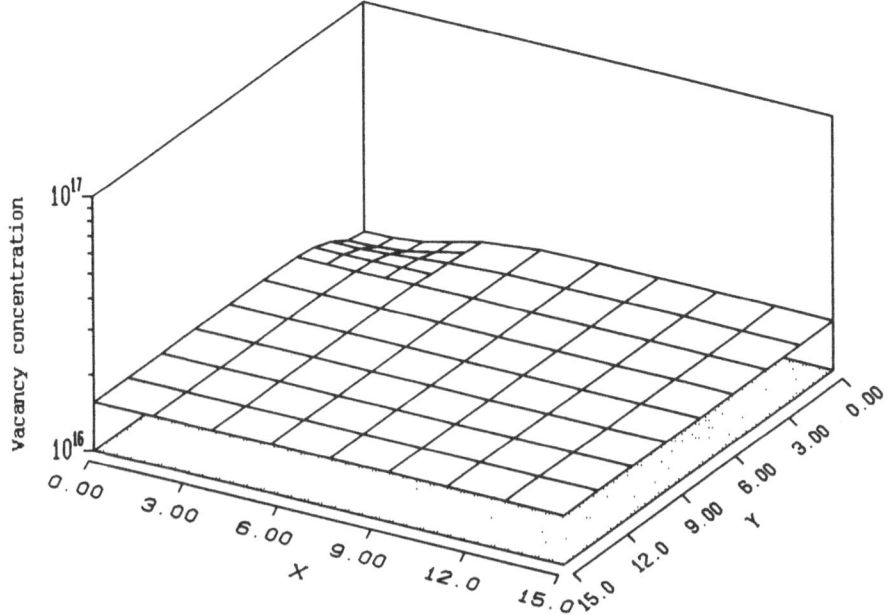

Fig. 5.25 Vacancy profile after 60s of the local oxidation process

significantly higher than N_I^*, the bulk recombination-generation term $k_R(N_I N_V - N_I^* N_V^*)$ depletes the vacancy concentration. This depletion then builds a gradient in vacancy concentration which transfers the vacancies from the bulk to the surface.

Composite grid surface plots of interstitial and vacancy profiles after $600s$ of the local oxidation process are shown in Figures 5.26 and 5.27.

Due to the large interstitial diffusion coefficient, the interstitial flux has reached the bottom semiconductor surface where the boundary recombination term $k_I(N_I - N_I^*)$ keeps the interstitial concentration at the equilibrium value. The abrupt change in the lateral interstitial concentration occurs all along the simulation in the vicinity of the local oxidation mask edge. The effect of the finite surface recombination term $k_V(N_V - N_V^*)$ on the top semiconductor surface is obvious by the well known vacancy-kink phenomenon which can be observed in Figure 5.27. The surface recombination mechanism counteracts the depletion vacancy front by generating vacancies at the surface. The result is the characteristic region of positive curvature, the kink in the vacancy profile.

The effect of the local grid decomposition strategy is obvious. The interstitial grid structure after 60s requires 5 levels while the corresponding vacancy profile requires only 3. The 5-th level in the interstitial discrete

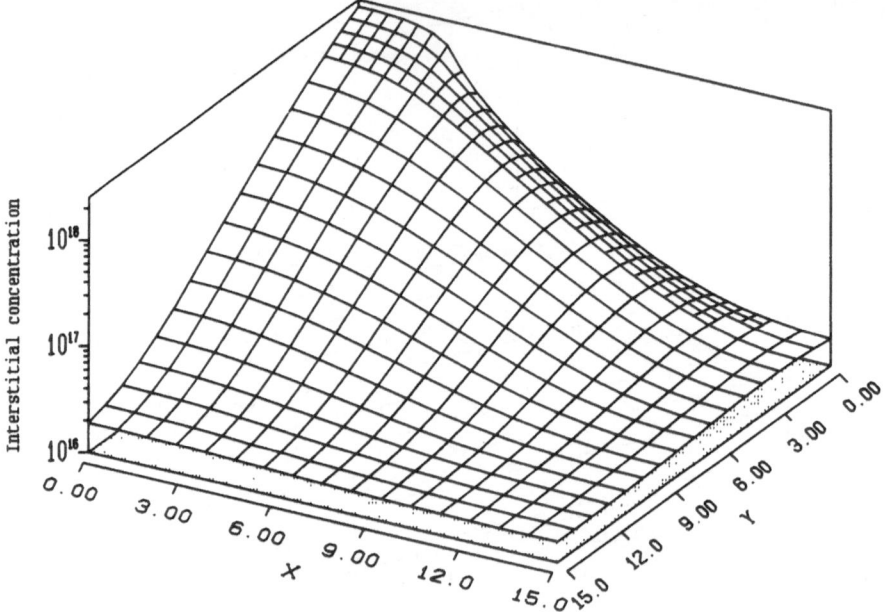

Fig. 5.26 Interstitial profile after 600s of the local oxidation process

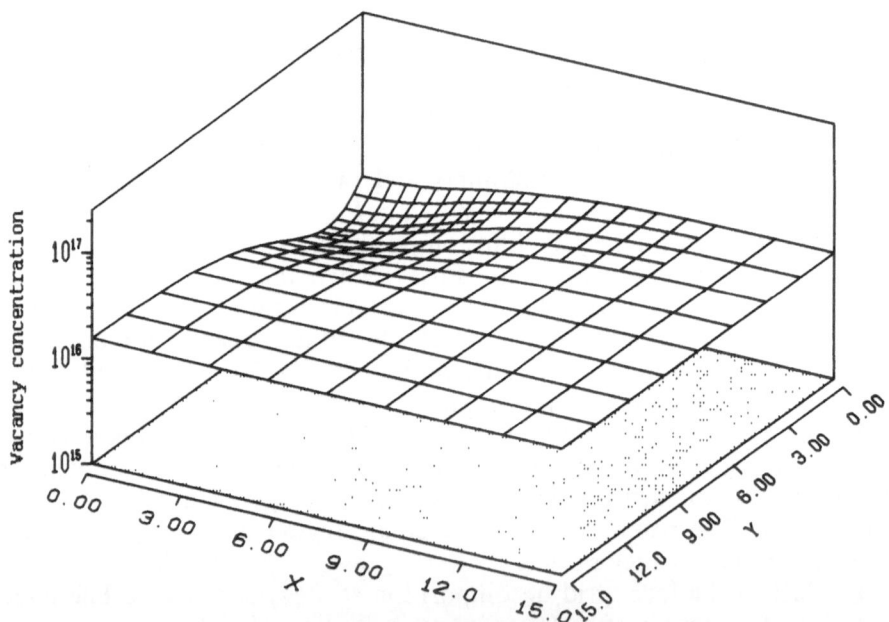

Fig. 5.27 Vacancy profile after 600s of the local oxidation process

approximation, which appeared near the edge of the local oxidation mask disappears with larger processing times. The appearance of the vacancy kink shape results in 4-th level of the vacancy grid structure at the end of the simulation.

5.5 Simulation of Critical Process Simulation Steps of BiCMOS Technology—Case Study

The basic algorithmic procedures for the simulation of evolution processes using the local grid-decomposition strategy are designed to handle general multiparticle processing sequences (evolution and intermediate auxiliary processes). A critical implantation and coupled impurity diffusion processes in the processing sequence for the fabrication of BiCMOS integrated circuits [51] serves as an example to demonstrate these capabilities. The BiCMOS technology which combines **Bi**polar and **C**MOS transistors in a single integrated circuit is attractive from the simulation point of view since it involves the simultaneous processing of three different impurity types in the implantation and the diffusion processes.

In the following, the process flows for realizing an NMOS, a bipolar and a PMOS transistor in BiCMOS technology are presented from a $0.8\,\mu m$ baseline CMOS transistor perspective. Critical processing steps for the fabrication of the NMOS transistor are used to describe the basic process flow while some of the NMOS processing steps are merged with the bipolar and the PMOS transistor process flows to realize their respective structures. The considered processing sequences are extracted from the complete process flow for BiCMOS integrated circuits and simplified to restrict the simulation to planar rectangular domains and constant substrate concentration. Because of the symmetric device structures of NMOS and PMOS bipolar transistors only one half of the device structures are simulated. The process parameters are selected from the literature [51] as typical but certainly not the optimal ones. Generally, the optimization of BiCMOS technology parameters is a multifaceted problem due to the conflicting requirements of the mixed bipolar and CMOS technology.

5.5.1 Simulation of the Lightly Doped Drain (LDD) Structure for the NMOS Transistor

Recently much attention has been paid to LDD NMOS devices. They have become the important VLSI device due to the reduction of the peak electric field at the drain which significantly improves the performance of the standard NMOS devices. This is usually accomplished by the introduction of a lightly doped drain extension along with the high doped source/drain

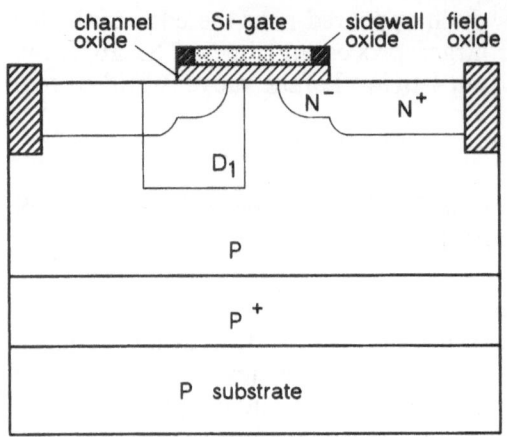

Fig. 5.28 Cross section of an NMOS transistor structure

profiles. The sidewall oxide at the gate edges acts as a mask against the high dose source/drain ion implant, preventing it from entering lightly doped source/drain regions. The cross-section of the NMOS transistor in BiCMOS technology as well as a rectangular simulation domain D_1 used for the two-dimensional simulation of the fabrication steps for the LDD source/drain structure are shown in Figure 5.28.

Process Flow 5.1 describes the basic implantation and diffusion steps used to realize the LDD source/drain impurity profile with corresponding processing parameters.

Process Flow 5.1. The LDD structure of the NMOS transistor

1.0 **Substrate**
> *Type*: *P-well*
> *Concentration*: $5 \cdot 10^{16} \, \text{cm}^{-3}$
> *Pad oxide*: $0.02 \, \mu\text{m}$

1.1 **Implantation** (*Channel regions*)
> *Dopant*: *boron*
> *Dose*: $1.5 \cdot 10^{12} \, \text{cm}^{-2}$
> *Energy*: 40 keV

1.2 **Diffusion** (*Redistribution of the channel profile*)
> *Temperature*: $1050 \, ^\circ\text{C}$
> *Time*: 15 min

1.3 **Implantation** (*Low doped NMOS source/drain region*)
> *Dopant*: *phosphorus*
> *Dose*: $10^{13} \, \text{cm}^{-2}$
> *Energy*: 30 keV
> *Masked region*: $x \geq 0.4 \, \mu\text{m}$

1.4 **Diffusion** (*The sidewall oxide formation*)
         Temperature: 900 °C
         Time: 20 min
1.5 **Implantation** (*High doped NMOS source/drain region*)
         Dopant: *phosphorus*
         Dose: $5 \cdot 10^{13}$ cm^{-2}
         Energy: 110 keV
         Masked region: $x \geq 0.2$ μm
1.6 **Implantation** (*High doped NMOS source/drain region*)
         Dopant: *arsenic*
         Dose: $5 \cdot 10^{15}$ cm^{-2}
         Energy: 80 keV
         Masked region: $x \geq 0.2$ μm
1.7 **Diffusion** (*Annealing*)
         Temperature: 900 °C
         Time: 20 min

Impurity profiles after processing steps 1.1–1.7 are shown in Figures 5.29–5.36. The total net impurity concentration of the NMOS LDD structure is shown in Figure 5.37.

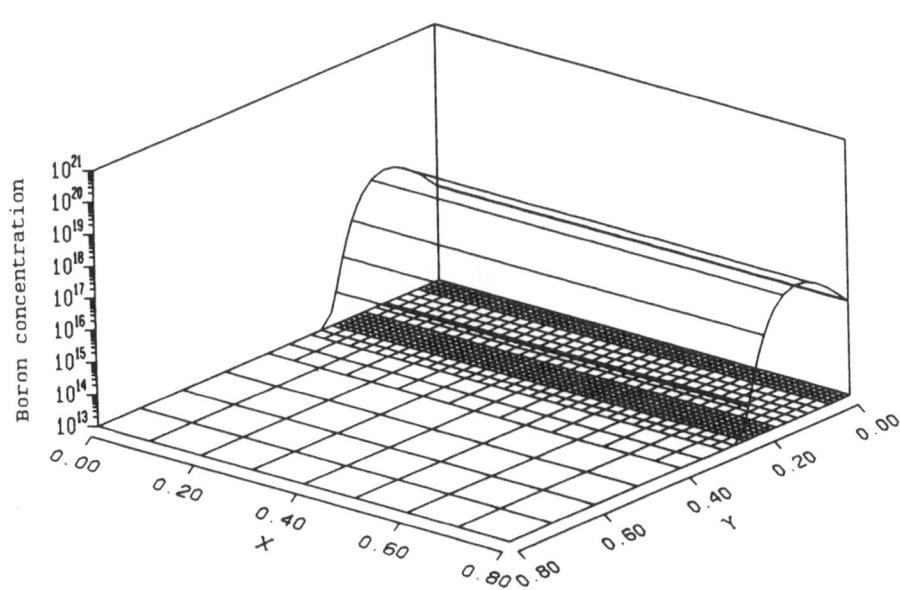

Fig. 5.29 A perspective isoconcentration plot of the boron profile after process step 1.1

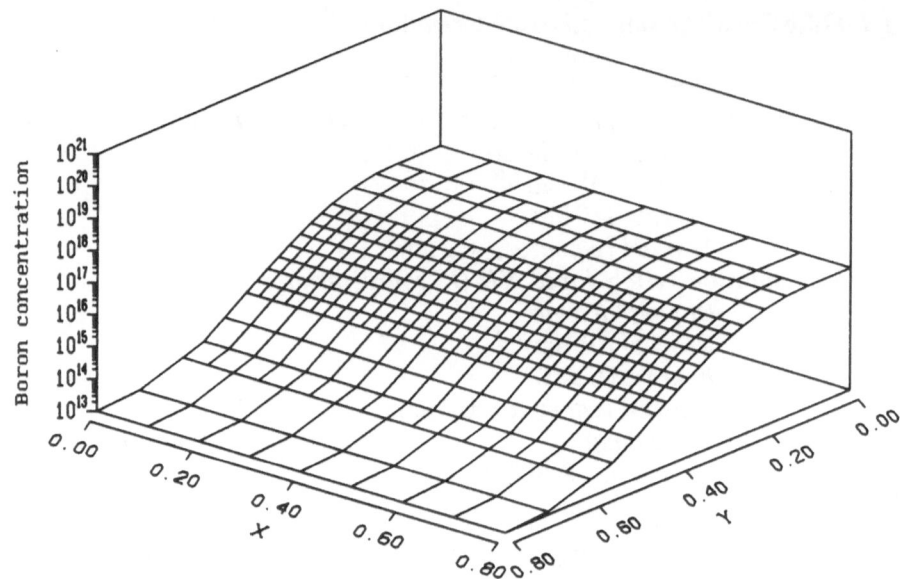

Fig. 5.30 Boron profile after process step 1.2

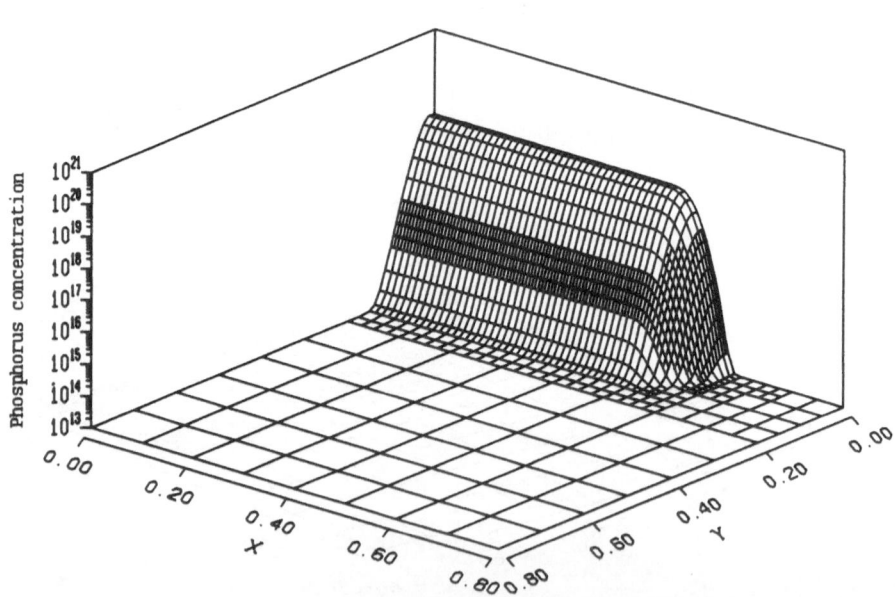

Fig. 5.31 Phosphorus profile after process step 1.3

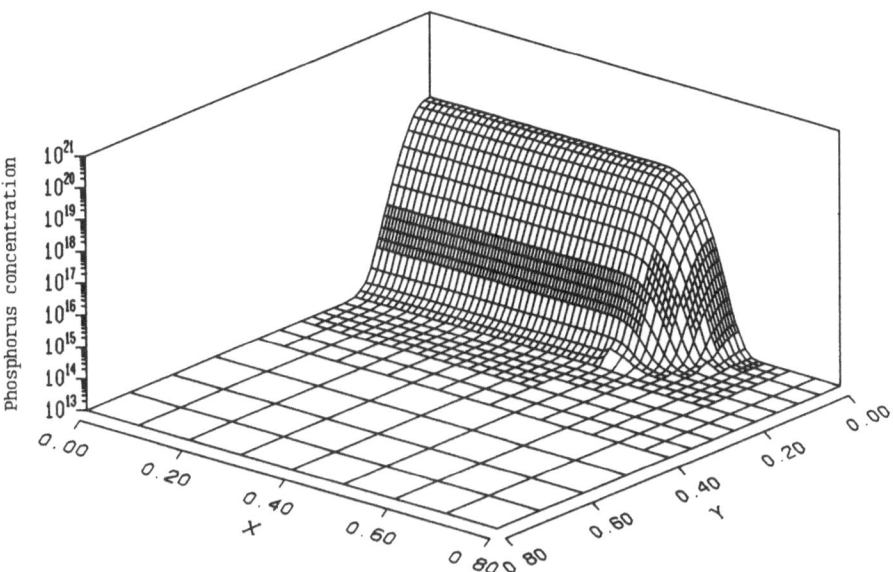

Fig. 5.32 Phosphorus profile after process step 1.4

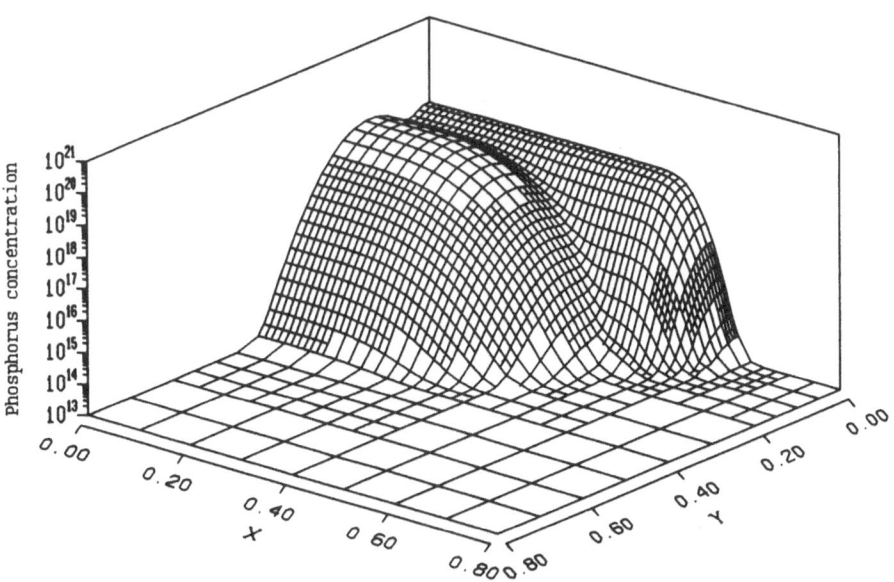

Fig. 5.33 Phosphorus profile after process step 1.5

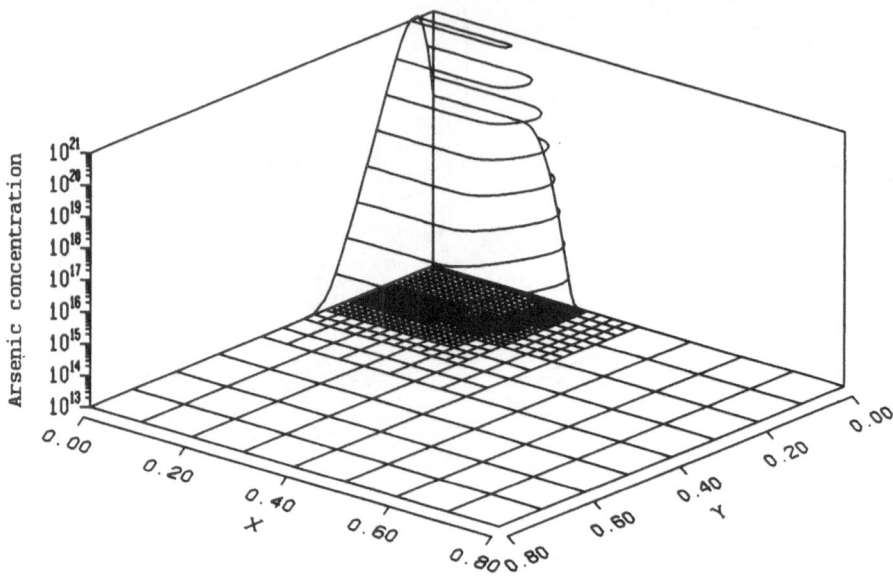

Fig. 5.34 A perspective isoconcentration plot over the composite grid of the arsenic profile
after process step 1.6

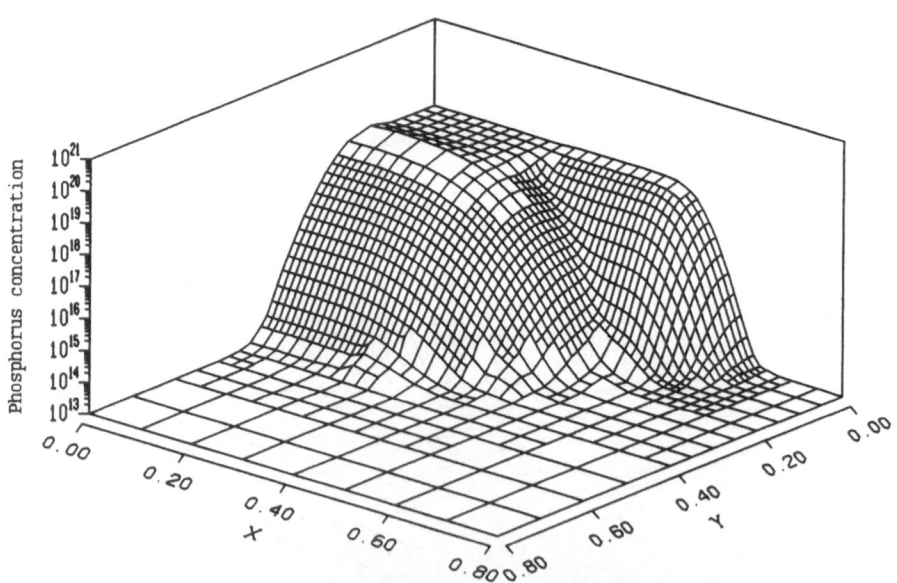

Fig. 5.35 Phosphorus profile after process step 1.7

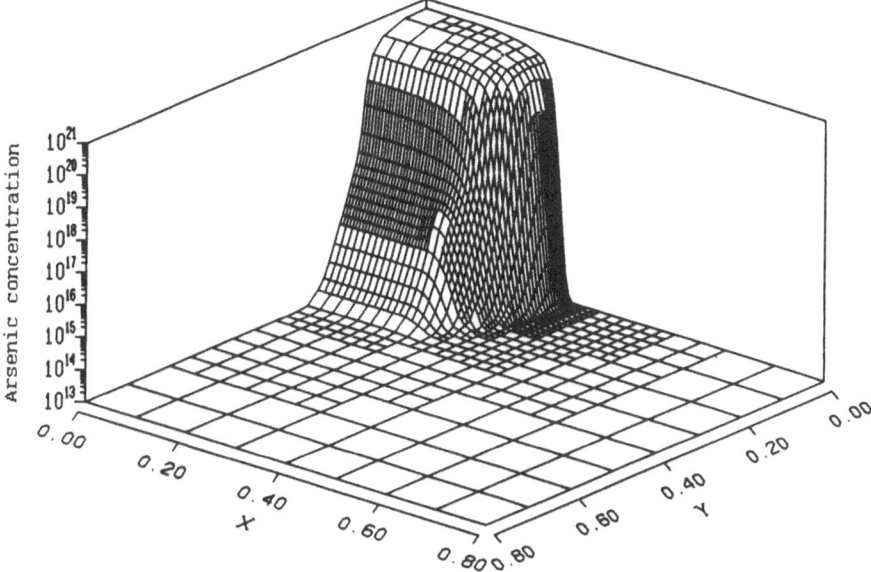

Fig. 5.36 Arsenic profile after process step 1.7

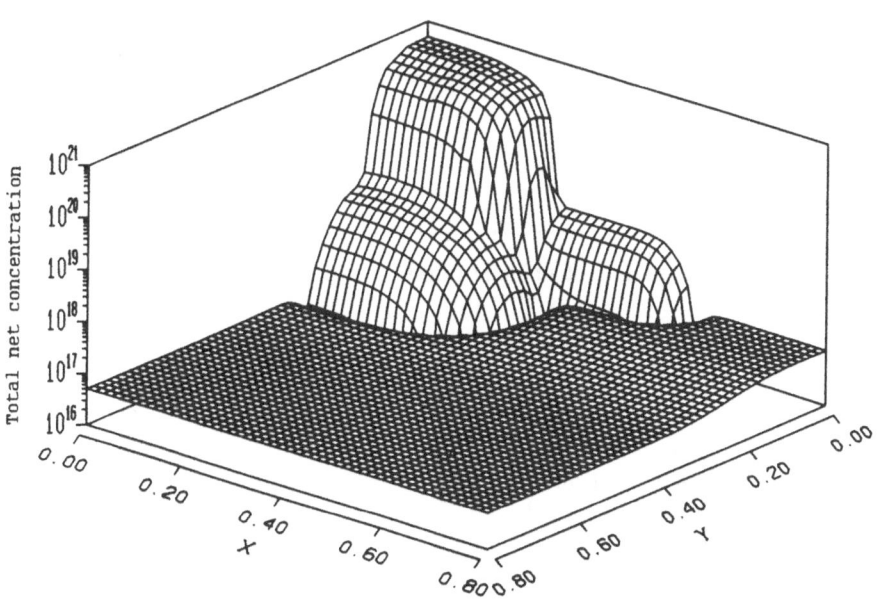

Fig. 5.37 Total net impurity profile of the NMOS transistor

5.5.2 Simulation of the Base and Emitter of the NPN Bipolar Transistor

The major challenge associated with the BiCMOS technology is how to obtain high performance bipolar and CMOS transistors with a minimal increase in process complexity compared to the conventional CMOS technology. The crucial process steps to create an NPN bipolar transistor are the ion implantation and the diffusion processes for base and emitter regions. It is common practice to implant the base region and minimize the subsequent thermal cycles in order to achieve a narrow base width. The formation of an extrinsic base region is necessary to minimize the base resistance. For simplicity, the present example is based on the technique where the emitter profile is directly implanted and annealed along with CMOS source/drain regions.

The cross-section of a bipolar structure and the chosen rectangular simulation domain D_2 are presented in Figure 5.38 while the Process Flow 5.2 includes the processing parameters.

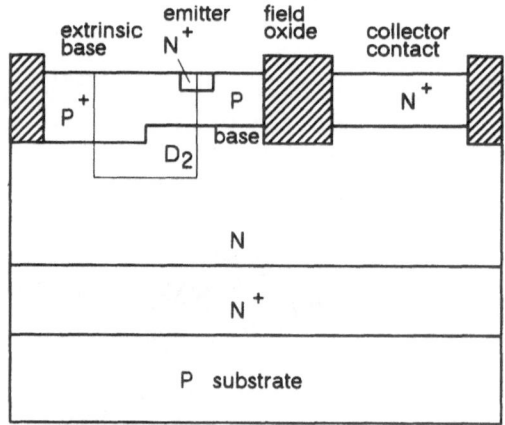

Fig. 5.38 A cross-section of a bipolar transistor with directly implanted emitter

Process Flow 5.2. The base and emitter of an NPN bipolar transistor

1.0 **Substrate**
>   *Type: N-well*
>   *Concentration:* 10^{16} cm^{-3}

2.1 **Implantation** *(Base region)*
>   *Dopant: boron*
>   *Dose:* 10^{13} cm^{-2}
>   *Energy:* 80 keV

2.2 **Diffusion** (*Step 1.2*)
2.3 **Implantation** (*Emitter region*)
    Dopant: arsenic
    Dose: 10^{16} cm^{-2}
    Energy: 50 keV
    Masked region: $x \leq 0.7\,\mu$m
2.4 **Diffusion** (*Step 1.4*)
2.5 **Implantation** (*Base contact region*)
    Dopant: boron
    Dose: $6 \cdot 10^{15}$ cm^{-2}
    Energy: 50 keV
    Masked region: $x \geq 0.1\,\mu$m
2.6 **Diffusion** (*Step 1.7*)

Impurity profiles after processing steps 2.1–2.6 are given in Figures 5.39–5.45. The total net impurity concentration of the bipolar transistor base and emitter structure is shown in Figure 5.46.

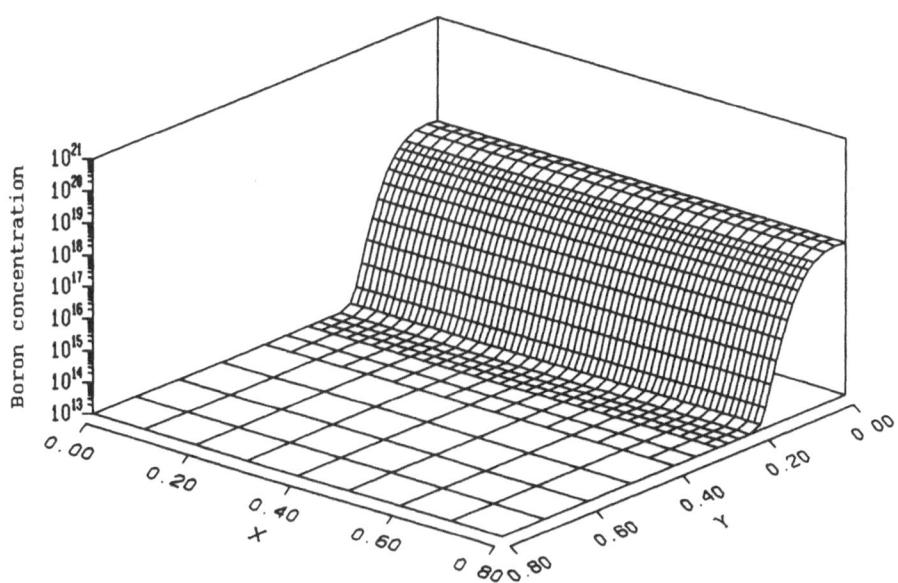

Fig. 5.39 Boron profile after process step 2.1

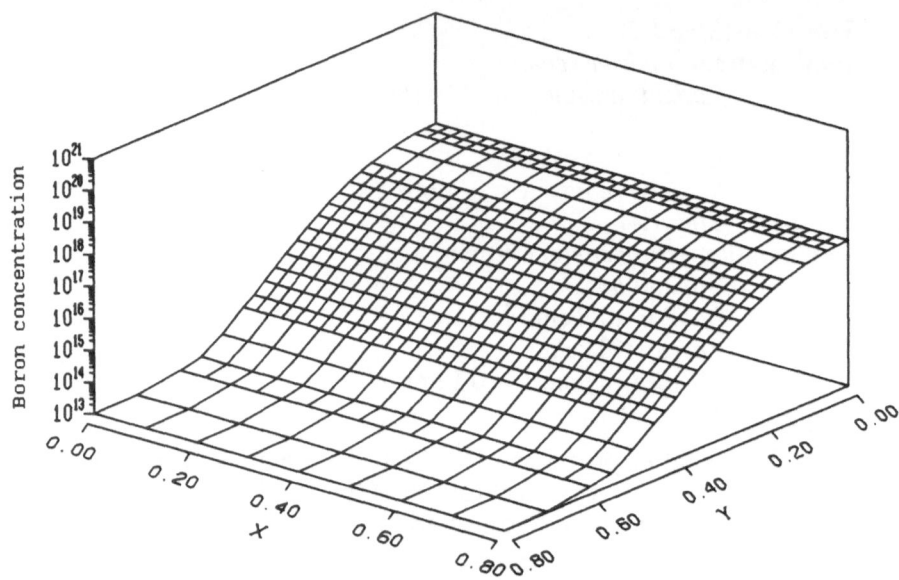

Fig. 5.40 Boron profile after process step 2.2

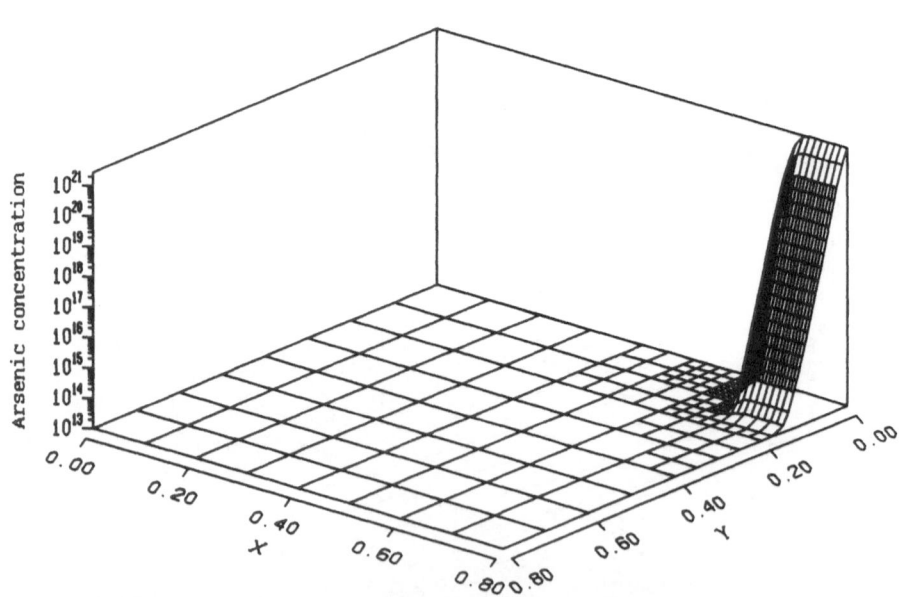

Fig. 5.41 Arsenic profile after process step 2.3

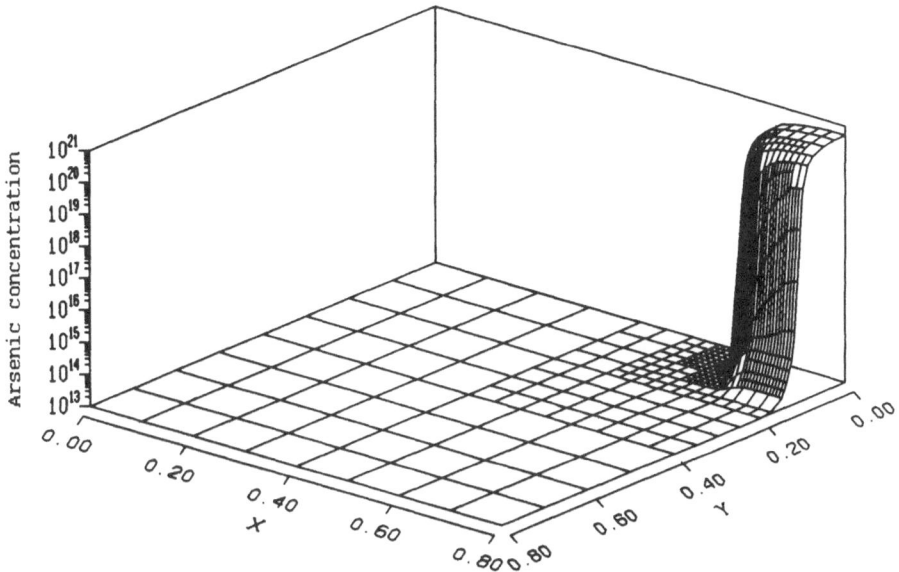

Fig. 5.42 Arsenic profile after process step 2.4

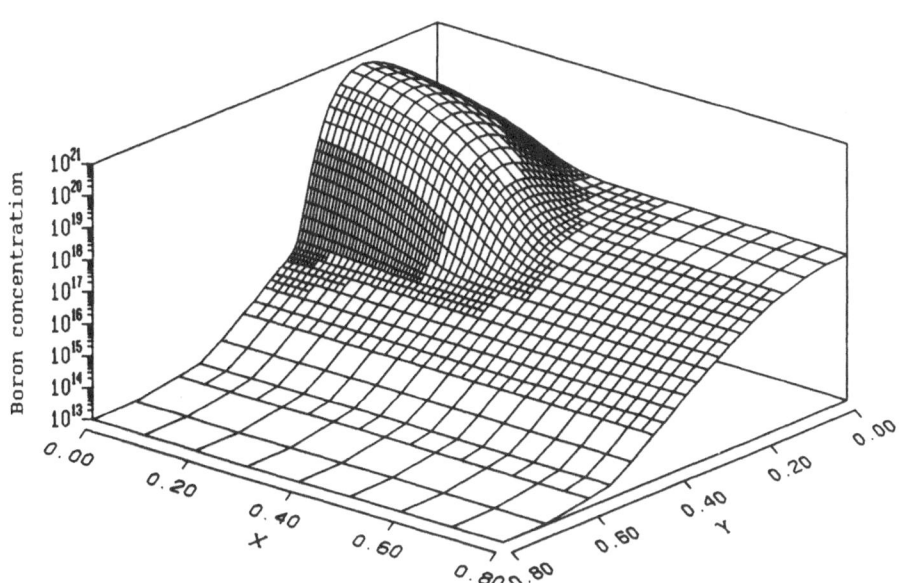

Fig. 5.43 Boron profile after process step 2.5

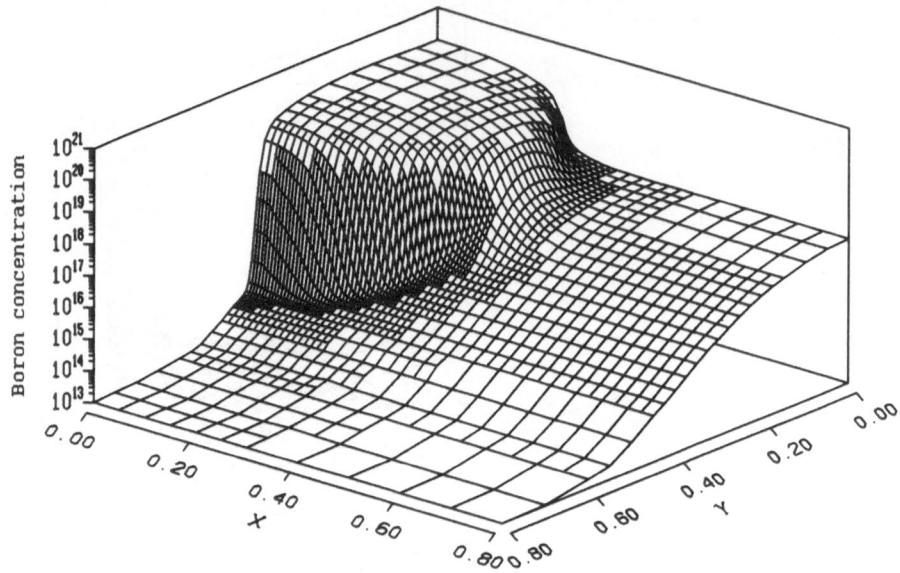

Fig. 5.44 Boron profile after process step 2.6

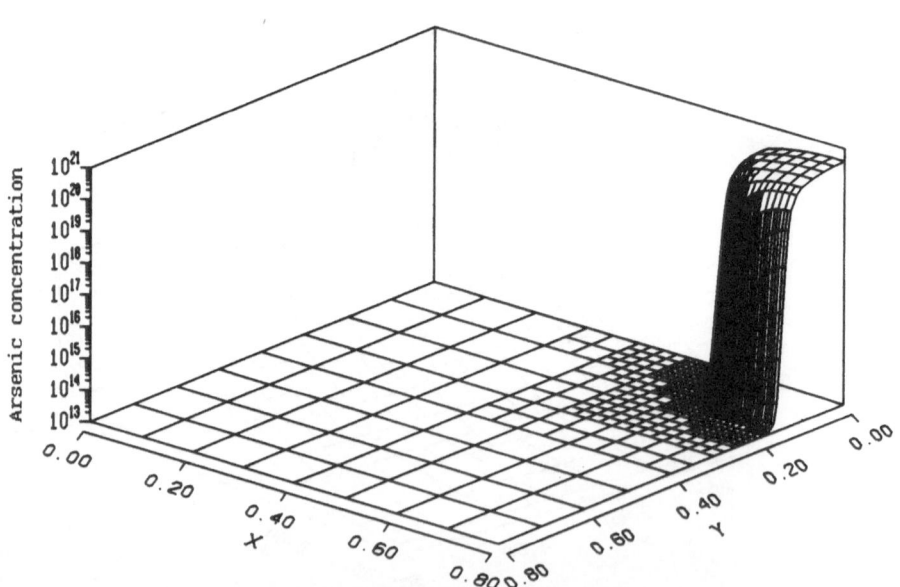

Fig. 5.45 Arsenic profile after process step 2.6

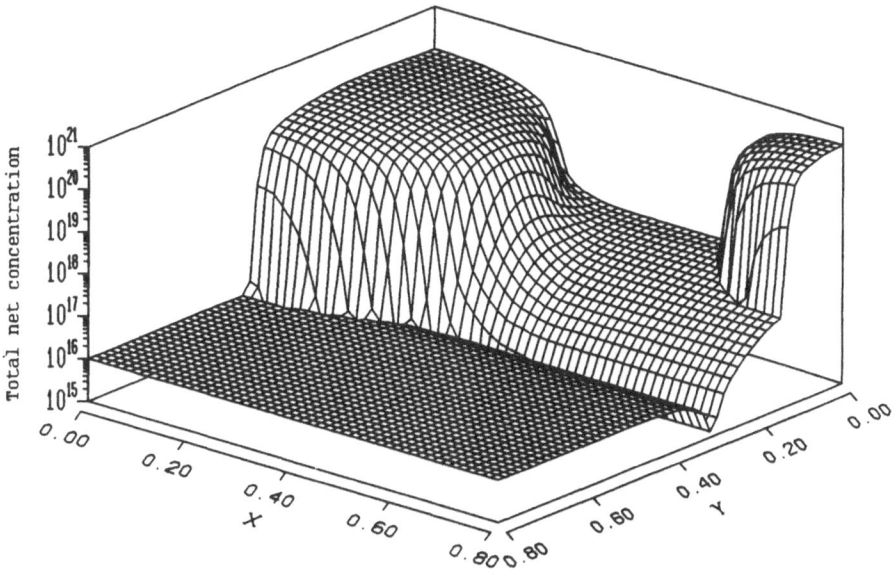

Fig. 5.46 Total net impurity profile of the bipolar transistor

5.5.3 Simulation of the Buried Channel Low Doped Drain (BCLDD) Structure of the PMOS Transistor

The realization of the PMOS transistor does not require implantation and annealing of the low-doped source/drain regions as required for the NMOS

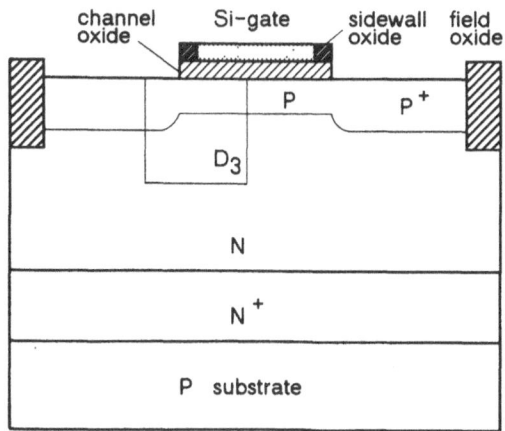

Fig. 5.47 Cross-section of the PMOS transistor

transistor. The boron ion implantation, which is used to adjust the CMOS threshold voltages, extends beyond the gate edge, underneath the sidewall oxide creating a buried channel LDD or BCLDD.

The cross section of the PMOS transistor with BCLDD structure and the rectangular simulation domain D_3 are shown in Figure 5.47. Process Flow 5.3 describes the technology parameters for the simulation.

Process Flow 5.3. The BCLDD structure of the PMOS transistor

3.0 **Substrate**

> *Type: N-well*
> *Concentration*: 10^{16} cm^{-3}
> *Pad oxide*: $0.02 \, \mu m$

3.1 **Implantation** *(Step 1.1)*

3.2 **Diffusion** *(Step 1.2)*

3.3 **Diffusion** *(Step 1.4)*

3.4 **Implantation** *(PMOS source/drain region)*

> *Dopant: boron*
> *Dose*: $6 \cdot 10^{15}$ cm^{-2}
> *Energy*: $60 \, keV$
> *Masked region*: $x \geq 0.2 \, \mu m$

3.5 **Diffusion** *(Step 1.7)*

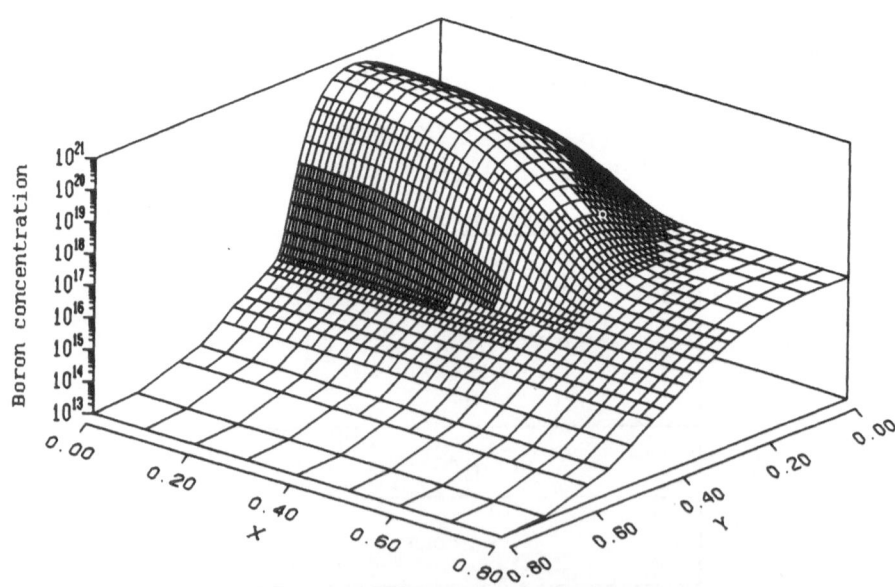

Fig. 5.48 Boron profile after process step 3.4

Impurity profiles after processing steps 3.4 and 3.5 are shown in Figures 5.48 and 5.49 while the total net impurity concentration of the PMOS BCLDD structure is shown in Figure 5.50.

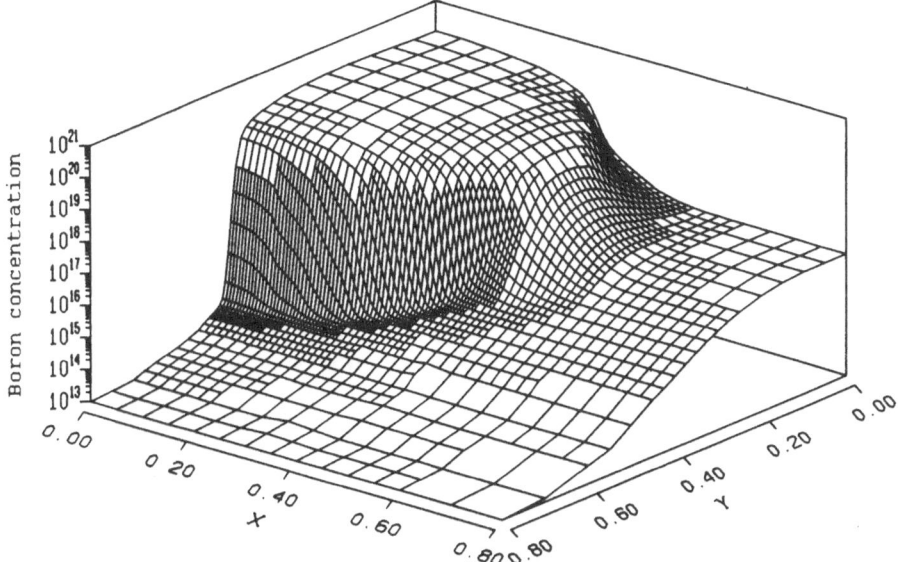

Fig. 5.49 Boron profile after process step 3.5

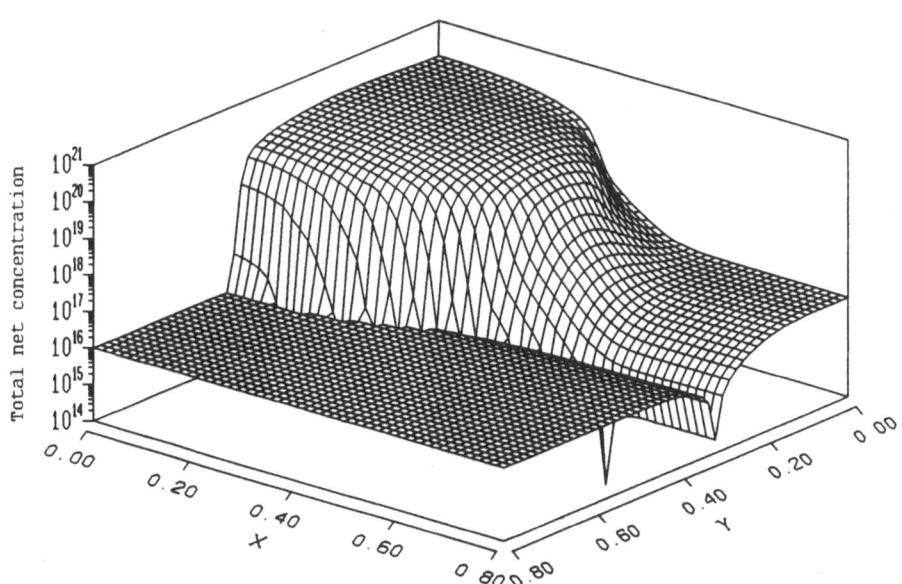

Fig. 5.50 Total net impurity profile of PMOS transistor

References

[1] Bank, R. E., Rose, D. J., *Parameter Selection for Newton-Like Methods to Nonlinear Partial Differential Equations*, SIAM J. Numer. Anal. 17, No. 6, pp. 806–822, 1980

[2] Bank, R. E., Rose, D. J., *Global Approximate Newton Methods*, Numer. Math. 37, pp. 279–295, 1981

[3] Bank, R. E., Coughran, jr., W. M., Fichtner, W., Grosse, E. H., Rose, D. J., Smith, R. K., *Transient Simulation of Silicon Devices and Circuits*, IEEE Trans. Electron Devices, Vol. ED-32, pp. 1992–2007, 1985

[4] Beatson, R. K., Wolkowicz, H., *Post-Processing Piecewise Cubics for Monotonicity*, SIAM J. Numer. 26, pp. 480–502, 1989

[5] Becker, K., *Mehrgitterverfahren zur Lösung der Helmholtzgleichung im Rechteck mit Neumannschen Randbedingungen*, Diplomarbeit, Univ. Bonn, 1981

[6] Becker, K., *Ein Mehrgitterverfahren zur Berechnung subsonischer Potential-strömungen um Tragflächenprofile*, GMD-Bericht Nr. 152, Oldenbourg Verlag, 1985

[7] Borucki, L., Hansen, H. H., Varahramyan, K., *FEDSS—A 2D Semiconductor Fabrication Process Simulator*, IBM J. Res. Develop., Vol. 29, No. 3, May 1985

[8] Börsch-Supan, *Über Stabilität und Schrittweitensteuerung bei der Lösung parabolischer Differentialgleichungen mit Differenzenverfahren*, Diplomarbeit, Univ. Bonn, 1979

[9] Braess, D., *The Convergence Rate of a Multigrid Method with Gauss-Seidel Relaxation for the Poisson Equation*, Lecture Notes in Mathematics, 960, pp. 368–386, 1982

[10] Brandt, A., *Multi-Level Adaptive Solutions to Boundary Value Problems*, Math. Comp., Vol. 31, pp. 333–390, 1977

[11] Brandt, A., *Multigrid Solvers on Parallel Computers*, in "Elliptic Problem Solvers (Schultz, M., ed.)", Academic Press, New York, 1981

[12] Brandt, A. *Multigrid Techniques: 1984 Guide with Applications to Fluid Dynamics*, GMD-Studie No. 85, GMD, Sankt Augustin, 1984

[13] Brandt, A., Ophir, D., *GRIDPACK: Toward Unification of General Grid Programming*, PDE Software: Modules, Interfaces and Systems, (Engquist, B., ed.), North-Holland, 1984

[14] Brandt, A., Greenwald, J., *Parabolic Multigrid Revisited*, in: Multigrid Methods III, (Hackbusch, W., Trottenberg, U., eds.), International Series of Numerical Mathematics, Volume 89, Birkhäuser Verlag, Basel, 1991

[15] Briggs, W. L., *A Multigrid Tutorial*, Society for Industrial and Applied Mathematics, Philadelphia, Pennsylvania, 1987

[16] Brodie, I., Muray, J. J., *The Physics of Microfabrication*, Plenum Press, New York, London, 1982

[17] Chan, T. F., Saad, Y., *Multigrid Algorithms on the Hypercube Multiprocessor*, IEEE Trans. Comput. 35, pp. 969–977, 1986

[18] Chan, T. F., Schreiber, R., *Parallel Networks for Multigrid Algorithms: Architecture and complexity*, SIAM J. Sci. Comput. 6, pp. 698–711, 1985

[19] Chin, D., Kump, M. R., Dutton, R. W., *SUPRA: Stanford University Process Analysis Program*, Stanford University Laboratories, Stanford University, Stanford, 1981

[20] Chin, D., Oh, S.-Y., Dutton, R. W., *A General Solution Method for Two-Dimensional Nonplanar Oxidation*, IEEE Trans. Electron Devices, 9/1983, pp. 993–998

[21] Collard, D., Baccus, B., Dubois, E., Morel, D., *IMPACT 1-2-3 AN INTEGRATED 2D PROCESS/DEVICE SIMULATOR FOR M.O.S. TECHNOLOGY.*, Lecture Notes and Digest of the NASECODE VI, Associated Short Course and Software Forum, (Crans, W., ed.), Boole Press Ltd., 1989

[22] Deal, B. E., Grove, A. S., *General Relationship for the Thermal Oxidation of Silicon*, J. Appl. Phys., 36, pp. 3770–3778, 1965

[23] Dekker, K., Verwer, J. G., *Stability of Runge-Kutta Methods for Stiff Nonlinear Differential Equations*, North-Holland, Amsterdam, 1984

[24] Deuflhard, P., *A Modified Newton Method for the Solution of Ill-Conditioned Systems of Nonlinear Equations with Application to Multiple Shooting*, Numer. Math. 22, pp. 289–315, 1974

[25] Dinar, N., *Fast Methods for the Numerical Solutions of Boundary Value Problems*, PhD-Thesis, Weizmann Institute of Science, Rehovot, Israel, 1979

[26] Dick, E., Linden, J., *A Multigrid Method for Solving the Steady Incompressible Navier-Stokes Equations Based on Flux-Difference Splitting*, Paper presented at "Numerical Methods in Laminar and Turbulent Flow", Sixth International Conference held at Swansea, 11th–15th July, 1989

[27] Dutton, R. W., Hansen, S. E., *Process Modeling of Integrated Circuit Device Technology*, Proceedings of the IEEE, No. 10, pp. 1305–1320, 1981

[28] Dutton, R. W., *Modeling of the Silicon Integrated-Circuit Design and Manufacturing Process*, IEEE Trans. Electron Devices, 9/1983, pp. 968–985

[29] Eisenstat, S. C., Jackson, K. R., Lewis, J. W., *The Order of Monotone Piecewise Cubic Interpolation*, SIAM J. Numer. Anal., Vol. 20, No. 6, pp. 1220–1238, 1985

[30] Fair, R. B., *Concentration Profiles of Diffused Dopants in Silicon*, in: Impurity Dopants in Silicon (Wang, F. F. Y, ed.), North-Holland, New York, 1981

[31] Fair, R. B., *Physics and Chemistry of Impurity Diffusion and Oxidation of Silicon*, in: Silicon Integrated Circuits–Part B (Kahng, D., ed.), Academic Press, New York, 1981

[32] Franz, A. F., Franz, G. A., Selberherr, S., Ringofer, C., Markowich, P., *Finite Boxes— A Generalization of the Finite-Difference Method Suitable for Semiconductor Device Simulation*, IEEE Trans. Electron Devices, pp. 1070–1082, 1983

[33] Fritsch, F. N., Carlson, R. E., *Monotone Piecewise Cubic Interpolation*, SIAM J. Numer. Anal. Vol. 17, No. 2, pp. 119–138, 1980

[34] Fritsch, F. N., Carlson, R. E., *Monotonicity Preserving Bicubic Interpolation: A Progress report*, Computer Aided Geometric Design, Vol. 2, pp. 117–121, 1985

[35] Fritsch, F. N., Carlson, R. E., *An Algorithm for Monotone Pieciecewise Bicubic Inter-polation*, SIAM J. Numer. Anal. Vol. 28, pp. 230–238, 1989

[36] Gannon, D., Van Rosendale, J., *On the Structure of Parallelism in a Highly Concurrent PDE Solver*, J. Parallel and Distributed Comput. 3, pp. 106–135, 1986

[37] Gärtel, U., *Parallel Multigrid Solver for 3D Anisotropic Elliptic Problems*, Arbeitspapiere der GMD Nr. 390, St. Augustin, 1989

[38] Gärtel, U., Krechel, A., Niestegge, A., Plum, H.-J., *Parallel Multigrid for 2D and 3D Elliptic Equations: Standard and Nonstandard Smoothing*, in: Multigrid Methods III (Hackbusch, W., Trottenberg, U., eds.), International Series of Numerical Mathematics 98, Birkhäuser Verlag, Basel, 1991

[39] Gärtel, U., Ressel, K., *Parallel Multigrid: Grid Partitioning Versus Domain Decom-position*, in: Computing Methods in Applied Sciences and Engineering (Glowinski, R., ed.), Nova Science Publishers, New York, 1991

[40] Gendler, E., *Multigrid Methods for Time-Dependent Parabolic Equations*, Masters Thesis, Weizmann Institute of Science, Rehovot, Israel, 1986

[41] Gerodolle, A., Corbex, C., Poncet, A., Pedron, T., Martin, S., *TITAN 5, A Two-Dimensional Process and Device Simulator*, Lecture Notes and Digest of the NASECODE VI, Associated Short Course and Software Forum (Crans, W., ed.), Boole Press Ltd., 1989

[42] Ghandi, S. K., *VLSI Fabrication Principles*, John Wiley & Sons, New York, 1983

[43] Giles, M. D., Gibbons, J. F., *Two-Dimensional Ion Implantation Profiles from One-Dimensional Projections*, J. Electrochem. Soc., Vol. 132, pp. 2476–2480, 1985

[44] Giles, M. D., *Calculation of Ion Implantation Profiles for Two-Dimensional Process Modeling*, Proc. 2nd International Conference on Simulation of Semiconductor Devices and Processes (SISDEP), (Board, K., Owen, D. R. J., eds.), Vol. 2, pp. 233–246, Pineridge Press, Swansea, 1986

[45] Greenbaum, A., *A Multigrid Method for Multiprocessors*, Appl. Math. Comput. 19, pp. 75–88, 1986

[46] Giebel, M., Schneider, M., Zenger, C., *A Combination Technique for the Solution of Sparse Grid Problems*, SFB-Bericht Nr. 342/19/90 A, Technische Universität München, 1990

[47] Hackbusch, W., *Multi-Grid Convergence Theory*, Lecture Notes in Mathematics, 960, pp. 177–219, 1982

[48] Hackbusch, W., *Parabolic Multi-Grid Methods*, in: Computing Methods in Applied Science and Engineering (Glowinski, R., Lion, J. L., eds.), North-Holland, Amsterdam, 1984

[49] Hackbusch, W., *Multi-Grid Methods and Applications*, Springer, Berlin, 1985

[50] Hackbusch, W., *Iterative Lösung großer schwachbesetzter Gleichungssysteme*, Studienbücher Mathematik, Teubner, Stuttgart, 1991

[51] Haken, R. A., Havemann, R. H., Eklund, R. H., Hutter, L. N., *BiCMOS Process Technology*, in: *BiCMOS Technology and Application* (Alvarez, A. R., ed.), Kluwer Academic Publishers, Boston, 1990

[52] Hemker, P. W., van der Maarel, H. T. M., Everaars, C. T. H., *BASIS: A Data Structure for Adaptive Multigrid Computations*, Report NM-R9014, Centre for Mathematics and Computer Science, Amsterdam, 1990

[53] Hempel, R., Schüller, A., *Vereinheitlichung und Portabilität paralleler Anwendersoftware durch Verwendung einer Kommunikationsbibliothek*, Arbeitspapiere der GMD Nr. 234, GMD, St. Augustin, 1986

[54] Hempel, R., Schüller, A., *Experiments with Parallel Multigrid Using the SUPRENUM Communications Library*, GMD-Studie Nr. 141, GMD, St. Augustin, 1988

[55] Hempel, R., Ritzdorf, H., *The GMD Communications Subroutine Library for Grid-Oriented Problems*, Arbeitspapiere der GMD Nr. 589, St. Augustin, 1991

[56] Herbin, R., Gerbi, S., Sonnad, V., *Parallel Implementation of a Multigrid Method on the Experimental ICAP Supercomputer*, Appl. Math. Comput. 27, pp. 281–312, 1988

[57] Ho, C. P., Plummer, J. D., Hansen, S. E., Dutton, R. W., *VLSI Process Modeling—SUPREM III*, IEEE Trans. Electron Devices, 11/1983, pp. 1438–1453

[58] Hobler, G., Langer, E., Selberherr, S., *Two-Dimensional Modeling of Ion Implantation with Spatial Moments*, Solid-State Electronics, No. 4, pp. 445–455, 1987

[59] Hobler, G., Selberherr, S., *Monte Carlo Simulation of Ion Implantation into Two- and Three-Dimensional Structures*, IEEE Trans. Computer-Aided Design, No. 5, pp. 450–459, 1989

[60] Hyman, J. M., *Accurate Monotonicity Preserving Interpolation*, SIAM J. Sci. Statist. Comput. Vol. 4, pp. 645–654, 1983

[61] Irvine, L. D., Marin, S. P., Smith, P. W., *Constrained Interpolation and Smoothing*, Constr. Approx. Vol. 2, pp. 129–151, 1986

[62] Ismail, R., Amaratunga, G., *Adaptive Meshing Schemes for Simulating Dopant Diffusion*, IEEE Trans. Computer-Aided Design, No. 3, pp. 276–289, 1990

[63] Joppich, W., *Mehrgitterverfahren für Diffusionsprobleme der Prozeßsimulation*, Dissertation Universität Bonn, 1990, and GMD-Bericht Nr. 189, Oldenbourg Verlag, 1991

[64] Joppich, W., *A Multigrid Algorithm with Time-Dependent, Locally Refined Grids for Solving the Nonlinear Diffusion Equation on a Non-Rectangular Geometry*, ELECTROSOFT, No. 4, 1990

[65] Joppich, W., *A Multigrid Algorithm with Time-Dependent, Locally Refined Grids for Solving the Nonlinear Diffusion Equation on a Nonrectangular Domain—Practical Aspects*, Transactions of the NASECODE VII Conference, Copper Mountain, USA, COMPEL, No. 4, Dez. 1991

[66] Joppich, W., *A Multigrid Algorithm with Time-Dependent, Locally Refined Grids for Solving the Nonlinear Diffusion Equation on a Nonrectangular Domain*, Multigrid Methods III, International Series of Numerical Mathematics, Birkhäuser Verlag, Vol. 98, 1991

[67] Joppich, W., Lorentz, R. A., *High Order Positive, Monotone and Convex Multigrid Interpolations*, GMD-Arbeitspapier 558, July 1991 and COMPEL (to appear)

[68] Joppich, W., Lorentz, R. A., *High Order Monotone and Convex Multigrid Interpolations for a VLSI-Problem*, in Multigrid Methods: Special Topics and Applications II, Papers presented at the 3rd European Conference on Multigrid Methods, Bonn, October 1–4, 1990, GMD-Studie 189, May 1991

[69] Keyes, D. E., Gropp, W. D., *A Comparison of Domain Decomposition Techniques for Elliptic Partial Differential Equations and Their Implementation*, SIAM J. Sci. Stat. Comp. 8, pp. 166–202, 1987

[70] Krechel, A., Plum, H.-J., Stüben, K., *Parallel Solution of Tridiagonal Linear Systems*, in: Hypercube and Distributed Computers (Andre, F., Versus, J. P., eds.), North-Holland, Rennes, 1989

[71] Krechel, A., Plum, H.-J., Stüben, K., *Parallelization and Vectorization Aspects of the Solution of Tridiagonal Linear Systems*, Parallel Computing 14, pp. 31–49, 1990

[72] Kump, M. R., Dutton, R. W., *The Efficient Simulation of Coupled Point Defect and Impurity Diffusion*, IEEE Trans. Computer-Aided Design, 2/1988, pp. 191–204

[73] Kroll, N., *Direkte Anwendung von Mehrgittertechniken auf parabolische Anfangsrandwertaufgaben*, Diplomarbeit, Univ. Bonn, 1981

[74] Lambert, J., *Computational Methods in Ordinary Differential Equations*, John Wiley & Sons, London, 1973

[75] Lemke, M., *Erfahrungen mit Mehrgitterverfahren für Helmholtzähnliche Probleme auf Vektorrechnern und Multiprozessor-Vektorrechnern*, Arbeitspapiere der GMD Nr. 278, GMD, St. Augustin, 1987

[76] Lemke, M., Quinlan, D., *Local Refinement Based Fast Adaptive Composite Grid Methods on SUPRENUM*, in: Multigrid Methods: Special Topics and Applications II, (Hackbusch, W., Trottenberg, U., eds.), GMD-Studien Nr. 189, 1991

[77] Lemke, M., Schüller, A., Solchenbach, K., Trottenberg, U., *Parallel Processing on Distributed Memory Multiprocessors*, GI—20. Jahrestagung (Reuter, A., ed.), Informatik-Fachberichte 257, Springer, 1990

[78] Linden, J., *Mehrgitterverfahren für die Poisson-Gleichung in Kreis und Ringgebiet unter Verwendung lokaler Koordinaten*, Diplomarbeit, Univ. Bonn, 1981

[79] Linden, J., *Mehrgitterverfahren für das erste Randwertproblem der biharmonischen Gleichung und Anwendung auf ein inkompressibles Strömungsproblem*, Dissertation, Univ. Bonn, 1985

[80] Linden, J., Steckel, B., Stüben, K., *Parallel Multigrid Solution of the Navier-Stokes Equations on General 2D-Domains*, GMD-Arbeitspapier No. 294, GMD, Sankt Augustin, 1988

[81] Linden, J., Steckel, B., Stüben, K., *Parallel Multigrid Solution of the Navier-Stokes Equations on General 2D Domains*, Parallel Computing 7, pp. 461–475, 1988

[82] Linden, J., Lonsdale, G., Steckel, B., Stüben, K., *Multigrid for the Steady-State Incompressible Navier-Stokes Equations: A Survey*, GMD-Arbeitspapier No. 322, GMD, Sankt Augustin, 1988

[83] Lindhard, J., Scharff, M., Schiott, H. E., *Range Concept and Heavy Ion Ranges*, Mat. Fys. Medd. K. Dan. Vidensk Selsk, No. 14, 1963

[84] Lonsdale, G., Schüller, A., *Parallel and Vector Aspects of a Multigrid Navier-Stokes Solver*, Arbeitspapiere der GMD Nr. 550, GMD, St. Augustin, 1991

[85] Lonsdale, G., Schüller, A., *Maintaining Multigrid and Parallel Efficiency for the Navier-Stokes Equations*, Proceedings of the Conference on "Parallel Computational Fluid Dynamics", Stuttgart, Germany, 10–12 June 1991 (Reinsch et al., eds.), Elsevier Science Publishers B.V., Amsterdam

[86] Lorenz, J., Svoboda, M., *ASWR—Method for the Simulation of Dopant Redistribution in Silicon*, Simulation of Semiconductor Devices and Processes Vol. 3 (Baccarani, G., Rudan, M., eds.), Bologna (Italy), Tecnoprint, 1988

[87] Lorenz, J., Pelka, J., Ryssel, H., Pichler, P., *Programs for VLSI Process Simulation*, Lecture Notes and Digest of the NASECODE VI, Associated Short Course and Software Forum (Crans, W., ed.), Boole Press Ltd., 1989

[88] Maldonado, C. D., Louie, S. A., Murphy, W. D., Hall, W. F., *MEMBRE: An Efficient Two-Dimensional Process Code for VLSI*, COMPEL, 4/1982, pp. 219–239

[89] Maldonado, C. D., *ROMANS II: A Two-Dimensional Process Simulator for Modeling and Simulation in the Design of VLSI Devices*, Applied Physics (Solids and Surfaces), A31, pp. 119–138, 1983

[90] Maldonado, C. D., Custode, F. Z., Louie, S. A., Pancholy, R., *Two-Dimensional Simulation of a 2-μm CMOS Process Using ROMANS II*, IEEE Trans. Electron Devices, 11/1983, pp. 1462–1469

[91] McAllister, D. F., Roulier, J. A., *Interpolation by Convex Quadratic Splines*, Math. Comp. Vol. 32, pp. 1154–1162, 1978

[92] Marchuk, G. I., *Methods of Computational Mathematics*, Nauka, Moscow, 1989

[93] Markowich, P. A., *The Stationary Semiconductor Device Equations*, Springer-Verlag, Wien, 1986

[94] Marsal, D., *Die numerische Lösung partieller Differentialgleichungen in Wissenschaft und Technik*, Springer, Berlin-Heidelberg-New York, 1978

[95] McBryan, O. A., Frederickson, P. O., Linden, J., Schüller, A., Solchenbach, K., Stüben, K., Thole, C.-A., Trottenberg, U., *Multigrid Methods on Parallel Computers—A Survey of Recent Developments*, Impact of Computing in Science and Engineering 3, pp. 1–75, 1991

[96] McCormick, S. F., *Multilevel Adaptive Methods for Partial Differential Equations*, Frontiers in Applied Mathematics 6, Siam, Philadelphia, 1989

[97] McCormick, S. F., *MLAT vs. FAC*, Preprint, University of Colorado at Denver.

[98] Mijalković, S., Stojadinović, N., *Multigrid Method: An Efficient Numerical Tool in VLSI Process Modeling*, Proc. 1st International Conf. on Computer Technology, Systems and Applications (COMPEURO'87), (Proebster, W. E., Reiner, H., eds.), Hamburg, pp. 508–509, May, 1987

[99] Mijalković, S., Stojadinović, N., *Solution of the Diffusion Equation in VLSI Process Modeling by a Nonlinear Multigrid Algorithm*, Numerical Methods and Approximation Theory III, Niš, August, 18–21, 1987

[100] Mijalković, S., Pantić, D., Stojadinović, N., *On Efficiency of Multigrid Methods in Two-Dimensional Impurity Redistribution Simulation*, Proc. 3rd International Conference on Simulation of Semiconductor Devices and Processes (SISDEP'88), (Baccarani, G., Rudan, M., eds.), Vol. 3, pp. 463–474, September, Bologna, 1988

[101] Mijalković, S., Stojadinović, N., *Efficient Simulation of Impurity Redistribution in*

VLSI Fabrication Processes, Solid-State Electronics, No. 12, pp. 1689–1693, 1988

[102] Mijalković, S., *An Adaptive Multigrid Algorithm for Simulation of Diffusion Processes in Semiconductor Device Fabrication*, Electrosoft, No. 4, pp. 277–290, 1990

[103] Mijalković, S., Pantić, D., Prijić, Z., Mitrović, S., Stojadinović, N., *MUSIC — A Multigrid Simulator for IC Fabrication Processes*, COMPEL, No. 4, pp. 599–610, 1991

[104] *MUSIC — User's Manual*, Faculty of Electronic Engineering, University of Niš, 18000 Niš, Beogradska 14, Yugoslavia

[105] Mijalković, S., *An Adaptive Multigrid Simulation of Coupled Multiparticle Diffusion in Semiconductor Device Fabrication*, Proc. 3rd European Conference on Multigrid Methods, in GMD-Studien No. 189, pp. 203–214, Bonn, May, 1991

[106] Mijalković, S., Pantić, D., Prijić, Z., Stojadinović, N., *Adaptive Multigrid Strategies for Simulation of Diffusion Processes*, Proc. 4th International Conference on Simulation of Semiconductor Devices and Processes (SISDEP'91), (Aemmer, D., Fichtner, W., eds.), Vol. 4, pp. 505–511, September, Zurich, 1991

[107] O'Brien, R. R., et al., *Two-Dimensional Process Modeling: A Description of the SAFEPRO program*, IBM J. Res. Develop., No. 3, May, 1985

[108] Orlowski, M., *Challenges for Process Modeling and Simulation in the 90's — An Industrial Perspective*, Proc. 4th International Conference on Simulation of Semiconductor Devices and Processes (SISDEP'91), (Aemmer, D., Fichtner, W., eds.), Vol. 4, pp. 505–511, September, Zurich, 1991

[109] Ortega, J. M., Rheinboldt, W. C., *Iterative Solution of Nonlinear Equations in Several Variables*, Academic Press, New York-London, 1970

[110] Ortega, J. M., Voigt, R. G., *Solution of Partial Differential Equations on Vector and Parallel Computers*, SIAM Rev. 27, pp. 149–240, 1985

[111] Pantić, D., Mijalković, S., Stojadinović, N., *A New Multi-Layer Ion Implantation Model for Process Simulation*, Microelectronics Journal, No. 6, pp. 5–10, 1989

[112] Penumalli, B. R., *A Comprehensive Two-Dimensional VLSI Process Simulation Program, BICEPS*, IEEE Trans. Electron Devices, 9/1983, pp. 986–992

[113] Pichler, P., Jüngling, W., Selberherr, S., *Simulation of Critical IC-Fabrication Steps*, IEEE Trans. Electron Devices, pp. 1940–1953, 1985

[114] Polak, S. J., Den Heijer, C., Schilders, W. H. A., Markowich, P., *Semiconductor Device Modelling from the Numerical Point of View*, Int. J. for Numerical Methods in Engineering 24, pp. 763–838, 1987

[115] Pollul, W., *Private Kommunikation über monotone und tendenzerhaltende Interpolation schwach oszillierender Daten*, Institut für Angewandte Mathematik der Universität Bonn, 1990

[116] Rheinboldt, W. C., *On a Theory of Mesh-Refinement Processes*, SIAM J. Numer. Anal., Vol. 17, No. 6, pp. 766–778, 1980

[117] Ressel, K., *Gebietszerlegung und Mehrgitterverfahren: Varianten, numerische Untersuchungen und Aspekte der Parallelisierung*, Diplomarbeit, Mathematisches Institut der Universität Köln, Wintersemester 1989/1990

[118] Ritzdorf, H., *Lokal verfeinerte Mehrgitter-Methoden für Gebiete mit einspringenden Ecken*, Diplomarbeit, Universität Bonn, 1984

[119] Russell, R. D., Christiansen, J., *Adaptive Mesh Selection Strategies for Solving Boundary Value Problems*, SIAM J. Numer. Anal., Vol. 15, No. 1, pp. 59–80, 1978

[120] Ryssel, H., Ruge, I., *Ion Implantation*, John Wiley & Sons, Chichester, 1986

[121] Ryssel, H., Biersack, J. P., *Ion Implantation Models for Process Simulation*, in: Process and Device Modeling (Engl, W. L., ed.), North-Holland, Amsterdam, 1986

[122] Schüller, A., *Mehrgitterverfahren für Schalenprobleme*, Dissertation, Universität Bonn, 1987

[123] Schüller, A., Solchenbach, K., Trottenberg, U., *Grid Partitioning for CFD Applications*, Proceedings of the Conference on "Parallel Computational Fluid Dynamics",

Stuttgart, Germany, 10–12 June 1991 (Reinsch et al., eds.), Elsevier Science Publishers B.V., Amsterdam.

[124] Seidl, A., *A Multigrid Method for Solution of the Diffusion Equation in VLSI Process Modeling*, IEEE Trans. Electron Devices, 9/1983, pp. 999–1004

[125] Selberherr, S., Guerrero, E., *Simple and Accurate Representations of Implantation Parameters by Low Order Polynomials*, Solid-State Electronics, Vol. 24, pp. 591–593, 1981

[126] Selberherr, S., *Analysis and Simulation of Semiconductor Devices*, Springer-Verlag, Wien, 1984

[127] Solchenbach, K., *Einsatz schneller elliptischer Löser zur Lösung nichtlinearer parabolischer Anfangsrandwertaufgaben*, Diplomarbeit, Univ. Bonn, 1980

[128] Solchenbach, K., Stüben, K., Trottenberg, U., Witsch, K., *Efficient Solution of a Nonlinear Heat Conduction Problem by Use of Fast Elliptic Reduction and Multigrid Methods*, Lecture Notes in Mathematics, 968, pp. 114–148, 1982

[129] Stoer, J., Bulirsch, R., *Einführung in die Numerische Mathematik II*, Heidelberger Taschenbücher, Springer, Berlin-Heidelberg-New York, 1973

[130] Stüben, K., *A Multi-Grid Program to Solve $\Delta U - c(x, y)U = f(x, y)$ (on Ω), $U = g(x, y)$ (on $\partial\Omega$) on Nonrectangular Bounded Domains Ω*, IMA-Report: 82.02.02, Gesellschaft für Mathematik und Datenverarbeitung mbH Bonn, 1982

[131] Stüben, K., Trottenberg, U., *On the Construction of Fast Solvers for Elliptic Equations*, Technical Report, IMA-Report 82.0201, 1982

[132] Stüben, K., Trottenberg, U., *Multigrid Methods: Fundamental Algorithms, Model Problem Analysis and Applications*, Lecture Notes in Mathematics, 960, pp. 1–176, 1982

[133] Sze, S. M., ed., *VLSI Technology*, McGraw-Hill Book Company, New York, 1983

[134] Thole, C.-A., Trottenberg, U., *Basic Smoothing Procedures for the Multigrid Treatment of Elliptic 3D-Operators*, in: Advances in Multigrid Methods (Braess, D., Hackbusch, W., Trottenberg, U., eds.), Notes on Numerical Fluid Mechanics 11, Vieweg, Braunschweig, 1985

[135] Vandewalle, S., *Waveform Relaxation Methods for Solving Parabolic Partial Differential Equations*, Proceedings of the Fifth Distributed Memory Computing Conference, Charleston, South-Carolina, April 9–12, 1990

[136] Vandewalle, S., Piessens, R., *A Parallel and Vectorizable Algorithm for Solving Parabolic Partial Differential Equations*, Proceedings of the Sixth GAMM-Seminar, Kiel, January 19–21, 1990 (Hackbusch, W., ed.), Notes on Numerical Fluid Mechanics, Volume 31, Vieweg, Braunschweig, 1991

[137] Winter, G., *Fourieranalyse zur Konstruktion schneller MGR-Verfahren*, Dissertation, Univ. Bonn, 1982

[138] Yan, Z., *Piecewise Cubic Curve Fitting Algorithm*, Math. Comp. Vol. 49, pp. 203–213, 1987

[139] Yeager, H. R., Dutton, R. W., *An Approach to Solving Multiparticle Diffusion Exhibiting Nonlinear Stiff Coupling*, IEEE Trans. Electron Devices, pp. 1964–1975, 1985

[140] Zenger, C., *Sparse Grids*, Proceedings of the Sixth GAMM-Seminar, Kiel, January 19–21, 1990 (Hackbusch, W., ed.), Notes on Numerical Fluid Mechanics, Volume 31, Vieweg, Braunschweig, 1991

Index

Narain D. Arora

MOSFET Models for VLSI Circuit Simulation
Theory and Practice

(Computational Microelectronics)
1993. Approx. 260 figures. Approx. 600 pages.
Cloth öS 2086,–, DM 298,–, US $ 198.00
ISBN 3-211-82395-6

Prices are subject to change without notice

The book covers the MOS transistor models and their parameters required for VLSI simulation of MOS integrated circuits. It gives the first detailed presentation of model parameter determination for MOS models. Various models are developed ranging from simple to more sophisticated models that take into account new physical effects observed in submicron devices used in today's MOS VLSI technology. The assumptions used to arrive at the models are emphasized so that the accuracy of the model in describing the device characteristics are clearly understood. Understanding these models is essential when designing circuits for the state of the art MOS IC's. Threshold voltage being the single most important MOSFET parameter, a full chapter is devoted to the development of the device threshold voltage model. Due to the importance of designing reliable circuits, the device reliability models as applied for circuit simulations are also covered. Since the device parameters vary due to inherent processing variations, how to arrive at worst case design parameters are covered.

Presentation of the material is such that even an undergraduate student not well familiar with semiconductor device physics can understand the intricacies of MOSFET modeling. The book serves as a technical source in the area of MOSFET modeling for state of the art MOSFET technology for both practicing device and circuit engineers and engineering students interested in the said area.

Springer-Verlag Wien New York

Sachsenplatz 4-6, P O. Box 89, A-1201 Wien · Heidelberger Platz 3, D-14197 Berlin
175 Fifth Avenue, New York, NY 10010, USA · 37-3, Hongo 3-chome, Bunkyo-ku, Tokyo, 113, Japan

Siegfried Selberherr, Hannes Stippel, Ernst Strasser (eds.)

Simulation of Semiconductor Devices and Processes
Vol. 5

(Computational Microelectronics)
1993. Approx. 530 pages
Cloth oS 1386,–, DM 198,–
ISBN 3-211-82504-5

The SISDEP 93 conference proceedings present outstanding research and development results in the area of numerical process and device simulation. The miniaturization of today's semiconductor devices, the usage of new materials and advanced process steps in the development of new semiconductor technologies suggests the design of new computer programs.
This trend towards more complex structures and increasingly sophisticated processes demands advanced simulators, such as fully three-dimensional tools for almost arbitrarily complicated geometries. With the increasing need for better models and improved understanding of physical effects, these proceedings support the simulation community and the process- and device engineers who need reliable numerical simulation tools for characterization, prediction, and development. This book covers the following topics: process simulation and equipment modeling, device modeling and simulation of complex structures, device simulation and parameter extraction for circuit models, integration of process, device and circuit simulation, practical applications of simulation, algorithms and software.

Franz Fasching, Stefan Halama, Siegfried Selberherr (eds.)

Technology CAD Systems

(Computational Microelectronics)
1993. Approx. 320 pages.
Cloth oS 896,–, DM 128,–
ISBN 3-211-82505-3

The proceedings of the first "Workshop on Technology CAD Systems" is an authoritative tutorial on CAD software systems for the physical design of semiconductor devices and manufacturing processes. Fourteen invited papers by academic and industrial representatives from USA, Europe, and Japan, as well as from commercial software vendors provide an excellent overview of the work in this area. This book covers the following topics: Coupling and integration of process simulation, device simulation, parameter extraction, and circuit simulation, technology CAD requirements, architectures and strategies of existing CAD systems, implementation and software aspects, practical applications and experiences with technology CAD, and directions of future work in this field.

Prices are subject to change without notice

Springer-Verlag Wien New York

Sachsenplatz 4-6, P.O. Box 89, A-1201 Wien · Heidelberger Platz 3, D-14197 Berlin
175 Fifth Avenue, New York, NY 10010, USA · 37-3, Hongo 3-chome, Bunkyo-ku, Tokyo, 113, Japan

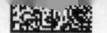